History of Physics

The Springer book series *History of Physics* publishes scholarly yet widely accessible books on all aspects of the history of physics. These cover the history and evolution of ideas and techniques, pioneers and their contributions, institutional history, as well as the interactions between physics research and society. Also included in the scope of the series are key historical works that are published or translated for the first time, or republished with annotation and analysis.

As a whole, the series helps to demonstrate the key role of physics in shaping the modern world, as well as revealing the often meandering path that led to our current understanding of physics and the cosmos. It upholds the notion expressed by Gerald Holton that "science should treasure its history, that historical scholarship should treasure science, and that the full understanding of each is deficient without the other." The series welcomes equally works by historians of science and contributions from practicing physicists.

These books are aimed primarily at researchers and students in the sciences, history of science, and science studies; but they also provide stimulating reading for philosophers, sociologists and a broader public eager to discover how physics research – and the laws of physics themselves – came to be what they are today.

All publications in the series are peer reviewed. Titles are published as both print and eBooks. Proposals for publication should be submitted to Marina Forlizzi (marina.forlizzi@springer.com) or one of the series editors.

Gianenrico Bernasconi · Ileana Chinnici ·
Marco Storni
Editors

Astronomical Observatories and Chronometry, 18th-20th Century

Studies in Honour of Paolo Brenni

Springer

Editors
Gianenrico Bernasconi
University of Neuchâtel
Neuchâtel, Switzerland

Ileana Chinnici
INAF Osservatorio Astronomico di Palermo
Palermo, Italy

Marco Storni
Université libre de Bruxelles
Bruxelles, Belgium

ISSN 2730-7549 ISSN 2730-7557 (electronic)
History of Physics
ISBN 978-3-031-95873-1 ISBN 978-3-031-95874-8 (eBook)
https://doi.org/10.1007/978-3-031-95874-8

This Springer imprint is published by the registered company Springer Nature Switzerland AG
The registered company address is: Gewerbestrasse 11, 6330 Cham, Switzerland

Paolo Brenni at the Museo della Fondazione Scienza e Tecnica in Florence, 2019. Photo by Anna Giatti

Preface

In recent decades, the history of astronomical observatories has been renewed by studies that emphasise the importance of these institutions as one of the first sites of highly specialised scientific research.[1] Recent scholarship has focused on the observing apparatus and practices within a long history of precision, on the contextual negotiations between astronomers and instrument makers, as well as on the social organisation of scientific work at the observatory and the professionalisation of the astronomer.[2] Historians of the modern and contemporary period, from the eighteenth to the twentieth century, have also studied the observatory in its architectural dimension and its place in the environment, the connections between science and the state—as in the case of the calculation of longitude to improve navigation—and the collective endeavours of positional astronomy (the *Carte du Ciel*) and practical astronomy (e.g. the applications of astronomy to geodesy or meteorology).[3]

In this new historiography of observatories, the relationship between astronomy and chronometry is crucial. Since the early modern period, pendulums have been fundamental tools in the practice of astronomical observation, since the measurement of time was essential for recording the transit of the stars and thus deduce the celestial coordinates for producing star catalogues and sky maps.[4] Clockmakers used their knowledge of mechanics to create astronomical models, which were not only technical masterpieces but also epistemic tools for understanding the movement of the stars. Conversely, it should be noted that astronomical observation was the very method used to determine time before the introduction of atomic clocks in the mid-twentieth century. Observatories served as central hubs for the dissemination of accurate time, a critical element in navigational practices as well as in urban space as a standard of public time.[5]

---

[1] See, for example, Lamy (2007); Aubin et al. (2010); Macdonald (2018); Nassim (2021).

[2] Key studies addressing these issues include Schaffer (1988); Boistel (2005); Saint Martin (2011); Gluch (2024).

[3] On these topics see Aubin (2003); Dunn and Higgitt (2014); Daston (2023).

[4] Lamy (2008).

[5] On these aspects, see McCrossen (2013); Rothauge (2023).

From the end of the eighteenth century, the role of the observatories as public institutions became more prominent, particularly as their activities in practical astronomy became aligned with the interests of states (navigation, cartography, geodesy, etc.). In this context, the work carried out in observatories on chronometry gained momentum, driven by the emergence of an international watchmaking market. As scientific institutions, the observatories were called upon to play a key role in the certification of chronometers, thus contributing to the creation of a quality-based economy. This was achieved by issuing performance certificates and awarding prizes to guide the market and encourage advances in the precision of timepieces.

Although the quest for astronomical precision led to a process of technical innovation in watchmaking, it should not be forgotten that the development of methods to quantify the astronomer's reaction time and elaborate the personal equation would not have been possible without the spread of chronographs and chronoscopes in the mid-nineteenth century.[6] Indeed, the accurate measurement of short intervals of time was essential in order to take account of individual differences in reaction times when recording observations. As well as refining the practice of astronomical observation, advances in horology had a profound impact on other scientific disciplines.[7] In physics, for example, precise timekeeping was essential for experiments in mechanics and electromagnetism, while in engineering the ability to measure time intervals with extreme precision was vital for the calibration of machinery and the development of early transport systems such as railways and telegraphs, which required synchronisation over long distances.

The growth of scholarship on observatories and chronometry has been encouraged by the intensified collaboration between historians of ideas and scientific practice on the one hand, and historians of scientific instruments on the other. The many conferences and publications sponsored by the Scientific Instrument Commission, and the publication in 2022 of a new *General History of Horology*, which presents a rich synthesis of the results of research in this domain, bear witness to the dynamism of this field of study.[8]

In this favourable context for the history of observatories, scientific instruments and chronometry, Paolo Brenni and Gianenrico Bernasconi organised a workshop to bring together, in an informal setting, a number of specialists in the field in order to promote collaboration and exchange on research perspectives and open questions. The aim of the event was to combine the presence of established scholars, who discussed their past achievements and the latest results of their work, with young scholars, who aimed to update the received views and contribute to writing the future of the discipline, promoting the collaboration between museum curators and historians of science. Following the event, which took place in Neuchâtel from 1 to 3 December 2021, Paolo and Gianenrico launched the project of publishing a collection of the texts presented at the workshop.

---

[6] On the personal equation, see Schmidgen (2009).

[7] Canales (2009).

[8] Turner et al. (2022).

On 4 December, we learned that Paolo had passed away at his home in Mendrisio. Deeply moved and saddened by the news of the sudden demise of such a great and generous *maestro*, Gianenrico—together with Ileana Chinnici and Marco Storni—decided to carry out the publication project that Paolo had so warmly supported. Over the past decades, as historical scholarship increasingly embraced the material dimensions of knowledge, Paolo made decisive contributions. He bridged different approaches, revitalised the study of scientific instruments and encouraged critical reflection on their conservation and restoration. With boundless curiosity, unwavering passion, deep expertise and generosity, Paolo brought dynamism to the international scientific community. This was evident even at the Neuchâtel workshop, where he strongly advocated the involvement of young researchers.

The volume resulting from the conference is a collection of fourteen papers, plus a general introduction written by the renowned historian of science David Aubin. The papers are divided into five sections: from the eighteenth century to the nineteenth and twentieth centuries, the relations between observatories and chronometry—and more generally between practices and materials—are analysed from the point of view of the quest for precision, the certification of timepieces, the acquisition and use of specific apparatus, and the circulation of knowledge and instruments on a global scale.

The intention to dedicate this publication to Paolo's memory is underlined by the inclusion of two final tributes, one written by Paolo Galluzzi, former director of the Galileo Museum in Florence and Paolo's main mentor, and the other by Jean-François Gauvin, professor of history and museum studies at Laval University, who knew Paolo for two decades and offers an assessment of Paolo's views on the conservation and restoration of scientific instruments.

Our thoughts and friendship are with his wife Andrea, his children Camilla and Oliver, and all the colleagues who worked with Paolo.

Neuchâtel, Switzerland — Gianenrico Bernasconi
Palermo, Italy — Ileana Chinnici
Bruxelles, Belgium — Marco Storni

References

Aubin, D. (2003). The Fading Star of the Paris Observatory in the Nineteenth Century: Astronomers' Urban Culture of Circulation and Observation. *Osiris* 18: 79–100.

Aubin, D., Bigg, C., & Sibum, H.O. (eds.). (2010). *The Heavens on Earth: Observatories and Astronomy in Nineteenth-Century Science and Culture.* Duke University Press.

Boistel, G. (ed.). (2005). *Observatoires et patrimoine astronomique français.* ENS Éditions.

Canales, J. (2001). Exit the Frog, Enter the Human: Physiology and Experimental Psychology in Nineteenth-Century Astronomy. *British Journal for the History of Science* 34(2): 173–197.

Daston, L. (2023). *Rivals: How Scientists Learned to Cooperate.* Columbia Global Reports.

Dunn, R., & Higgitt, R. (2014). *Ships, Clocks, and Stars: The Quest for Longitudes.* HarperCollins.

Gluch, S. (2024). Promises of Precision: Questioning Precision in "Precision" Instruments. *Annals of Science* 81(1–2): 1–9.

Lamy, J. (2007). *L'observatoire de Toulouse aux XVIIIe et XIXe siècles: archéologie d'un espace savant*. Presses universitaires de Rennes.

Lamy, J. (ed.) (2008). *La Carte du Ciel. Histoire et actualité d'un projet scientifique international.* EDP Sciences.

Macdonald, L. (2018). *Kew Observatory and the Evolution of Victorian Science: 1840–1910.* University of Pittsburgh Press.

McCrossen, A. (2013). *Marking Modern Times: A History of Clocks, Watches, and Other Timekeepers in American Life.* The University of Chicago Press.

Nasim, O. (2021). *The Astronomer's Chair: A Design History of Observing and Sitting*. MIT Press.

Rothauge, C. (2023). *Zeiten in Deutschland 1879–1919: Konzepte, Kodizes, Konflikte.* Brill.

Saint-Martin, A. (2011). L'astronomie française à la Belle Époque: professionnalisation d'une activité scientifique. *Sociologie du travail* 53: 253–272.

Schaffer, S. (1988). Astronomer Mark Time: Discipline and the Personal Equation. *Science in Context* 2(1): 115–145.

Schmidgen, H. (2009). *Hirn und Zeit: Die Geschichte eines Experiments, 1800–1950.* Matthes & Seitz.

Turner, A. J., Nye, J., & Betts, J. (eds.). (2022). *A General History of Horology*. Oxford University Press.

Acknowledgments

The present volume is the result of the international workshop *Astronomical Observatories and Chronometry: Time, Science and Instruments (18th–20th Century)*, held at the University of Neuchâtel and at the International Museum of Horology of La Chaux-de-Fonds from 1 to 3 December 2021. The event was supported by the Swiss National Science Foundation (SNSF) as part of the research project *L'observatoire cantonal de Neuchâtel (1858–1948): cultures de la précision, économie de la qualité et "marchandisation" de l'heure*, funded by the Swiss National Science Foundation (grant number 100012_175949), by the University of Neuchâtel, the Fondation Watch Academy, the Bureau de contrôle des ouvrages en métaux précieux, and the Musée international d'horlogerie of La Chaux-de-Fonds.

We would like to thank the Dr. Martin Othmar Winterhalter Foundation of Mendrisio for its support in funding the preparation of this volume. We are grateful to Mark F. Rogers for revising the English texts of non-native authors, and to Aude Monié for her assistance in proofreading all the chapters.

February 2025

Gianenrico Bernasconi
Ileana Chinnici
Marco Storni

Contents

Editors and Contributors

About the Editors

Gianenrico Bernasconi is associate professor and *directeur de recherche* in history of technology at University of Neuchâtel. He was the director of the project *L'observatoire cantonal de Neuchâtel (1858–1948): cultures de la précision, économie de la qualité et "marchandisation" de l'heure*. He is currently continuing his research on the observatory in the project *Atomic Clocks at the Neuchâtel Observatory: Time, Quantum Technologies and Innovation (1948–2001)*, which he co-directs with Gaetano Mileti. He recently co-edited with Marco Storni the book *Early Modern Fire: Science, Technology and Urban Space* (Brill, 2025).

Ileana Chinnici is an astronomer at the INAF Osservatorio Astronomico di Palermo, where she works mainly in the field of history of nineteenth-century astronomy. She has been curator of the historical collections of the Observatory for many years and is currently scientific advisor for these collections, with a unified approach to instruments, books, and archives. Co-author of the catalogue of the old scientific instruments of the Observatory and author of many publications in specialist journals, she was awarded the 2021 Osterbrock Book Prize. Recently involved in projects for the preventive conservation of the scientific heritage of the historical observatories, in 2024 she was elected President of the Scientific Instrument Commission for the triennium 2025–2028.

Marco Storni is a postdoctoral fellow at the Université libre de Bruxelles. After completing his Ph.D. at the École Normale Supérieure de Paris and the University of Bologna, he was a postdoctoral fellow at the Ca' Foscari University of Venice and at the University of Neuchâtel. His research focuses on the history and philosophy of science in the early modern period, especially in the eighteenth century. He is the author of *Maupertuis. Le philosophe, l'académicien, le polémiste* (Champion, 2022; winner of the Prize of the Fondation Del Duca-Institut de France 2023) and co-editor, with Gianenrico Bernasconi, of *Early Modern Fire: Science, Technology, and the Urban Space* (Brill, 2025).

Contributors

Emily Akkermans National Maritime Museum (trading as Royal Museums Greenwich), Greenwich, London, England

David Aubin Institut de mathématiques de Jussieu-Paris rive gauche, Sorbonne Université, Paris, France

Trudy E. Bell Rocky River, OH, USA

Amandine Cabrio Manuscript and Private Archives Service, Bibliothèque publique de Neuchâtel, Neuchâtel, Switzerland

Ileana Chinnici INAF—Istituto Nazionale di Astrofisica, Osservatorio Astronomico di Palermo, Palermo, Italy

Dalia Deias Sciences, Philosophie, Humanités SPH, Bordeaux Montaigne, Bordeaux, France;
Ph.D. in History of Science, Centre Alexandre Koyré, Paris, France

Bart Fried New York, USA

Paolo Galluzzi Florence, Italy

Jean-François Gauvin Université Laval, Quebec city, Canada

Louis Georges Université Paris-Nanterre, Paris, France

Julien Gressot Institut d'histoire, University of Neuchâtel, Neuchâtel, Switzerland

Romain Jeanneret Institut d'histoire, University of Neuchâtel, Neuchâtel, Switzerland;
Haute Ecole Arc Conservation-Restauration, HES-SO University of Applied Sciences and Arts Western Switzerland, Neuchâtel, Switzerland

Ion Mihailescu Institut d'histoire, Université de Neuchâtel, Neuchâtel, Switzerland

Yolanda Muñoz Rey Universidad de Cádiz, Cadiz, Spain

Günther Oestmann Technischen Universität, Berlin, Germany

S. Prashant Kumar Department of History, University of Chicago, Chicago, USA;
Archives at NCBS, Tata Institute of Fundamental Research, Bengaluru, India

Christiaan Sterken Vrije Universiteit Brussel, Brussels, Belgium

Gudrun Wolfschmidt GNT, Hamburg Observatory, University of Hamburg, Hamburg, Germany

Chapter 1
Introduction: Observatory Time

David Aubin

Abstract The introduction surveys the various ways in which time was construed in astronomical observatories from the early modern period to the twentieth century.

Observatory Time—The Lovers was the name the artist Man Ray gave to a large canvas he worked on for two years between 1932 and 1934 (Fig. 1.1).[1] A main feature of Surrealist exhibitions in Paris, London, and New York over the following years, the painting is a striking representation of a magnificent pair of bright red lips—believed to have been those of an ex-lover Ray had come to obsess about, the photographer Lee Miller. Almost filling up the visible part of a cloudy sky, Miller's lips hovered over a dark forest from which emerged the silhouette of an observatory whose two domes, Ray wrote in his autobiography, were "like breasts dimly indicated on the horizon."[2] This, he went on to say, was "an impression of [his] daily walks through the Luxembourg Gardens," from which the sight of the Paris Observatory above the horizon is still indeed conspicuous.

At the time of its unveiling, however, Ray put forward a more interesting, properly Surrealist interpretation of his work of art. Contemporary critics sometimes failed to swoon over the painting's "sad mediocrity," one seeing in it "a strange mixture of the facile poetry that emanates from old-fashioned things and the rigor that comes from modernist ones."[3] But, as André Breton argued in his preface to the London 1936 exhibition catalogue, the true success of Surrealism precisely lay in its ability at

[1] The painting was unveiled at the Galerie Cahiers d'Art in Paris in 1935 and presented at London's *International Surrealist Exhibition* in the summer of 1936 and at the New York *Fantastic Art, Dada, Surrealism* show at the end of the year. See Ray ([1935] 2015, p. 140).

[2] Ray (1963, p. 255). About the relationship between Miller and Ray, see Mileaf (2004, pp. 4–23).

[3] The first quote comes from *Les Beaux-Arts* (6 décembre 1935, p. 8); the second is taken from *L'Intransigeant* (9 décembre 1935, p. 4).

D. Aubin (✉)
Institut de mathématiques de Jussieu-Paris rive gauche, Sorbonne Université, 4 place Jussieu, 75005 Paris, France
e-mail: david.aubin@imj-prg.fr

G. Bernasconi et al. (eds.), *Astronomical Observatories and Chronometry, 18th-20th Century*, History of Physics, https://doi.org/10.1007/978-3-031-95874-8_1

Fig. 1.1 *Àl'heure de l'observatoire—Les amoureux* [observatory time—the lovers], 1968, serigraph; reproduction of an oil painting from 1932–1934

bridging the gap between mental representation and exterior perception.[4] In a short companion piece to *Observatory Time—The Lovers*, Man Ray fully subscribed to this ambition by contrasting two realities: the mundane one, represented by "neutral light and empty space"—the sort of light he called "the astronomical accomplice of time"—versus the "sole reality" that only he could perceive—his lover's lips. At seven o'clock in the morning, he noted with precision, he found himself in a dreamlike state irresistibly attracted to them.[5] Time was both absolutely defined by modern science and mentally perceived. Trying to reconcile this age-old confrontation in his painting, Ray pointed to the observatory as the place where technological modernity merged with the cold reality from which lovers hoped however fleetingly to escape.

How had the observatory come to embody the representation of mundane reality Surrealists artists wished to surpass? A striking coincidence of time suggests still another way of reading *Observatory Time—The Lovers*, which as far as I know has never been noted before. At the very moment Ray was working on his painting, observatory time was for the first time endowed with a mechanical voice. In the spring of 1932, the Paris Observatory director Ernest Esclangon announced at the Academy of Sciences his idea of having a speaking clock connected to the telephonic system. Launched on Valentine's Day, 1933, the speaking clock was an immediate success and, on that day, the busiest phone number in Paris was *Odéon 8400*, so much so that the system was overwhelmed and most callers did not even get to hear a busy signal, let alone the time.[6] With the speaking clock, observatory time had become an object of mass consumption. It is therefore tempting to see in Man Ray's giant lips the symbol of the overpowering seduction exerted by the automated voice that had taken the Sun's place as a beacon of time and made the ultraprecise product of contemporary science accessible to all.

---

[4] Breton (1936, n.p.).

[5] Ray (1935, pp. 127–129).

[6] Esclangon (1932, pp. 921–924).

> Every one the least versed in *Astronomy*, does know of what great Concernment the *Observation* of the exact Moment of Time is therein. Without this no celestial *Observation* can be performed *Accurately*, and *Astronomy* is like to receive but little advancements.[7]

As explained by Irish natural philosopher William Molyneux, time was a central concern for astronomers. The collection of essays assembled in this book certainly portrays astronomical observatories as places that were utterly preoccupied with it. Following Man Ray's cue, we may call "observatory time" the particular ways in which time was construed, measured and handled by this special culture. It is good to recall that observatories owed their origins in part to the imperial powers' desire to exert control over space (via maps) and time (via the calendar). While the incessant ballet of clocks and stars was put to use for the drawing of accurate maps of the remotest places on Earth, the lives of generations of astronomers chained to their eyepieces were ruled by special routines. Demands for ever higher standards of precision led them to dive deeper into the fabric of time and astronomers perceived that the transfer of information along human nerves was not instantaneous and could even be measured.[8] Observatories became major purveyors of accurate time to clockmakers, seamen, as well as the general public. Beyond the observatory, time was transported by hand, announced by firing cannons, signaled by falling balls, electrically wired through the telegraph, wirelessly beamed on radio waves—and even spoken over the phone by melodious voices. By the 1930s, the observatory had come to stand as a symbol of the mechanical time that, according to a famous thesis put forward by E.P. Thompson, now seemed to be governing most aspects of life in modern industrial societies.[9]

Observatory time was the set of techniques designed by a certain scientific culture in order to tackle timeless problems.[10] "What is time?" the early Christian philosopher Augustine famously asked in his *Confessions*, only to share his perplexity: "If nobody asks me, I know: but if I were desirous to explain it to one that should ask me, plainly I know not." Like Aristotle and Lucretius before him, Augustine emphasized the subjective nature of a concept that, in their views, was intimately linked with their different understandings of the human soul.[11] In the early modern period, Isaac Newton forcefully voiced an opposite, purely objective conception of time. "Absolute, true and mathematical time in and of its nature, without reference to anything external," he wrote in the *Principia*, "flows uniformly."[12]

---

[7] Molyneux (1686, p. 1).

[8] Canales (2009). On the history of the personal equation, see also Schaffer (1988, pp. 115–145); Schmidgen (2003, pp. 237–275); Hoffmann (2007, pp. 333–365).

[9] Thompson (1967, pp. 56–97). See also Mumford ([1934] 2010); Landes (1983); Dorn-van Rossum (1996); Martineau (2015).

[10] On "observatory techniques," see Aubin et al. (2010).

[11] Augustine (1912, p. 239). On Augustine, see Doody et al. (2021). On Aristotle, see Coope (2016, pp. 125–151).

[12] Newton (1999, p. 408). See also Čapek (1987, pp. 595–608); Edwards (2013); Gorham and Slowik (2014, pp. 119–137); Morfouli (2022, pp. 36–66).

As this book shows, scientists working in astronomical observatories were not concerned with philosophical quandaries regarding the nature of time, so much as they were with the mundane practice of time measurement. A wide range of ancient cultures of chronometry relied on celestial motions. One of the oldest Indian texts on astronomy clearly stated that this was "the science of time."[13] In Egypt, we find sundials dating from as early as around 1500 BCE.[14] But, as Augustine went on to say, time is *not* the motion of a body. If the Sun were to move twice as fast, days, that is, the amount of time separating two consecutive sunsets, would be shorter, but the flow of time would remain the same. In the early seventeenth century, the astronomer Pierre Gassendi concurred with Augustine, noting that the use of the Sun as a timekeeper was no more than mere convenience:

> Since there is no motion more general, more regular and more known than that of the sun, so we accept its own motion as a sort of general clock for measuring the flow of time. Not that time would flow faster or slower if the sun would move faster or slower; whatever motion the sun may have, we took it as a measure of time; if this motion were twice as fast, time would not be faster because of it.[15]

As a timekeeper, the Sun was not only conventional; it also was inaccurate. The distinction between the flow of time and the Sun's motion was reinforced by the discovery of the so-called "equation of time," already identified in the second century CE by Ptolemy who noticed that days were not of equal duration throughout the year.[16] Early modern astronomers therefore defined an abstract notion of "mean time" that differed from "true time" observed by using the Sun. Since days were unequal, Antoine de Villon for example explained in 1524, it was "necessary for Astrologers [*i.e.* astronomers] to use regular and equal time."[17] Until the establishment of modern observatories, however, there was no consensus on the exact values of these discrepancies. "Although it be apparent and well known to all," the English astronomer Vincent Wing remarked in 1656, "that the days natural are unequal, yet in finding the true quantity and proportion of the inæquality, there hath been ingendred great variety of opinions amongst the best learned and most cunning Astronomers in all ages."[18] In the 1670s, the Astronomer Royal John Flamsteed settled the matter with the help of some of the most accurate pendulum clocks of the day made by Thomas Tompion.[19]

Like the Royal Observatory set up by Flamsteed in Greenwich in 1675, the *Observatoire royal* in Paris was the place where Ancient astral chronometry met with the pendulum clock. By adopting this name for the building arising from the ground on

---

[13] In published translations, astronomy is variously called the "knowledge of time" (Shamasastry (1936)) or "the lore of time" (Kuppanna Sastry (1985)). I thank Aditya Kolachana for pointing out this quotation to me and suggesting the "science of time" as an equally valid translation.

[14] Bickel and Gautschy (2014, pp. 3–14); Vodolazhskaya (2014, pp. 1–18).

[15] Gassendi (1658, p. 224), cited in Čapek (1987, p. 603).

[16] Jones (2022, pp. 12–35).

[17] De Villon (1624, p. 67).

[18] Wing (1656, p. 48).

[19] Cook (2001, pp. 9–27).

the outskirts of Paris starting in 1667, the group of scientists organized as the Royal Academy of Sciences signaled their wish to invent a special place for the pursuit of knowledge, the like of which was difficult to find before.[20] In Uraniborg, which was explicitly claimed as a model by the Observatory, the clocks had for example "never played an important part."[21] As Dalia Deias underscores here, the situation was completely different at the Observatoire. With Giovanni Domenico Cassini's arrival, four years later, and his stay as the figurehead of the Paris Observatory for more than four decades, the modern concept of the observatory as a building specially dedicated to the precise observation of the sky with the help of telescopes and clocks was fixed.

The pendulum clock was particularly well suited to embody observatory time: "All Watches that go true, measure the *equal* or *mean* time, consequently ought to differ from the apparent time shewn by a *Sun-dial*."[22] Deias underscores here that the bas-relief carved on the Southern façade of the Paris Observatory depicted, among other things, a pendulum clock paid for by the king, a clear symbol of the importance that was attributed to the device. She might also have emphasized that the sculpture now stood in the place where an early embodiment of the Observatory had prominently displayed sundials.[23] As a symbol of the Sun king's earthly powers, the ornamentation of the Paris Observatory was obviously chosen with great care. Observatories marked time and they did it with mechanical clocks. Pendulums and, later, chronometers and chronographs henceforth became a significant part of any observatory's equipment. Whether they wished to improve the astronomical clocks in use or to test instruments for the Navy, observatories from Palermo to Cádiz, as recalled here by Ileana Chinnici and Yolanda Muñoz Rey, were replete with them.

Yet, one should be wary of too quickly equating this modern conception of linear time, for which the spread of the mechanical clock usually takes the blame, with observatory time. At the Paris Observatory, not all time measurements were made using clocks and, as evoked here by Louis Georges, Cassini had a meridian line built to follow the sun's motion from within a building that was filled with clocks.[24] Even in the modern period, clock time coexisted with other types of time experience that felt no less modern.[25] In terms of discourse and material practice, changes brought about by the diffusion of clock technologies started to be felt much before the advent of the modern observatory and impacted society, at least at first, quite independently from it.[26] In addition, the idea that time measuring devices before the invention of

---

[20] Deias (2020). See also Deias (2017, pp. 101–130); Deias (2019, pp. 138–173); and Deias (2022, pp. 74–97). On clocks in the Paris Observatory and other observatories modeled on it, see Deias's contribution to this volume.

[21] Dreyer (1890, p. 324).

[22] Molyneux (1686, p. 43).

[23] "Elevation perspective de l'observatoire." Bibliothèque nationale de France, département Estampes et photographie, RESERVE HA-18 (9)-FOL; https://gallica.bnf.fr/ark:/12148/btv1b5250 53525 (accessed on 11 December 2024).

[24] Heilbron (1999).

[25] Wersan (2017, pp. 282–310).

[26] Scattergood (2003, pp. 453–474); Bernasconi and Thürigen (2020).

the mechanical clock only had a marginal impact on science and society is now disputed.[27]

By adopting the mean time indicated by their pendulum clocks, observatories thus relied on time reckoning that differed from public time. Paradoxically perhaps, time in civil society remained aligned with true solar time, while observatory time was the result of conventions. Watchmakers as a consequence needed to print pamphlets explaining the difference between clock and solar time, although the actual difference was most often small. In *An explanation of the Nature of the Equation of Time*, one of them underscored that such knowledge was necessary "to distinguish betwixt Good and Bad Workmanship […] or at any time to know the True Moment of the Day or Night by the best Watch or Clock in the World."[28] Makers developed pendulum clocks in which a curve cut according to solar inequalities allowed true time to be shown on the dial, only to see them shunned by both civil consumers and astronomers. While the former were deterred by the costs of such delicate devices, the latter feared that added gears would introduce further irregularities in the running of the clock, thus undermining their precision.[29]

Indeed, observatory time ought above all to be precise. At the Paris Observatory, Charles Perrault remarked that one found "pendulum clocks of unconceivable precision [*justesse*]."[30] In the Quarrel of the Ancients and the Moderns, observatory clocks were held as beacons of modernity of such precision that daily inequalities in the motion of the Sun could be corrected by them. In the French equivalent to the British *Nautical Almanac*, Philippe de La Hire concurred that the cycloid pendulum clocks invented by Christiaan Huygens were so perfect that "one could not imagine anything more precise [*juste*]." They were maintained at the Observatory with "an exact regularity of one second by the heavenly bodies, whose motion is now reciprocally determined by the help of clocks with more precision than ever before."[31]

Like time, precision is a complex notion whose modern definition is in significant part due to the observatory culture.[32] Precision lay in the care with which instruments were made by artisans and intimately known and cared for by their users.[33] In the observatory, as recalled by Bart Fried, the quest for more accurate timekeeping also involved the development of high-precision optical instruments. Sometimes, precision was a personal quality characterizing some exceptional astronomer's skill. In this volume, Emily Akkermans shows how much faith was invested in officers who were, as she put it, "highly skilled in astronomical observations and often trained by

[27] Jones (2016); J. Miller (2023).

[28] Clay (1731, p. iii).

[29] Lepaute (Paris, 1760).

[30] Perrault (1688–1697, t. 4, p. 101).

[31] *Connaissance des temps* (1681, pp. 73–74).

[32] Aubin (2024, pp. 139–159). See also Gluch (2024a, 2024b, pp. 1–9) and the various articles assembled in the same special issue.

[33] Gluch (2024a, 2024b, pp. 329–365). On the history of precision more generally, see Wise (1995).

members of the Royal Society to follow instruction and record instrumental practices." More and more, precision became this very ability to follow labor-intensive procedures, involving many participants as well as delicate instruments, which Julien Gressot aptly calls "operative chains."[34]

Observatory time was precise in two distinct ways. Adopted in 1994, the ISO 5725 norm specifies the meanings of *trueness* as the closeness of agreement between the arithmetic mean of a large number of test results and the true or accepted reference value and *precision* specifies the closeness of agreement between test results. Taken together, both terms describe the *accuracy* of a measurement method.[35] Using this vocabulary, we can say that observatory time therefore was *precise* due to its great regularity and *true* in its ability to be in exact correspondence with mean astronomical time.

The precision of observatory time was essential to navigation. In her chapter about chronometric expeditions and telegraphic longitude determination in the continental USA, Trudy E. Bell reminds us that, as early as 1530, Dutch mathematician and cartographer Gemma Frisius explained that longitude could be determined by transporting a "very exquisite" timekeeper to distant locations.[36] Huygens was driven to develop the pendulum clock in the first place, due to this need for clocks capable of reliably keeping time over long periods, that is, clocks that were *precise* in the sense just given, or extremely regular. The controversial way in which the British Board of Longitude awarded a prize to the clockmaker John Harrison for developing his timekeepers is well known by now.[37] Observatory scientists played a key role in the testing and validation of proposed methods for finding longitude. It was quite natural therefore for observatories to became places of reference to test the regular running of chronometers used by the navies around the world. In his contribution, Günther Oestmann shows the importance of the task for naval observatories in imperial Germany, underscoring that, despite some resistance from the merchant navy, the *Deutsche Seewarte* in Hamburg checked more than a thousand timepieces for various customers over a period of fifteen years. The level of precision achieved in an active professional astronomical observatory setting is nicely illustrated here by Christiaan Sterken's study of Munich observatory scientist Felix Schmeidler who made it his lifelong pursuit to achieve unequalled clock precision.

Observatory time was extremely regular and, more and more, as discussed here by Ion Mihailescu, the observatory scientists' own time was spent in minute, routine, and repetitive tasks.[38] Astronomers' taxing time schedules had a profound impact on the

[34] See also Gressot and Jeanneret (2022, pp. 27–48).

[35] "Accuracy (trueness and precision) of measurement methods and results—Part 1: General principles and definitions", https://www.iso.org/obp/ui/#iso:std:iso:5725:-1:ed-1:v1:en (accessed on 11 December 2024). For a further discussion of this, see Aubin (2024).

[36] Quoted from the French translation: Frisius (1557, p. 52). See also Pogo (1530) and his treatise on triangulation (1533, pp. 469–506); Hallyn (2008, p. 29).

[37] Andrewes (1996); Sobel, (1995); Dunn and Higgitt (2014).

[38] On work regimes in observatories, see also Saint-Martin (2011, pp. 253–272); Donnelly (2014, pp. 479–503).

interior design of observatories and even their furniture.[39] This official, bureaucratic regime also entailed legal components as observatories were increasingly drawn into legislative frameworks to enforce chronological regimes put in place by modern states in the service of industry. In her chapter, Amandine Cabrio for example underscores how starting in 1858, in Switzerland, a "technical and regulatory framework" was set up around the Neuchâtel observatory in order to certify chronometers.

With accurate timekeepers, such as the replica of Harrison's made by Larcum Kendall, which was taken by Captain James Cook on his voyages across the globe, it was sometimes easier to know Greenwich time on faraway beaches in the Pacific than in the heart of London. For many, this state of affairs was unacceptable. The clockmaker Ferdinand Berthoud may have been among the first to suggest, as early as 1754, that public clocks should follow mean time.[40] But such recommendations were only slowly implemented. In 1823, the Viennese astronomer Joseph Johann Littrow thought that "the only place in which the time of the day can be known with certainty at every instant was the observatory."[41] Thus, the second aspect of observatory time's accuracy, not so much its precise regularity, but its trueness with respect to agreed-upon standards, was emphasized. After Geneva, London, and Amsterdam, Paris required its public clocks to drop true time and adopt mean time on 24 December, 1826.[42] Only observatory time was now legal, public, and official.

The official adoption of mean time, although well suited to the regularity of precise chronometers, created new problems for clockmakers and users who could no longer rely on the Sun to set their timepieces to the correct time. Previously, they would simply have used various devices such as meridians, sundials, or even automatic gun firing fuses set up by sunlight. But, no longer. The equation of time had been used to transform the time observed on clocks into the time featured in astronomical tables. Now, its function was reversed: the equation of time was needed in order to transform true solar time into the mean artificial time to which clocks were set. Henceforth, true solar time was increasingly called "apparent time," underscoring the need to redefine what trueness meant for clocks. A true clock from then on showed the mean time that observatories computed.

In most places, however, the use of the equation of time (or its graphic representation on sundials like the analemma) proved too cumbersome for most users. In ports, as examined here by Gudrun Wolfschmidt, time-balls provided an early technical solution to the problem. In her account, Wolfschmidt finds that more than 250 examples of this technology were deployed on all continents. More often than not, however, time-balls told Greenwich mean time rather than local time. In cities where local mean time was usually preferred, time had to be conveyed from the observatory to the workshop using a pocket watch (Fig. 1.2). With the spread of telegraphic

[39] Nasim (2021).

[40] "Lettre à M. L'abbé Raynal, sur quelques parties de l'horlogerie, et sur les moyens de profiter de toute la justesse de ces machines" (Novembre 1754, pp. 142–153). See Dequidt (2016, pp. 20–28); Bianchi (2021, pp. 337–344).

[41] Littrow (27 February 1823, pp. 193–201); English translation in Carey (1825, p. 29).

[42] Hartmann (1827, p. 5).

Fig. 1.2 Claverie, "Paris clockmakers taking time at the Paris observatory's regulator," *Journal illustré* (1877)

technologies, electrical systems were developed for the distribution of time to major clocks.[43] It is fitting to recall here that the importance of scientific instrument makers in the implementation of such projects was of course not missed by the late lamented Paolo Brenni who noted that the observatory's master clock was built by Breguet.[44] Innovations however did not bring about an early end to the practice of conveying time with pocket watches.[45]

"The time has arrived," the English railway engineer Henry Booth insisted in 1847, to uniformize time across the country.[46] A year earlier, to make scheduling easier, Booth from the Liverpool and Manchester Railway had decided that the clocks in both stations would be set to Greenwich mean time. The U.S. story of railway time and its relation to the development of the time zones has received great attention by historians.[47]

---

[43] For Paris, see the map of the project from 1878 repr. In Aubin (2003, p. 98). See also Galison (2003); Souchier, (2019, pp. 219–237). The British is well-documented, see Gay (2003, pp. 107–140); Rooney and James (2009, pp. 5–30); Ishibashi (2020, pp. 25–46). On Berlin, see Sauter (2007, pp. 685–709).

[44] Brenni (1996, pp. 19–24).

[45] Rooney (2008).

[46] Booth (1847). See Schivelbush (1987, pp. 55–64).

[47] See, especially, Bartky (2000). See also, Dick (1983, p. 190).

Observatory time became global in October 1884, when an international conference was held at Washington, D.C., for the purpose of fixing a prime meridian and a universal day.[48] In 1912, the diffusion of radio technology increased the precision with which time could be broadcast over long distances. This led astronomers to convene another international conference in Paris.[49] Set up in 1913 to establish "definitive time," the International Time Bureau was hosted by the Paris Observatory until 1987, when it was transferred to the International Bureau of Weights and Measurements. By going global, observatory time of course clashed with other conceptions of time. The case of Ragoonathachary discussed in this volume by S. Prashant Kumar is especially enlightening. While his own understanding of time was pushed aside, Ragoonathachary himself became "a time machine," that is, both "a well-regulated chronometric assistant and also the means by which modernity would be conveyed to the colony."

"Time is man-made," claimed the amateur astronomer Newton Mayall in a radio broadcast aired from Harvard University on May 31, 1941.[50] And, by and large, Mayal might have added, it was made by observatories across the globe. By focusing on the history of observatory time, this book shows the extent to which the reverse has also been true. A certain conception of time and the responsibilities that came with it helped to build the modern observatory in a myriad of interesting ways. However with the development of atomic clocks in the second half of the twentieth century, the official definition of time was taken away from observatories.[51] But the history of the most recent avatars of observatory time remains to be written.

References

Andrewes, W. J. (Ed.). (1996). *The Quest for Longitude*. Harvard University Press.

Arias, E. F., Panfilo, G., & Petit, G. (2018). The 50th anniversary of the atomic second. *IEEE Transactions on Ultrasonics, Ferroelectrics, and Frequency Control, 65*(5), 898–903.

Aubin, D. (2003). The fading star of the Paris observatory: astronomers' urban culture of circulation and observation. *Osiris, 18*, 79–100.

Aubin, D. (2024). Popularizing precision: cultures of exactness at the Paris observatory, 1667–1742. *Annals of Science, 81*, 139–159.

Aubin, D., Bigg, C., & Sibum, H. O. (2010). *The Heavens on Earth: Observatories and Astronomy in Nineteenth-Century Science and Culture*. Duke University Press.

Augustine. (1912). *Confessions* (W. Watts, Trans.). Loeb Classical Library. Harvard University Press.

Bartky, I. (2000). *Selling the True Time: Nineteenth-Century Timekeeping in America*. Stanford University Press.

[48] Howse (1989, pp. 138–151); Zerubavel, (1982, pp. 1–23); Palmer (2002, pp. 7–24); Withers (2017); Oggle (2015).

[49] Bureau des longitudes, Conférence international de l'heure, Paris, octobre 1912 (1912); Guinot (2000, pp. 175–184).

[50] Newton Mayall (1941, pp. 524–530).

[51] Leschiutta (2005, S10–S19); Guinot and Arias (2005, S20–S30); Arias et al., (2018, pp. 898–903).

Bernasconi, G., & Thürigen, S. (Eds.). (2020). *Material Histories of Time: Objects and Practices, 14th–19th Centuries*. De Gruyter.
Bianchi, S. (2021). Where was mean solar time first adopted? *Journal of Astronomical History and Heritage, 24*(2), 337–344.
Bibliothèque nationale de France. *Elevation perspective de l'observatoire*. Département Estampes et photographie.
Bickel, S., & Gautschy, R. (2014). Eine ramessidische Sonnenuhr im Tal der Könige. *Zeitschrift Für Ägyptische Sprache und Altertumskunde, 141*(1), 3–14.
Booth, H. (1847). *Uniformity of Time, Considered Especially in Reference to Railway Transit and the Operations of the Electric Telegraph, in a Letter to the Right Hon. Edward Strutt, M.P.* John Weale; Thomas Baine.
Brenni, P. (1996). 19th century French scientific instrument makers. XII: Louis Clement Breguet and Antoine Louis Breguet. *Bulletin of the Scientific Instrument Society, 50*, 19–24.
Breton, A. (1936). Preface (Transated by D. Gascoyne). In *The International Surrealist Exhibition: Catalogue*. New Burlington Galleries.
Bureau des longitudes. (1912). *Conférence internationale de l'heure (Paris, octobre 1912)*. Gauthier-Villars.
Canales, J. (2009). *A Tenth of a Second: A History*. The University of Chicago Press.
Čapek, M. (1987). The conflict between the absolutist and the relational theory of time before Newton. *Journal for the History of Ideas, 48*, 595–608.
Carey, G. G. (1825). *Astronomy as It Is Known in the Present Day*. William Cole.
Clay, F. (1731). *An Explanation of the Nature of the Equation of Time for Adjusting Watches and Clocks to the Motion of the Sun, also the Description of a Time-Keeper for Astronomical and Other Uses* (Printed for F. Clay).
Connaissance des temps, ou calendrier et éphémérides du lever et du coucher du Soleil, de la Lune, et des autres Planètes, avec les éclipses pour l'an MDCLXXXI.(1681). Jean Baptiste Coignard, pp. 73–74.
Cook, A. (2001). Time and the royal society. *Notes and Records of the Royal Society of London, 55*, 9–27.
Coope, U. (2005). *Time for Aristotle*. Oxford University Press.
de Villon, A. (1624). *L'Usage des éphémérides avec la méthode de dresser et corriger toute sorte de figues cœlestes, et juger par le moyen d'icelles des diuerses constitutions des temps et saisons de l'années, et de toutes les autres choses qui en dépendent, comme sont guerres, pestes, famines, mortalitez et autres*. Jean Moreau.
Deias, D. (2017). Un observatoire parmi les lieux d'observation du ciel: Le projet de l'Observatoire royal de Louis XIV au prisme de son nom. *Cahiers D'histoire. Revue D'histoire Critique, 136*, 101–130.
Deias, D. (2019). The Paris Observatory in the Early Modern Ecosystem of Knowledge (1667–1712). In M. Feingold & G. Giannini (Eds.), *The Institutionalization of Science in Early Modern Europe* (pp. 138–173). Brill.
Deias, D. (2022). A vanished curiosity from the observatoire of Louis XIV: The Parterre géographique, a tool both of precision and of representation. *Archives Internationales D'histoire des Sciences, 72*, 74–97.
Deias, D. (2020). *Inventer l'Observatoire: sciences et politiques sous Giovanni Domenico Cassini (1625–1712)*. Ph.D. Dissertation, EHESS.
Dequidt, M.-A. (2016). Les horlogers parisiens et le temps: Afficher l'heure officielle ou mesurer les durées? *Romantisme, 174*, 20–28.
Dick, S. J. (1983). *Sky and Ocean Joined: The U.S. Naval Observatory, 1830–2000*. Cambridge University Press.
Donnelly, K. (2014). On the boredom of science: positional astronomy in the nineteenth century. *British Journal for the History of Science, 47*, 479–503.
Doody, J., Hannan, S., & Paffenroth, K. (Eds.). (2021). *Augustine and time*. Lexington Books.

Dorn-van Rossum, G. (1996). *History of the Hour: Clocks and Modern Temporal Orders.* The University of Chicago Press.
Dreyer, J. L. E. (1890). *Tycho Brahe: A Picture of Scientific Life and Work in the Sixteenth Century.* Adam and Charles Black.
Dunn, R., & Higgitt, R. (2014). *Finding Longitude: How Clocks and Stars Helped Solve the Longitude Problem.* Collins.
Edwards, M. (2013). *Time and the Science of the Soul in Early Modern Philosophy.* Brill.
Esclangon, E. (1932). Sur les horloges parlantes destinées à la distribution téléphonique de l'heure. *Comptes-Rendus De L'académie des Sciences, 184*, 921–924.
Frisius, G. (1557). *Les Principes d'astronomie et cosmographie, avec l'usage du globe* (Claude de Boissière, Trans). Guillaume Cavellat.
Galison, P. (2003). *Einstein's Clocks, Poincaré's Maps: Empires of Time.* Norton.
Gassendi, P. (1658). *Syntagma philosophicum*, vol. 1. Anisson.
Gay, H. (2003). Clock synchrony, time distribution and electrical timekeeping in Britain 1880–1925. *Past & Present, 181*, 107–140.
Gluch, S. (2024a). Promises of precision: questioning precision in "precision" instruments. *Annals of Science, 81*, 1–9.
Gluch, S. (2024b). Timing the stars: clocks and complexities of precision in eighteenth-century observatories. *History of Science, 62*, 329–365.
Gorham, G., & Slowik, E. (2014). Locke and Newton on Space and Time and Their Sensible Measures. In Z. Biener & E. Schliesser (Eds.), *Newton and Empiricism* (pp. 119–137). Oxford University Press.
Gressot, J., & Jeanneret, R. (2022). Determining the right time, or the establishment of a culture of astronomical precision at neuchâtel observatory in the mid-19th century. *Journal for the History of Astronomy, 53*, 27–48.
Guinot, B. (2000). History of the Bureau international de l'Heure. *International Astronomical Union Colloquium, 178*, 175–184.
Guinot, B., & Arias, E. F. (2005). Atomic time-keeping from 1955 to the present. *Metrologia, 42*(3), S20–S30.
Hallyn, F. (2008). *Gemma Frisius, arpenteur de la terre et du ciel.* Champion.
Hartmann. (1827). *Le Temps vrai et le temps moyen, ou résumé de la cause de l'irrégularité apparente des montres et pendules réglées sur le temps vrai, suivi d'une table d'équation, indiquant jour par jour l'instant du passage du Soleil au méridien, et la variation qu'il faut faire éprouver aux montres et pendules pour les régler sur le Temps moyen à Midi vrai.* Chez l'auteur.
Heilbron, J. L. (1999). *The Sun in the Church: Cathedrals as Solar Observatories.* Harvard University Press.
Hoffmann, C. (2007). Constant differences: Friedrich Wilhelm Bessel, the concept of the observer in early nineteenth-century practical astronomy and the history of the personal equation. *British Journal for the History of Science, 40*, 333–365.
Howse, D. (1989). *Greenwich Time and the Discovery of Longitude.* Oxford University Press.
Ishibashi, Y. (2020). Constructing the 'automatic' greenwich time system: George Biddell Airy and the telegraphic distribution of time, c. 1852–1880. *British Journal for the History of Science, 53*, 25–46.
Jones, A. (2022). Precision of time observation in Greco-Roman astrology and astronomy. *Almagest, 13*, 12–35.
Jones, A. (Ed.). (2016). *Time and Cosmos in Greco-Roman Antiquity.* Princeton University Press.
Kuppanna Sastry, T. S. (Ed.). (1985). *Vedānga Jyotisa of Lagadha.* International Science Academy.
L'Intransigeant. (9 December 1935).
Landes, D. S. (1983). *Revolution in Time: Clocks and the Making of the Modern World.* Belknap Press of the Harvard University Press.
Lepaute, J. -A. (1760). *Description d'une pendule à secondes, qui marque le temps moyen et le temps vrai, sans être exposée aux inconvéniens qu'on a remarqué jusqu'à présent dans les Pendules d'équation.* J. Chardon.

Les Beaux-Arts. (6 December 1935).
Leschiutta, S. (2005). The definition of the 'atomic' second. *Metrologia, 42*(3), S10–S19.
Lettre, à M. (November 1754). L'abbé Raynal, sur quelques parties de l'horlogerie, et sur les moyens de profiter de toute la justesse de ces machines. *Mercure de France*, pp. 142–153.
Littrow, J. J. (1823). Über die Regierung der öffentlichen Urhen. *Wiener Zeitschrift Für Kunst, Literatur, Theater und Mode, 25*, 193–201.
Martineau, J. (2015). *Time, Capitalism and Alienation: A Socio-Historical Inquiry into the Making of Modern Time*. Brill.
Mayall, R. N. (1941). Time and timekeeping. *Popular Astronomy, 49*, 524–530.
Mileaf, J. (2004). Between you and me: man Ray's object to be destroyed. *Art Journal, 63*, 4–23.
Miller, K. J. (2023). *Time and Ancient Medicine: How Sundials and Water Clocks Changed Medical Science*. Oxford University Press.
Molyneux, W. (1686). *Sciothericum Telescopium; Or, A New Contrivence of Adapting a Telescope to an Horizontal Dial for Observing the Moment of Time by Day or Night, Useful in Astronomical Observations, and for Regulating and Adjusting Curious Pendulum-Watches and Other Time-Keepers with Proper Tables Requisite Thereto*. Andrew Crook and Samuel Helsham.
Morfouli, M. (2022). La précision dans la mesure du temps et la Nouvelle Philosophie. Étude de cas (Alexandre Koyré et Richard Westfall). *Almagest, 13*, 36–66.
Mumford, L. ([1934] 2010). *Technics and Civilization*. Chicago University Press.
Nasim, O. (2021). *The Astronomer's Chair: A Visual and Cultural History*. MIT Press.
Newton, I. (1999). *The Principia: Mathematical Principles of Natural Philosophy* (I.B. Cohen, A. Whitman, & J. Budenz, Trans.). University of California Press.
Oggle, V. (2015). *The Global Transformation of Time, 1870–1950*. Harvard University Press.
Palmer, A. W. (2002). Negotiation and resistance in global networks: the 1884 international meridian conference. *Mass Communication and Society, 5*(1), 7–24.
Perrault, C. (1688–1697). *Parallèle des Anciens et des Modernes*, vol. 4. J.-B. Coignard.
Pogo, A. (1933). Gemma Frisius, his method of determining differences of longitude by transporting timepieces (1530) and his treatise on triangulation (1533). *Isis, 22*(193), 469–506.
Ray, M. (1963). *Self Portrait*. Little, Brown.
Ray, M. ([1935] 2015). At Observatory Time… The Lovers. In J. Mundy (Ed.), *Writings on Art* (pp. 128–129). The Getty Research Institute.
Rooney, D., & Nye, J. (2009). "Greenwich observatory time for the public benefit": standard time and victorian networks of regulation. *British Journal for the History of Science, 42*, 5–30.
Rooney, D. (2008). *Ruth Belville: The Greenwich Time Lady*. National Maritime Museum.
Saint-Martin, A. (2011). L'astronomie française à la Belle Époque. Professionalization d'une activité scientifique. *Sociologie Du Travail, 53*, 253–272.
Sauter, M. J. (2007). Clockwatchers and stargazers: time discipline in early modern Berlin. *The American Historical Review, 112*, 685–709.
Scattergood, J. (2003). Writing the clock: the reconstruction of time in the late middle ages. *European Review, 11*(4), 453–474.
Schaffer, S. (1988). Astronomers mark time: discipline and the personal equation. *Science in Context, 2*, 115–145.
Schivelbusch, W. (1987). *The Railway Journey: The Industrialization and Perception of Time and Space in the Nineteenth Century*. University of California Press.
Schmidgen, H. (2003). Time and noise: the stable surroundings of reaction experiments, 1860–1890. *Studies in History and Philosophy of Science Part C: Studies in History and Philosophy of Biological and Biomedical Sciences, 34*, 237–275.
Shamasastry, R. (Ed.). (1936). *Vedangajyautisha*. Government Branch Press.
Sobel, D. (1995). *Longitude: The True Story of a Lone Genius Who Solved the Greatest Scientific Problem of His Time*. Walker.
Souchier, C. (2019). Sciences et techniques au service du public. La synchronisation des horloges parisiennes au XIXe siècle. *Politix, 125*, 219–237.
Thompson, E. P. (1967). Time, work-discipline, and industrial capitalism. *Past & Present, 38*, 56–97.

Vodolazhskaya, L. N. (2014). Reconstruction of ancient Egyptian sundials. *Archaeoastronomy and Ancient Technologies, 2*(2), 1–18.
Wersan, K. (2017). The early Melon and the mechanical gardener: toward an environmental history of timekeeping in the long eighteenth century. *Environmental History, 22*(2), 282–310.
Wing, V. (1656). *Astronomia Instaurata: Or A New and Compendious Restauration of Astronomie in Four Parts.* R. and W. Leybourn.
Wise, M. N. (Ed.). (1995). *The Values of Precision.* Princeton University Press.
Withers, C. W. J. (2017). *Zero Degrees: Geographies of the Prime Meridian.* Harvard University Press.
Zerubavel, E. (1982). The standardization of time: a sociohistorical perspective. *American Journal of Sociology, 88*(1), 1–23.
Zinn, P. (2016). Lucretius on time and its perception. *Kriterion, 30*, 125–151.

David Aubin is Professor of the History of Science at Sorbonne Université and a member of the Institut de Mathématiques de Jussieu-Paris Rive Gauche, Paris. He is the author of *L'Élite sous la mitraille. Les normaliens, les mathématiques et la Grande Guerre, 1900–1925* (Rue d'Ulm, 2018) and *Femmes, vulgarisation et pratique des sciences au siècle des Lumières: Les Dialogues sur l'astronomie et la Lettre sur la figure de la Terre de César-François Cassini de Thury* (Brepols, 2020).

Part I
Observatories, Time Measurement, and the Public Space in the Eighteenth Century

Chapter 2
Synchronising Paris in the Eighteenth Century? Henry Sully's Meridian Line at Saint-Sulpice: Astronomy, Clockmaking, and Timekeeping

Louis Georges

> *"Houses there have no other clock than the course of the Sun; these people are three centuries behind in terms of skills and customs."*
> L-S. Mercier, *Tableau de Paris,* 1782, pp. 269–270.

Abstract This study reevaluates the role of the 1728 meridian line in Paris' Saint-Sulpice church, as both a technical instrument and a discursive project. Designed by the English clockmaker Henry Sully, purportedly to synchronise Parisian public clocks via precise solar noon, the project aimed at repurposing the horizontal church *meridiana,* used in the 16th and 17th centuries for astronomical observation, into a civic timekeeping tool. Unfinished at Sully's death, the instrument is primarily documented through a bold and programmatic 1728 text, which advocated for astronomy's role in everyday horology. Through archival analysis and material evidence, this article contextualises Sully's social identity (an outsider seeking legitimacy beyond guild structures) and argues that, rather than a realistic and functional device, the Saint-Sulpice line embodied a « technology »: a performative discourse promoting a new conception of the clockmaker-astronomer as civic synchronizer. By highlighting Sully's editorial strategies and the meridian's hybrid design, it complicates John Heilbron's model of meridian lines transitioning belatedly to timekeeping uses. Instead, it reveals the coexistence and negotiation of astronomical precision, mechanical time, and social authority in Enlightenment Paris, demonstrating how gnomonics contributed to emerging ideals of urban temporal uniformity.

L. Georges (✉)
Université Paris-Nanterre, Paris, France
e-mail: louis.h.georges@gmail.com

G. Bernasconi et al. (eds.), *Astronomical Observatories and Chronometry, 18th-20th Century*, History of Physics, https://doi.org/10.1007/978-3-031-95874-8_2

Describing the Faubourg Saint-Marcel as "the Parisian neighbourhood where the poorest, most restless, and unruly populace lives," Louis-Sébastien Mercier associates timekeeping with social status: keeping time on a sundial—or without instrumentation altogether, as suggested by his turn of phrase—is here a form of social discredit. Lowering gnomonics to a dishonourable anteriority, he bears witness to the social stratification of timekeeping, but also to the influence of a technical rhetoric constructed by clockmakers from the beginning of the eighteenth century onwards. Through extensive editorial activity,[1] clockmakers built their historicization of technology on a break between the modern, reliable mechanical orientation, and another, "pre-mechanical," defined in opposition to the first.[2] Such a framework proved influential: although more nuanced, the prolific historiography of clockmaking has sometimes perpetuated the relegation of gnomonic orientation, associating technical modernity with mechanisation.[3] Sundials have often been studied as relative and dependent instruments: of fragile use due to meteorological and nocturnal discontinuities, and imprecise compared to the intra-hour accuracy of mechanical clocks.[4] As a result, gnomonic and mechanical timekeeping were disjointed on the opposite sides of a chronological, social, and economic break: the latter constituting "the key machine of the industrial era,"[5] the former merely a delay on the path towards modernity, here symbolised by the "slow men"[6] of Saint-Marcel.

A specific type of sundial has, nevertheless, been widely studied through a different historiography than that of timekeeping. The *meridiana*, or meridian line, is at the centre of European early-modern astronomy from the 1650s to the 1750s. These dials, limited to the exact indication of solar noon, were first used as a precision device for celestial and calendarical calculation, and proved valuable for cartography. John L. Heilbron famously studied the emergence of meridian lines in Italian churches between 1650 and 1750.[7] He emphasised a triple evolution of the "oracle of Apollo": architectural context (from churches to royal observatories), patronage (from the Catholic Church to rival European monarchies), and function (from calendarical computus to geographical determination). However, *meridiane* were not planned as timekeeping instruments: they were not introduced as clocks, nor as a complements to clocks. In fact, timing would be, according to Heilbron, a secondary function in

---

[1] About the link between historicization and economic practices in early modern horology, see Graf (2020).

[2] The historical framework created by clockmakers in the eighteenth century elevates Huygens and Coster's pendulum clock of 1656 and 1657 as the sudden eruption of modern horology. See Sully and Le Roy (1737); Lepaute (1755); and above all Berthoud (1802).

[3] Drawing upon this technical chronology, classical economic studies of timekeeping by Cipolla, Landes, or Aveni, tended to consider mechanical clocks as the exclusive component of the "horological revolution." See Cipolla (1967); Landes (1983); Aveni (1989).

[4] Schechner (2001).

[5] Mumford (1934, pp. 12–18, 470).

[6] Vidal (2020).

[7] Heilbron (1999).

chronology and value[8]: only once overshadowed by mural quadrants with achromatic telescopes at the end of the eighteenth century were astronomical meridians converted to horological use. To serve as the reference for mechanical timepieces, meridian lines had to be converted: to mean time by the equation of time; and to different systems of civil hour counting, especially in Italy.[9]

In 1728, a clockmaker of some repute, Henry Sully, began drawing a monumental *meridiana* in the newly rebuilt church of Saint-Sulpice of Paris. He was no astronomer, and would describe the purpose of his project as such:

> For several years, I have been thinking of a practicable means, accessible to everyone, to know with precision, every day of the year, the true time of the Sun, to which all the instruments used for time measurement must be related, and especially public clocks.
>
> It is as inconvenient as it is bizarre that in a city like Paris, one is exposed to such great discordance and uncertainty about the true time of day, as has always been the case until now […].
>
> So that the reader no longer finds any difficulty in knowing the true noon, on all days when the sun is not visible, and to notify all of Paris at the instant of noon, every day of the year, it is only necessary to add in two words, that by means of a well-regulated pendulum, and a signal whose manner is already conceived and established, we will achieve uniformity in time throughout the city, and there is reason to believe that everyone will take pleasure in conforming to it.[10]

Focusing on the social identity of Henry Sully, the context, the project, and the writings surrounding the 1728 *meridiana* at Saint-Sulpice, this paper aims to challenge the received image of discontinuity and radical difference between sundials and meridian lines, as well as the chronological hierarchy between the astronomical purpose and secondary timekeeping use of the latter. Based on recent historiography of horology and gnomonics, which takes into account the social context of timekeeping practice,[11] highlights the coexistence and interdependence of different modes of reference to time,[12] and emphasises the role of gnomonics in the construction of modern categories of precision and synchronisation,[13] this study considers the case of Sully as an opportunity to contextualise the link between astronomy and horology in early eighteenth-century Paris. An expression of the "quantifying spirit"[14] of the eighteenth century, this instrument echoes many issues within the history of measurement: the epistemological and social boundary between scientists and artisans; the diversity of the notion of "precision" (solar and mean); the circulation of knowledge across national and scientific borders; the emergence of quantified measurements and construction of shared norms. While these boundaries may seem firm and defined from a technical standpoint, social reality appears more complex

[8] Heilbron (1999, pp. 265–289).

[9] Heilbron (1999, pp. 265–268).

[10] Sully (1728a, pp. 1591–1593).

[11] Bernasconi and Thürigen (2020, pp. 7–16); Georges (2023).

[12] Turner (1982, 2002); Glennie and Thrift (2009); Nye et al. (2022).

[13] Sauter (2007); Savoie (2001, 2021).

[14] Frängsmyr et al. (1990).

than a simple top-down historical model, from science to civil society, from observatories to cities, from astronomers to artisans. More accurately, conflicts in usage and conception surround *meridiane* throughout the eighteenth century.

Paris constitutes in this regard a prime example: the city experienced a so-called "golden age of meridian lines"[15] in the early eighteenth century, expressed by a wide diffusion in the urban landscape. Unfinished and replaced in 1743 by a second and better-known line, Sully's *meridiana* allows for a nuanced outlook on the "church observatories" studied by Heilbron, and on astronomy's ties to timekeeping. Sully was not an astronomer, but a clockmaker seeking to offer the notability and social legitimacy of science to the mechanical arts. The line's purpose was therefore not astronomic: it was supposedly a precise timekeeper used to adjust the church's clock. The centre of an ambitious network of temporal cues, it was also to serve as a horological reference to synchronize all of Paris's clocks with Saint-Sulpice's local solar time for the benefit of watch users. Finally, the project was supplemented by a piece of writing, published in the *Mercure de France* in July 1728, which is both a theoretical essay expressing the role of the clockmaker and gnomonics in the life of the city, and an editorial gesture aimed at reviving Sully's career. As he tried to establish a civic observatory, Sully contributed to the advent of the meridian line as an urban object.

Henry Sully: A Clockmaker in Search of Social Legitimacy

The social persona of the clockmaker is obviously neither an eighteenth century innovation, nor a Parisian one. However, it then and there experienced a gradual shift: from patronage, royal privilege and the clockmakers' guild of Paris founded in 1544, emerged a valorisation of clockmakers as individuals, driven by the notion of "talent."[16] Clockmaking, a practical knowledge based on oral, often safeguarded guild transmission, developed into a theoretical knowledge, claiming scientific validity and conveyed through an extensive literature. Clockmakers became writers, public figures of knowledge and debates,[17] although such an image sometimes proved rhetorical, as most clockmakers remained either masters or privileged.[18]

In Paris, Henry Sully assumed the archetype of a free-thinking technician, carefully constructed throughout his writings, and, after his death in 1728, throughout those of his followers. He is a celebrated and well-studied figure,[19] but the precise steps of his career remain elusive. This is not simply a case of an historical figure and his myth being out of sync: the categories used to understand eighteenth century clockmaking and mechanical arts were merely emerging during Sully's life, and

[15] Gotteland (2002, p. 74).

[16] Vérin (1993); Pansini (2008). On the social evolution of mechanical arts, see Bertucci (2017).

[17] Augarde (1996, pp. 72–82).

[18] Dequidt (2014, pp. 139–173).

[19] Hilaire-Pérez (1997); Turner (2015); Bertucci (2017).

therefore fail to grasp the complexity and indeterminacy of his social and professional identity. His biography is part of the larger narrative of the history of early-modern mechanical arts, of a discourse on technique produced by technicians. A prolific author, he was also the subject of discourses, a celebrated life, elevated as a model for the mechanical arts milieu. Shortly after his demise on October 13, 1728, the *Mercure de France* published a long obituary that provided an initial narrative framework:

> M. Henry Sully, an Englishman from the county of Somerset and clockmaker of H.S.H. the Duke of Orleans, established in France for a long time, died in Paris on the 13th of this month, at the age of 48, after having abjured the Anglican Religion in the presence of the priest of Saint-Sulpice and having received the sacraments with the sentiments of exemplary faith and piety. He was not only very skilled in his profession but also distinguished in everything related to mechanics, navigation and even astronomy (…). His last work, which he did not have the pleasure of seeing in its perfection, is the meridian line of the church of Saint-Sulpice, of which we gave the description in the Mercure of July; but everything is traced, and there are only 7 degrees 30 minutes left to be at the equinoxes, that is to say, at the signs of ^ and d d.[20]

Several narrative tropes are brought to make of the deceased a remarkable member of Paris's intellectual society: the indefatigable pursuit of knowledge at the expense of the body, the elevation from mechanical arts to science, a conversion to Catholicism - symbolizing a desirable transfer of scientific expertise from England to France. This first portrait was later expanded upon: the main narrative was given, in panegyric mode, by Julien Le Roy as an appendix to the 1737 reissue of Sully's masterwork, the *Règle artificielle du temps*.[21] Friend, technical and editorial collaborator of the clockmaker, he lauded Sully as a "martyr" of mechanical arts:

> The arts have their martyrs as well as religions, although the motives of the former cannot be compared to those of the latter. There are some who would have lived in abundance and rest, if they had had more taste for their fortune than for the perfection of their works: the late Mr. Sully was one of them; I have seen him, full of zeal, going from door to door preaching to clockmakers the perfection of their art, and encouraging them with his speeches and advice to perfect it more and more.[22]

He also gave a striking narrative of the sudden demise of his friend, reminiscent of the seventeenth century "*belle mort*" genre, without any other sources to attest to its accuracy:

> Towards the end of the same week, having learned that someone in the St. Marceau district intended to show something curious to the Society of Arts, he took note of the address, which turned out to be false. He went back and forth in this large district without being able to find it, becoming heated and eventually dying four or five days later from a chest infection.
>
> His illustrious pastor, the *curé* of Saint Sulpice, after providing him with spiritual support during his illness, ordered his funeral to be held in his church, near the doors of the sanctuary of the high altar, and just west of the meridian line upon which he had traced the degrees of the signs just a few days before his death.

---

[20] Mercure de France (1728, pp. 2271–2274).

[21] Sully and Le Roy (1737, pp. 381–414).

[22] Sully and Le Roy (1737, pp. 381–382).

It was in October 1728 that the days of one of the most skilled clockmakers in Europe came to an end.[23]

The image and the main elements of his biography of Sully thus came from a later writing effort,[24] carrying less of a value of truth than a performative identity within the Parisian mechanical arts. From a simple craftsman, he became "a brilliant English clockmaker, friend of Newton, Leibniz, and Bernoulli,"[25] a "happy genius" who had "a very large share in the revolution that took place in European clockmaking and which resulted in definitively putting the sceptre of the art in the hands of French clockmakers."[26] He is a persona created by writing, constructed by posterity to embody an early Parisian occurrence of the ideal technician of the Enlightenment, uniting technical, economic, and scholarly practices.

Behind this written facade, the reality of Sully's social trajectory is opaque. Although a lack of precise documentation forbids any certainty on the subject, it is to be noted that his British origins are first expressed by the author himself in his works and seem ambiguous. His first book, the *Abrégé de quelques règles pour faire un bon usage des montres*, published in French in Leiden in 1710[27] and 1711, in Frankfurt in 1711 and 1712 and translated into German in 1713, was signed "*Henry Sully, horloger de Londres à Leide.*" A collection of parish and notarial documents shows that he did not simply pass through Leiden, but was there established for many years within a specific community. His son Henry, from his marriage to Anne Horton, was baptised on July 15, 1708 in the Walloon church of the city,[28] and a second son, Jean, on July 21, 1709 in the same context.[29] Both certificates are written in French, as is the entire register. On November 30, 1710, a daughter, Henrietta, was baptised, this time in a more central church, the Pieterskerk.[30] It appears that Anne Horton died in childbirth, the children being entrusted to two guardians in Leiden on September 18, 1711,[31] a date consistent with the aforementioned German editions. The presence of other occurrences of his surname amongst witnesses of these certificates suggests a lasting settlement of Sully's extended family in Leiden. Most importantly, the social and religious milieu of the documents strongly suggests a Francophone origin: the Vrouwekerk, a Walloon church since 1584 in Leiden, is

[23] Sully and Le Roy (1737, pp. 408–409).

[24] See also Alexandre (1734); Moréri (1735, p. 340).

[25] Bléchet (2011, p. 100).

[26] Dubois (1849, p. 51).

[27] *L'Europe savante* (janvier 1718, pp. 299–300).

[28] "Acte de baptême de Henry Sully, fils de Henry Sully et Anne Horton, le 15 juillet 1708," Erfgoeg Leiden en omstreken, Parish Registers of Baptisms, Dopen Waalse Kerk, Part 275, 1696–1748, Leiden, 1004, July 15, 1708, Dopen Vrouwekerk.

[29] "Acte de baptême de Jean Sully, fils de Henry Sully et Anne Horton, le 21 juillet 1709," Erfgoeg Leiden en omstreken), Parish Registers of Baptisms, Dopen Waalse Kerk, Part 275, 1696–1748, Leiden, 1004, July 21, 1709, Dopen Vrouwekerk.

[30] "Acte de baptême de Henrietta Sully, fille de Henrik Sullij et Anne Horton, le 30 novembre 1710," Erfgoeg Leiden en omstreken), Parish Registers of Baptisms, Dopen NH Pieterskerk, Part 227, 1710–1724, Leiden, 1004, numéro d'inventaire 227, November 30, 1710, Dopen Pieterskerk.

[31] See note "520 ligne 13" in Huygens (1934, p. 701).

the centre of the Huguenot community. The Sully family appears as members of it by language,[32] culture, and lasting involvement in this parish. It is not absolutely excluded, however, that Henry or his wife were English religious refugees, a group also familiar with the Vrouwekerk, although significantly earlier in the seventeenth century.[33]

Similarly, Sully's apprenticeship with Charles Gretton, his encounters with Newton and Leibniz,[34] seem rhetorical. In reality, his social position was often fragile and did not benefit from real support from a scientific elite; he relied on individual political patronage, such as that of the Duke of Arenberg[35] and the Duke of Orléans, and only obtained a royal pension, albeit somewhat substantial, in 1726.[36] Despite his requests and the interest of Jacques Cassini,[37] he was unable to gain entry or support from the Academy of Sciences and was refused membership in the clockmakers' guild of Paris, leaving him without his own workshop.[38] Sully's unique path cannot be understood without considering his efforts to create new forms of social legitimacy. Through various means, he sought to rectify the fact that "most of those who practice [the art of horology] have only been placed amongst the common mechanics,"[39] by distancing this art from a corporatist system he abhorred and by giving it a structure which would have allowed talent to flourish.[40] Sully's role in the foundation of the Society of Arts in Paris is well studied[41] and is consistent with his celebration of the mechanical arts as a body of knowledge: he there found an institution to promote his talents outside of the corporatist framework.

[32] In the preface to his *Règle artificielle du temps*, Sully "ask[s] the reader to consider that [he has] written in a language that is foreign to [him]", Sully (1714, p. 9). However, his entire bibliography, including works published in Leiden, Vienna, and Frankfurt, were primarily written and published in French, with a very few cases of later translations.

[33] Sully's familiarity with English technical literature could support this idea.

[34] The idea originates from a reaction by Leibniz to the *Regle artificielle*, which Sully included in and used to promote the second edition.

[35] "Épître dédicatoire au duc d'Arenberg," in Sully (1714).

[36] "Brevet de 500 livres de pension à Sully, horloger anglais, pour son travail qui peut être utile à la navigation," Archives nationales (AN), O/1/69, fol. 27.

[37] *Registre des procès-verbaux des séances de l'Académie royale des sciences*, tome 35, année 1716, fol. 185–186.

[38] Upon his arrival in Paris, his mastery was allegedly refused Bertucci (2017, p. 164).

[39] Sully (1711b).

[40] The hostility towards the guild system is a constant theme in his work. See Sully (1711b, pp. 14–16); Sully (1717, pp. 9–11, 69).

[41] Hahn (1971, p. 109); Bertucci and Courcelle (2015); Bertucci (2017).

The Equation of Time: A Social and Urban Issue in the 1720s

Saint-Sulpice's meridian line plays a central role within Sully's life narrative. The indirect cause of his death, it symbolizes the junction of clockmaking and astronomy, the ascent of the mechanical arts towards science. However, for this specific instrument, the form of the church *meridiana* is seemingly diverted towards a new use. The line appears as a testimony to the rise of clockmaking and clockmakers. According to its author, its intended purpose was to give precision to and synchronise clocks on local solar noon for the whole city. It is uncertain whether this ambition was a viable project or belonged to the realm of technical desires. However, it was then understood as genuine: in 1742, a description of this instrument by Piganiol de la Force echoed Sully's own words from the *Mercure*:

> Sir Sully, an Englishman who was a famous clockmaker and a good astronomer, was struck by the fact that in a city like Paris, people were exposed to a constant uncertainty about the true time of day. He conceived the idea of drawing a meridian on the pavement of this church, facing true north and south.[42]

The device's idiosyncrasy makes it difficult to fit in usual gnomonic classifications. Heilbron considers it an unclassifiable exception and also interprets the horologist's writing literally:

> Sully had an agenda of his own. It annoyed him that the clocks of Paris did not ring the hour in unison. The near completion of Saint-Sulpice suggested to him how to preserve direction from the inconvenience of multiple noons; he needed only to lay down a *meridiana* in the church and attach its attendant to a noisemaker to alert Languet's parishioners that midday had arrived.[43]

However, a scientific and technical context accounts for originality of this line. It was an attempt to combine two gnomonic genres: the horizontal astronomical *meridiana*, and the vertical noon mark erected on facades for the purpose of watch setting, particularly in vogue in the 1720s. At the heart of this project lies the new social role of clockmakers, as synchronisers of urban rhythms.

The equation of time had been a concern for clockmakers since the 1710s. The development of pendulum clocks and the anchor escapement allowed for an isochronous day, strictly equal to 24 h. This novelty caused a desynchronisation between two temporal units, true solar time and mean time, visible. Causes and calculations of the equation of time were subjects of debate amongst astronomers.[44] The professional stakes were significant: noticing a time difference between their watches and the sundials, users often turned against clockmakers and accused them of faulty workmanship or charlatanism.[45]

---

[42] Piganiol De la Force (1742, tome 6, p. 382).

[43] Heilbron (1999, p. 220).

[44] Savoie (2021).

[45] Sully repeatedly complained about the lack of knowledge and "unreasonable expectations" of the public, which tarnished the reputation of clockmakers. See for example Sully (1717, pp. 15–16).

In the late seventeenth century, English clockmakers attempted to theorise and popularise this discrepancy for the benefit of new users of mechanical horology. One of them, John Smith, published two influential works aimed at constructing the categories of "solar" (or "apparent") time and "mean" time, *The Equation of Time* (1678), and the *Horological Disquisitions* (1694).[46] These are not hierarchical in Smith's view: he supported, like most clockmakers before the mid-eighteenth century, a hybrid system adapting the regular movement of the pendulum to irregular solar time, chiefly through resetting hands alone, without considering mean time as an end in itself. This constituted a double adaptation of the machine: it requires assiduous human involvement through measurement and adjustment, but also rendered mechanical horology dependent on external devices. Two of these devices were then favoured: tables of the equation of time, and meridian sundials, serving as references for clocks.

Henry Sully attempted to promote both of them successively, first facing rivalry and criticisms when publishing tables. Sully built upon Smith's theoretical framework[47] to develop a pedagogy intended for Parisian users of the 1710s. He published his own equation tables as early as 1714,[48] but came into conflict with other scholars regarding their calculation and use. Jakub Kresa voiced criticisms of the *Règle artificielle* tables' methodology and accuracy, blaming Sully for excessively simplifying astronomical calculations for civic purposes. Building upon this controversy in the second edition of the book, Sully published Kresa's letter and his own response, in which he defended a purely horological conception of astronomical tables, convenience taking precedence over accuracy. In the same fashion, he fiercely attacked what he deemed were overly complex astronomical ephemerides, particularly the *Connaissances des temps* published by the *Académie des Sciences* from 1679 onwards.[49] Sully thus faced intense competition from other table publishers and, as a result, came to favour sundials. Critical of common gnomonics, he stressed the need for a precise *meridiana*, creating a ritualisation of gestures, engaging users in daily "ceremonies of measurement"[50] raising noon as a referential centre. As with Smith's, his adjustment system was hybrid, "the hands pointing on the dials as close as possible to the apparent time, while the movements are regulated on mean time."[51] Dials and clocks were placed in a complementary relation, a complementarity made evident in the frontispiece of the clockmaker's first work, showing the sun and a clock in unison at true noon, above the formula *Metior et emendo*.[52]

[46] Another example, although late, is Clay (1731).

[47] Sully mentioned the *Horological Disquisitions* as an inspiration, although he pointed out its limitations. Sully (1714, pp. 3–4).

[48] Sully (1714, pp. 53–54; 1728b, p. 1).

[49] "Réponses aux observations et remarques du R. P. Kresa S.J. sur le livre qui a pour titre, Regle artificielle du tems," Sully (1717).

[50] Schaffer (2015).

[51] Schaffer (2015, p. 36).

[52] Sully (1711a).

In the 1720s, Sully contributed to a general movement of debates and pedagogy in Paris regarding the equation of time, which was embodied by the emergence of meridian lines as an urban and social object. The first *meridiana* of the Paris Observatory, built in the late seventeenth century by Jean-Dominique Cassini obviously provided a formative context; however, the resumption of the line by Jacques Cassini from 1728 onwards seems to have been the determining influence for many urban *meridiane* in the 1720s and 1730s, including Sully's. A shift from seventeenth century churches or observatories towards private residences and public spaces and diversification of the device are unambiguously noticeable.[53] Alongside horizontal dials, known as "floor" lines and placed in *hôtels particuliers* (such as the Petit-Luxembourg in 1711)[54] vertical *meridiane*, known as "garden" lines, appeared more frequently outside private residences (such as the Hôtel de Taranne in 1710)[55] or in public spaces, although usually later on (on the Pont-au-Change in 1737, the Palais-Royal from the 1740s onwards.[56] These public vertical noon marks became famous for their social value: Casanova, for example, wondered in 1750 about the success of the Palais-Royal line: "Aren't there sundials everywhere? Yes, but the one at the Palais-Royal is the most accurate."[57] Finally, the hybrid adjustment system advocated by Smith and Sully, which always favoured true noon, competed from the 1730s onwards with mean time *meridiane*, carved with an eight-shaped curve allowing for direct reading of mean noon and which was placed, debatably for the first time, by Grandjean de Fouchy in a salon of the Petit-Luxembourg between 1728 and 1730.[58] The meridian sundial thus became a tool for verifying the clock rather than a reference for it.[59]

The 1728 Meridian at Saint-Sulpice: A Horological Observatory?

The excitement surrounding precision horological sundials was most spectacularly expressed by Sully's project at the Saint-Sulpice church in 1728. Replaced by the subsequent *meridiana* built in 1743 by Le Monnier, it is solely known through descriptions: that given by Sully in the *Mercure* of July 1728, which had programmatic and theoretical aims,[60] and later writings falling under the genres of city guides (Piganiol de la Force, 1742) or technical curiosities Le Monnier (1743), Lalande

[53] See graphs by Savoie (2021, pp. 198–199).

[54] Gotteland (2002, p. 81).

[55] Gotteland (2002, pp. 79–81).

[56] Gotteland and Camus (1997, pp. 193–194); Savoie (2021, p. 169).

[57] Casanova (1968, pp. 636–638).

[58] Savoie (2008).

[59] Landes (1983, pp. 118–124).

[60] Sully (1728a).

(d'Alembert, 1785). While a text by Boucher d'Argis (1752) has also been cited,[61] it is not clear whether it focuses on the 1728 meridian alone or mixes elements of both successive lines: it describes an obelisk, which, through its inscriptions, is unambiguously the one still standing at Saint-Sulpice, but also uses the exact words from the *Mercure*'s description. It is to be understood that the second line may have preserved and reused elements of the first, and that parts of the current obelisk at Saint-Sulpice—at least the pedestal—were first built in 1728—the winter solstice "column" described by Sully,[62] and later readapted and heightened.

This *meridiana* supposedly served a public utility purpose that Sully had long envisioned: rationalising the chiming of clock towers in Paris. The resumption of work on the new Saint-Sulpice church provided an opportunity, and Sully claimed to have himself suggested the construction of a meridian line to the priest of Saint-Sulpice, Jean-Baptiste Languet de Gergy:

> The frequent sight of the beautiful and spacious church of Saint-Sulpice, located in one of the most beautiful neighbourhoods of Paris, suggested to me an expedient that seemed at first quite fortunate and sufficient for this purpose. I imagined tracing a meridian there, and on the same day that this thought came to me, I communicated it to the parish priest of Saint-Sulpice, convinced that he would find it agreeable. The first idea of public utility left no time for hesitation to someone who had done so much for it. At that very moment, the priest of Saint-Sulpice approved of my proposal, wished for its execution, and honoured me by asking me to take care of it. I joyfully accepted this pleasant commission, and carried it out with extreme pleasure, flattered to have been able to contribute in some small way to such noble aims and to the embellishment and splendour of such a magnificent building.[63]

The gratuitous gesture of the clockmaker is clearly part of a narrative. According to Le Monnier, Languet had conceived the project of a meridian from the beginning of the construction of the church, himself delineating the space for it near the southern portal.[64] His idea was to draw inspiration from Italian churches both in architecture and astronomy. If we are to believe that Boucher d'Argis's text described Sully's line,[65] it could even be that Easter *computus* was indeed the original function of the line, regardless of the clockmaker's writing. The obelisk indeed displays the inscription "*Gnomon Astronomicus, ad certam Paschalis aequinoctii Explorationem.*" These words are still visible on the current gnomon; however, since Boucher d'Argis did not mention the long inscription that now follows and celebrates Le Monnier's work, it is plausible that the first gnomon only displayed this formula, later expanded upon.

The line seemed particularly ambitious. In a church of monumental dimensions, offset by 11° from a north/south axis, it covered the entire transept ("176 feet[66] on

[61] Gotteland et al., (1990, p. 198).

[62] Sully (1728a, p. 1606).

[63] Sully (1728a, pp. 1592–1593).

[64] Anonymous (1743, p. 146); Heilbron (1999, p. 220).

[65] The text's division between "description" and "supplement" makes its chronological interpretation difficult. However, the described line does bear the mark of the "summer solstice of June 21, 1728" (p. 340).

[66] After 1668, one "pied-de-roi" equates to 32.48 cm.

Fig. 2.1 Frontispiece engraving of the *Abrégé de quelques règles pour faire un bon usage des montres*, 1711

the pavement," or 57.16 m). The eyepiece, supported by a plate on the stained glass of the southern portal and one inch in diameter (2.70 cm), is located 75 feet above ground (24.36 m).[67] It was the largest meridian in Paris, far longer than that of the observatory, but shorter than Italian lines, particularly San Petronio of Bologna's (68.20 m). Its measuring inscriptions, represented in Fig. 2.1, were equally complex. On either side of the summer solstice point, engraved "Summer Solstice of June 21, 1728," it shows, between the southern portal and this point, two columns, the "division of the line into thousandths of the vertical height" on the left, and the "degrees and minutes of the distance from the zenith to the Tropic of Cancer." After the solstice, six columns measure, on the left, the "hours and minutes of the sunrise from 5 to 5 min," the "position of the sun in the ecliptic, in descending signs, characters and names of these signs and their degrees," and the "tangents to thousandths of the radius"; on the right, the "degrees of the declination of the sun, both northern and southern on either side of the equator, up to the other tropic," the "position of the sun in the ecliptic, in ascending signs, characters and names of these signs and their degrees," and the "hours and minutes of the sunset from 5 to 5 min."[68] All of these were described by Sully as pedagogical tools to cultivate astronomic accuracy within Parisian learned society. An innovation compared to Italian meridians, the line is both horizontal and vertical. Limited by the dimensions of the transept, the line, theoretically 240 feet, or 77.95 m, long is extended by a 25 foot, or 8.12-m, gnomon allowing observation up to the winter solstice.[69]

[67] Sully (1728a, p. 1594).

[68] Sully (1728a, pp. 1601–1602).

[69] Sully (1728a, p. 1606).

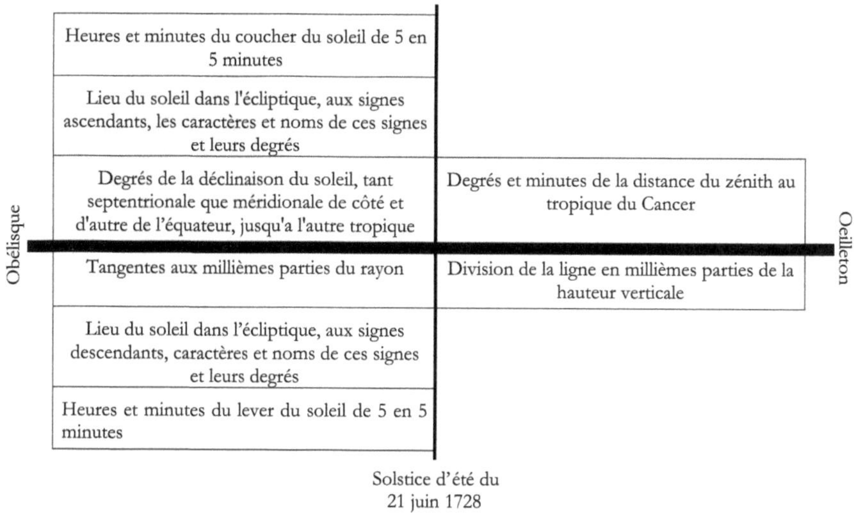

Fig. 2.2 Schematic measuring inscriptions of Henry Sully's meridian line at Saint-Sulpice, 1728[70]

It proves difficult to assess the extent to which Sully's project had progressed, based on the vestigial traces in the current church. These appear to indicate a well-established and carefully engraved line, subsequently mutilated by changes to and wear and tear of the flooring, resulting in its current fragmented appearance. Spaced about 45 cm east of the 1743 line, Sully's meridian is also significantly longer, extending all the way to the southern portal (Fig. 2.2). It is marked by two inscriptions (Fig. 2.3) now fragmentary due to a change in flooring, which match those described by Sully and summarised in Fig. 1.2. Only the base of the line still visibly displays a numbered measuring apparatus (Fig. 2.4). It is uncertain whether the line was intended to be decorated with marble and brass, or to be limited to a simple engraved line in the pavement (Fig. 2.5).

As cited in introduction, the *Mercure*'s article described at length a horological project aimed at producing an astronomy-based clock time for the parish, and at spreading this reliable information throughout the city. Out of sync bells and clocks represented a nuisance for Sully. Such a grievance is common in his work and amongst clockmakers, eager to distance themselves from the negative image of church clock conductors:

> Public clocks are most often poorly governed; the experience with them, even in large cities, sufficiently proves this by comparing the clocks in distant neighborhoods, and we need not give reasons that can otherwise account for such a strange effect.[71]

Sully claimed his precise gnomonic device, ensuring accurate time signaled visually and aurally at the church, to be the response to this incongruity. Yet, the most

[70] Sully (1728a); Boucher d'Argis (1752). This diagram develops that of Gotteland et al. (1990).

[71] Sully (1728b, p. 8). See also Sully (1711a, p. 4; 1717, p. 81).

Fig. 2.3 Fragment of the 1728 meridian line (foreground) and the 1743 meridian (background) in the current Saint-Sulpice church (photograph by the author)

Fig. 2.4 Fragments of inscriptions from the 1728 meridian line: "[...] en Millieme[s parties de] la Haut[eur verticale]...s " on top (west), "[...] Minutes du [zeni]th au trop[iq]ue du Cancer" on the bottom (east) (photograph by the author)

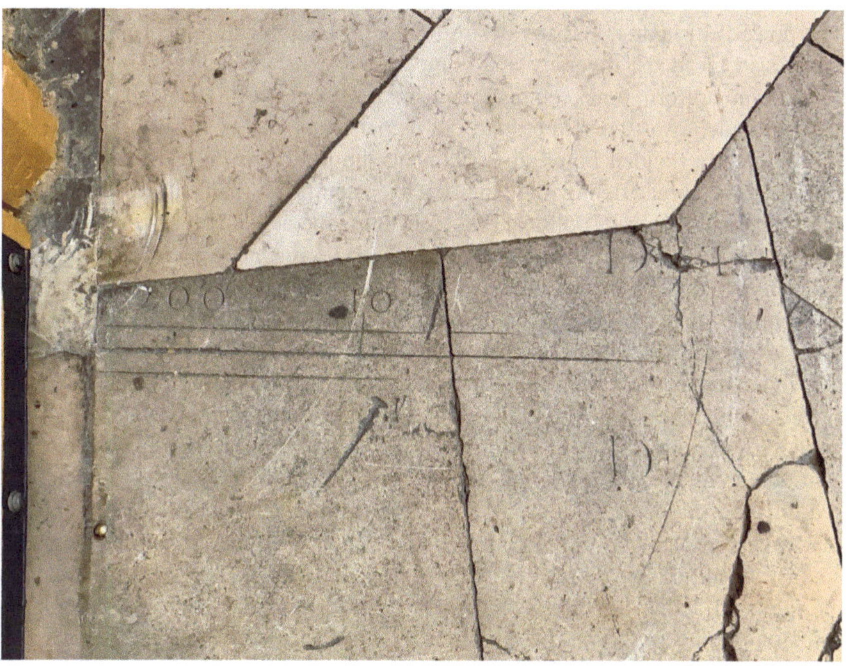

Fig. 2.5 Base and south end of the 1728 meridian line, along with its numbered measuring inscriptions. The words "Divis[ion]" and "De[grés]" can also be deciphered, matching the beginning of the inscriptions transcribed in Fig. 2.4 (photograph by the author)

innovative aspect of the *Mercure*'s text was to envision the circulation of precise solar time throughout the city, so that Parisians will "no longer encounter any difficulty in knowing the true noon every day that the sun is not visible, and to inform all of Paris at the stroke of noon every day of the year," thus "achieving uniformity in time throughout the city."[72] While the intention is clear, the means prove mysterious: a "well-regulated pendulum" in the church was to be connected to a "signal whose manner is already conceived and established." Without further information, it is impossible to deduce the nature of this signal, except, teleologically, under later popular synchronizers, such as cannons or time balls.

Sully showed notable assurance in proclaiming his goal of creating a civil observatory for Parisian horology; however, its feasibility was questionable. This made it necessary to precisely adjust the church's clock on solar noon as observed on the line; yet, clock management was a fragile function within the churchwardens' councils,[73] fulfilled by lowly beadles rather than technical personnel capable of adjustment, measurement and maintenance. Timekeeping was an irregular task, with visits happening several months apart: parish archives show that church clocks could remain faulty or defective for months or years on end without repair. It also proved a poorly compensated position, neglected as such by beadles and councils alike. The device envisioned by Sully would have required a fundamental shift in the churchwardens' system: yet it is unlikely that Saint-Sulpice's parish, one of the largest and most powerful in Paris, had any desire for economic reform. Synchronising public clocks also faced the diversity and competition of urban powers. Up until 1826, there was no uniform timekeeping policy in Paris: each street clock and the information it provided, was the sole property of its particular owner. In such a polycentric city, politically divided between Parliament, the police of the Châtelet, the royal domains, the "bureau de la ville", the "enclos", and fifty urban parishes, clocks and bells were instruments of political rivalry. Ringing the bell louder or before the hour, displaying a monumental dial were ways to assert dominance over the sensory space of the city.[74] Parishes thus would have found no benefit in clock synchronisation, as desynchronisation was blatantly a political tool.

It is even doubtful whether Sully truly had faith in the completion of his system: he seemed more eager to introduce ideas than implementing them. The 1728 *meridiana* appears in many respects as a tool of horological pedagogy, both visual and textual. The description given by Sully in the *Mercure* was not merely an informative document, but a complement to, if not an integral part of, the meridian line, conceived as a didactic project. Building upon horological manuals and including tables of solar time variations on the line, its purpose was to give the key to understanding this instrument to the general, albeit learned and curious readership of the *Mercure*. The intended audience is explicit: "We will not discuss here the uses of this line which only concern those who are familiar with the principles of Astronomy [...] We only aimed here for a clear and concise Description, which is accessible and

[72] Sully (1728a, p. 1593).

[73] Regarding the urban role and management of parish clocks, see Georges (forthcoming).

[74] Garrioch (2003).

useful to everyone."[75] As Heilbron did, we may postulate a connection between this *meridiana*, the text, and the creation of the Society of Arts, which would find in this instrument a first application of technical innovations for the benefit of the public good, an idea they advocated from 1728 onwards.

Conclusion

The meridian line designed by Henry Sully at Saint-Sulpice in 1728 offers an early example of a technicist "embellishment" of early-modern urban space. It reveals the influence of a clock-based conception of time and technology on landscapes, discourses and behaviours. Although the *Mercure* description imagined an early "metronomic city"[76] defined by a unified temporal framework, synchronized time cues, and large-scale timekeeping governance, it is reasonable to raise doubts about the reality of this project.

Yet, this discourse proves important purely as a discourse. It reveals a fundamental element of the technicist rhetoric emerging in the eighteenth century: a didactics of technology, based on the construction of a conceptual framework and its social diffusion. If a current trend in the history of temporalities is indeed to denaturalise temporal conventions, it must be considered that clock time does not constitute an innate body of knowledge, but arises from a slow learning process, both theoretical and practical.[77] Such an habituation creates social tensions: if users are proactive, influencing the reputation of clockmakers, their mastery of technology is affected by an intense editorial initiative from clockmakers. In this sense, such a movement marked the emergence of a new *techno-logy*, a discourse on technique, its place and role in society, and practically aimed at generalising its use.[78]

Furthermore, the 1728 meridian line is an important part of a transition that takes precision gnomonics out of churches and observatories to introduce it as a new urban object. Although close in form to classical church lines, its hybrid nature made it an early occurrence of urban noon marks built for the purpose of watch setting. It contributed to the eighteenth century "readable city" based on legible orientation tools in public spaces.[79] Finally, it expressed a horological temporalisation of activities which then tended to supersede the diversity of time references, especially those based on a local scale.[80] It would be easy to take Sully, teleologically, as a pioneer of the eighteenth century triumph of mechanical horology; but he appeared as a multifaceted craftsman trying to create new ways to acquire social legitimacy. The social identity and professional network he constructed in the 1710s and 1720s later served

[75] Sully (1728a, pp. 1606–1607).

[76] Young (1988).

[77] On this subject, see Glennie and Thrift (2009, pp. 93–96; 225–231).

[78] About the conceptual history of this notion, see Beaune (1980 particularly pp. 253–263).

[79] Bowlby (2007); Marin and Folin (2018).

[80] Grossin (2018).

as the foundation to develop the figure of the intellectual and scholar clockmaker, as expressed in the writings of Julien le Roy, Jean-André Lepaute and Ferdinand Berthoud.

References

Alexandre, J. (1734). *Traité général des horloges*. Hippolyte-Louis Guérin.

Anonymous. (1728). Nécrologie de Henry Sully. *Mercure de France*, pp. 2271–2274.

Anonymous. (1743). Sur le gnomon et l'obélisque de la méridienne de Saint Sulpice. *Histoire de l'Académie royale des sciences*. Imprimerie royale, pp. 142–147.

Augarde, J. -D. (1996). *Les ouvriers du temps: la pendule à Paris de Louis XIV à Napoléon 1er*. Antiquorum.

Aveni, A. (1989). *Empires of Time: Calendars, Clocks, and Cultures*. Basic Books.

Beaune, J. -C. (1980). *La technologie introuvable. Recherche sur la définition et l'unité de la technologie à partir de quelques modèles du XVIIIe et XIXe siècles*. Vrin.

Bernasconi, G., & Thürigen, S. (Eds.). (2020). *Material Histories of Time: Objects and Practices, 14th–19th Centuries*. De Gruyter.

Berthoud, F. (1802). *Histoire de la mesure du temps par les horloges*. Imprimerie de la République.

Bertucci, P., & Courcelle, O. (2015). Artisanal knowledge, expertise, and patronage in early eighteenth-century Paris: the Société des arts (1728–36). *Eighteenth-Century Studies, 48*, 159–179.

Bertucci, P. (2017). *Artisanal Enlightenment: Science and the Mechanical Arts in Old Regime France*. Yale University Press.

Bléchet, F. (2011). L'art de mesurer le temps à l'aube des Lumières. In Mauerhan, J. (Ed.), *Pratique et mesure du temps* (pp. 97–108). Éditions du Comité des travaux historiques et scientifiques.

Boucher d'Argis, A. -G. (1752). *Variétés historiques, physiques ou littéraires, ou recherches d'un sçavant*, tome 2. Chez Nyon et chez Guillyn.

Bowlby, R. (2007). Readable city. *PMLA, 122*(1), 306–309.

Casanova, G. (1968). *Mémoires*. Gallimard.

Cipolla, C. (1967). *Clocks and Culture: 1300–1700*. Collins.

Clay, F. (1731). *An Explanation of the Nature of Equation of Time*. Clay.

d'Alembert, J. le Rond. (1785). *Encyclopédie méthodique*, tome 2. chez Panckoucke.

Dequidt, M. -A. (2014). *Horlogers des Lumières: temps et société à Paris au XVIIIe siècle*. Éditions du Comité des travaux historiques et scientifiques.

Dubois, P. (1849). *Histoire de l'horlogerie depuis son origine jusqu'à nos jours*. Le Moyen-âge et la Renaissance, et chez l'Auteur.

Frängsmyr, T., et al. (Eds.). (1990). *The Quantifying Spirit in the 18th Century*. The University of California Press.

Garrioch, D. (2003). Sounds of the city: the soundscape of early modern European towns. *Urban History, 30*(1), 5–25.

Georges, L. (2023). Les heures locales: Horloges de rue et espace urbain à l'époque moderne (Paris, XVIIe-XVIIIe siècles). *Histoire Urbaine, 68*(3), 145–162.

Glennie, P., & Thrift, N. (2009). *Shaping the Day: A History of Timekeeping in England and Wales 1300–1800*. Oxford University Press.

Gotteland, A., et al. (1990). Les méridiennes de l'église Saint-Sulpice à Paris. *L'astronomie, 104*, 195–214.

Gotteland, A., & Camus, G. (1997). *Cadrans solaires de Paris*. CNRS Éds.

Gotteland, A. (2002). *Les cadrans solaires et méridiennes disparus de Paris*. CNRS Éds.

Graf, J. (2020). The First Black Forest Clock: How a Fake Impacted our Notion of Clock History. In G. Bernasconi & S. Thürigen (Eds.), *Material Histories of Time: Objects and Practices, 14th–19th Centuries* (pp. 211–224). De Gruyter.

Grossin, W. (2018). *Les temps de la vie quotidienne*. De Gruyter.

Hahn, R. (1971). *The Anatomy of a Scientific Institution: The Paris Academy of Sciences, 1666–1803*. University of California Press.

Heilbron, J. L. (1999). *The Sun in the Church: Cathedrals as Solar Observatories*. Harvard University Press.

Hilaire-Pérez, L. (1997). Transferts technologiques, droit et territoire: Le cas franco-anglais au XVIIIe siècle. *Revue D'histoire Moderne Et Contemporaine, 44*(4), 547–579.

Huygens, C. (1934). *Œuvres complètes de Christiaan Huygens.* Tome 18: *L'horloge à pendule ou à balancier de 1666 à 1695*. Martinus Nijhoff.

L'Europe savante. (1718). Tome 1. Chez A. de Rogissart.

Landes, D. S. (1983). *Revolution in Time: Clocks and the Making of the Modern World*. Belknap Press of Harvard University Press.

Lepaute, J. A. (1755). *Traité d'horlogerie contenant tout ce qui est nécessaire pour bien connoitre et pour régler les pendules et les montres*. Jacques Chardon.

Marin, B., & Folin, M. (2018). Comment s'orienter dans la ville modern. Introduction. *Histoire Urbaine, 53*(3), 5–16.

Mercier, L. -S. (1782). Chapitre LXXXV. Le faubourg Saint-Marcel. In *Tableau de Paris*. Tome I (pp. 268–272). Amsterdam.

Le Monnier, P. -C. (1743). Construction d'un obélisque à l'extrémité septentrionale de la Méridienne de l'église de Saint Sulpice. In *Histoire de l'Académie royale des sciences* (pp. 361–366). Imprimerie royale.

Moréri, L. (1735). *Supplément au grand dictionnaire historique, généalogique*. Jean-Baptiste Coignard.

Mumford, L. (1934). *Technics and Civilization*. Harcourt.

Nye, J., Betts, J., & Turner, A. J. (2022). *A General History of Horology*. Oxford University Press.

Pansini, V. (2008). Pour une histoire concrète du "talent": Les sélections méritocratiques et le coup d'œil du topographe. *Annales Historiques De La Révolution Française, 354*, 5–27.

Piganiol de La Force, J. -A. (1742). *Description de Paris, de Versailles, de Marly, de Meudon, de Saint-Cloud, de Fontainebleau....* Charles-Nicolas Poirion.

Sauter, M. (2007). Clockwatchers and stargazers: time discipline in early modern Berlin. *American Historical Review, 112*(3), 685–709.

Savoie, D. (2008). L'aspect gnomonique de l'œuvre de Fouchy: La méridienne de temps moyen. *Revue D'histoire des Sciences, 61*(1), 41–62.

Savoie, D. (2001). *La gnomonique*. Les Belles Lettres.

Savoie. D. (2021). *Une histoire des cadrans solaires en Occident: la gnomonique du Moyen Âge au XXe siècle*. Les Belles Lettres.

Schaffer, S. (2015). Les cérémonies de la mesure. *Annales. Histoire, Sciences Sociales*, 70(2), 409–435.

Schechner, S. J. (2001). The material culture of astronomy in daily life: sundials, science, and social change. *Journal for the History of Astronomy, 32*, 189–222.

Smith, J. (1678). *Equation of Time*. E. Tyler & R. Holt.

Smith, J. (1694). *Horological Disquisitions*. John Nicholson at the King's-Arms, and John Sprint at the Bell, in Little-Britain.

Sully, H., & Le Roy, J. (1737). *Regle artificielle du temps*. Grégoire Dupuis.

Sully, H. (1711a). *Abrégé de quelques règles pour faire un bon usage des montres*. chez l'auteur.

Sully, H. (1711b). *Court Essai sur l'utilité & l'excellence de l'art de l'Horlogerie*.

Sully, H. (1714). *Regle artificielle du tems*. André Heyinger.

Sully, H. (1717). *Regle artificielle du tems*. Grégoire Dupuis.

Sully, H. (1728a). Description de la ligne méridienne de l'Église Saint-Sulpice, avec l'explication de ses usages. *Mercure de France*, pp. 1591–1607.

Sully, H. (1728b). *Méthode pour régler les montres et les pendules.* Grégoire Dupuis.
Turner, A. J. (1982). "The accomplishment of many years": three notes towards a history of the sand-glass. *Annals of Science, 39*, 161–172.
Turner, A. J. (2015). The eclipse of the sun: sun-dials, clocks and natural time in the late seventeenth century. *Early Science and Medicine, 20*(2), 169–186.
Turner, A. J. (2002). Essential Complementarity: The Sundial and the Clock. In *Sundials at Greenwich*, pp.15–24.
Vérin, H. (1993). *La gloire des ingénieurs: l'intelligence technique du XVIe au XVIIIe siècle.* Albin Michel.
Vidal, L. (2020). *Les hommes lents : Résister à la modernité.* Flammarion.
Young, M. D. (1988). *The Metronomic Society: Natural Rhythms and Human Timetables.* Harvard University Press.

Louis Georges is a doctoral student at Université Paris-Nanterre (MéMo - Centre d'histoire des sociétés médiévales et modernes). His research focuses on the history of timekeeping in early modern Paris, especially on the connection between urban space and time cues: acquisition practices of city dwellers, production of "local time zones" by competing authorities, interdependence of mechanical and non-mechanical timepieces, or instrument craftsmanship. His doctoral work led to several publications in scholarly journals (*Dix-septième siècle, Histoire urbaine*).

Chapter 3
Instructions and Time Measurement at the Paris Observatory (1667–1789)

Dalia Deias

Abstract This chapter focuses on the presence, characteristics and uses of pendulum clocks at the Royal Observatory of Paris in the early eighteenth century. Through the analysis of unpublished manuscripts, I argue that such clocks were crucial to the organisation of work at the Observatory and to the performance of common astronomical observations. I also consider the European circulation of knowledge about these clocks, as evidenced by exchanges of letters and other publications. In fact, the study of the history of the Observatory's pendulum clocks makes it possible to illustrate the concrete functioning of the institution, its place in the European scene, as well as the ambitious cultural-political projects behind its activities.

Introduction

In a letter dated February 4, 1702, addressed to the astronomer Francesco Bianchini, Eustachio Manfredi, who held the chair of astronomy in Bologna, introduced the idea of constructing an "*osservatorio*" in his city. Taking inspiration from the Paris Royal Observatory, Manfredi informed Bianchini of his intention to use the methods of Louis XIV's "illustrious Academy" ("*insigne Accademia*"). Speaking of a French clock he was expecting from Paris, Manfredi stressed that his aim was to imitate as

D. Deias (✉)
Sciences, Philosophie, Humanités SPH, Bordeaux Montaigne, Bordeaux, France
e-mail: dalia.deias@u-bordeaux.fr

Ph.D. in History of Science, Centre Alexandre Koyré, Paris, France

G. Bernasconi et al. (eds.), *Astronomical Observatories and Chronometry, 18th-20th Century*, History of Physics, https://doi.org/10.1007/978-3-031-95874-8_3

closely as possible "the Observatory, at least in the accuracy and quality of the instruments."[1] To achieve this goal, Manfredi emphasized the importance of following the established rules for constructing such a building. This required seeking the advice of the French royal astronomers, becoming familiar with their techniques, obtaining certain instruments, particularly pendulums, to be sent from France to Bologna.

At the time, Manfredi and Bianchini knew that they could rely on their connections at the Academy of Sciences and the Paris Observatory, particularly with the Cassini family (as both had been corresponding for some years with Giovanni Domenico) and the Maraldi family.[2] Manfredi's wishes were indeed fulfilled: the astronomers of the Observatory were among the most important intermediaries for the circulation of the Observatory's observation techniques and objects in the Italian peninsula and played a fundamental role in the foundation of the Istituto delle scienze in Bologna.[3] Particularly important in these discussions, as we shall see, were the French pendulums, indispensable objects for astronomical observations.

In the dedication to Louis XIV of the *Horologium Oscillatorium* (1673), Christiaan Huygens spoke of the Royal Observatory as the "house of Urania."[4] A good number of clocks, he wrote, had found a home there, to be improved and used. Given the weight of these objects and their importance to the development of cartography, it

---

[1] Letter from Eustachio Manfredi to Francesco Bianchini (Bologna, February 4, 1702). Vallicelliana Library, Rome, Ms U 17, 868r–869r. This chapter summarises two parts of my doctoral dissertation. In my PhD thesis, I have referred to Giovanni Domenico Cassini's (Cassini I) entire conception of celestial observation sites as "accuracy (*esattezza*)," where the accuracy of observations and measurements is embodied first of all in the material building itself (its walls, its floors, its structure). As we shall see in this chapter, Cassini, in organising the building, wanted to make it a measuring instrument, with the aim to reduce the measurement errors inevitable in small instruments. On the concept of "accuracy (*esattezza*)," see Deias (2020a and 2020b).

[2] In 1702, Manfredi was busy founding the Institute of Sciences (*Istituto delle scienze*) in Bologna following the journey to Italy of Giovanni Domenico Cassini and his son Jacques (Cassini II), in 1695. Manfredi is known today for his weight in Bolognese astronomy of the time, in particular his astronomical observations (practised together with his sisters, whose contribution was fundamental but less visible via publications), published for instance in one of his major works, the posthumous *Istituzioni Astronomiche* (1749). Since 1699 Francesco Bianchini was an official correspondent of Giovanni Domenico and in 1706 he was appointed Foreign associate (*associé étranger*) of the Paris Academy. Manfredi received the same title in 1726. For information about the Istituto delle scienze founded by Manfredi in Bologna, see Cavazza (1990, 2002) and about letters between Manfredi and Cassini, see Giuntini (2001).

[3] On the circulation of objects and the importance of intermediaries in the Age of Enlightenment, see Bernasconi (2015). For the circulation of French pendulums specifically see Augarde (1996), in particular the section on the export routes of these objects, starting on p. 207. This author emphasises how individuals played an important role in the export of French pendulums at the time. For the circulation of knowledge more generally, see the work of Simon Schaffer and Kapil Raj.

[4] Huygens's contribution to pendulum clock technology is well known. Following Galileo's work, the mathematician relied on the construction techniques of the Dutchman Salomon Coster, particularly with regard to the choice of using a weight instead of a leaflet. Huygens carefully studied the functioning of the pendulum, and in particular the phenomenon of isochronism, which led the Paris Observatory, and later the Greenwich Observatory, to use long pendulums. On Huygens's life and work, see Andryesse (1998), the papers presented at the roundtable du CNRS *Huygens et la France* (1982), and Wolf (1902).

seemed natural to store them there. Between 1673 and 1681, Huygens lived at the Observatory. While the Observatory has been often associated with pendulum clock technology, less attention has been paid to the fact that the it also played a leading role in the development of clocks—particularly pendulums—into real work instruments, without which observation of the sky would have been impossible and the use of telescopes very limited.

In this chapter, I will focus on the pendulum clocks at the Paris Observatory between the second half of the seventeenth century and the eighteenth century.[5] The period I am referring to begins with the earliest evidence of pendulum-use techniques found in astronomers' manuscripts, from the 1680s, and extends to the eve of the French Revolution. We will see the main characteristics of these objects, how the royal academicians used them within the walls of the Observatory and the key role they used to have in the Academy's missions abroad, in Asia and Africa. To better understand the role of these objects and their techniques of use, it is necessary to trace their circulation from France to the Italian peninsula, and ultimately to Baghdad in the second half of the following century, using a global approach. We will see why Parisian pendulums were considered better than others and we will focus on the fundamental link between this type of instrument and the work and programme of the Observatory.[6]

After a description of the material characteristics and uses of the pendulum clocks at the Observatory (drawing on a variety of sources) in the first section of the chapter, in the second I will examine manuscript descriptions of the fundamental use of pendulums during the Academy's missions to Asia and Africa. The chapter concludes with a shift to Baghdad on the eve of the revolutionary French events. A geographical leap that is indispensable to show how techniques born and maturated at the *Observatoire royal* between the seventeenth and eighteenth centuries spread through institutional and privileged channels to reach the Iraqi city.

---

[5] In the following pages, unless otherwise stated, I will use the terms clock, pendulum clock, and pendulum synonymously. The term telescope always translated the French term *lunette*.

[6] In 1666 and 1667 respectively, the Paris Academy and the Observatory were founded. The latter's initial plan encompassed all of the Academy's fields of study, observation and experiment, as well as all of its collections and instruments. The problem of longitudes (initially both at sea and on land) held a central place in the research conducted at the Observatory, with in view the improvement of French geographical knowledge and mapmaking (of France, but also of distant places such as Asia or the coasts of Africa). The main method used by the Observatory members was the simultaneous observation from distant places on Earth of the eclipses of the satellites of the planet Jupiter—an idea that Galileo but could not put into practice, and that was only adopted by Cassini I in 1668. Since a century before Louis XIV, clocks had formed the core of the method to deduce longitude proposed by mathematician and cartographer Gemma Frisius, by timing the distances to be covered and estimated on ships. Given the importance of the question of longitude in the Observatory's initial project, it is hardly surprising that clocks received so much attention from the outset. On the quest for longitude, see Andrewes (1998), Sobel (1998) and Grillot (1976), and for the eighteenth century in particular Boistel (2006). On time and space in the seventeenth century, see Blamont (2001) and Costabel (1978). On scientific journeys coordinated by the Observatory astronomers, see Débarbat (1997, 1999), Dew (2008, 2010a, 2010b, 2011, 2015) and on these satellites, see Wilson and Débarbat (1989); on time measurements see Landes (1987).

Material Characteristics and Main Uses of Pendulums at the Observatory

When admiring the south facade of the Observatory attention should be paid to the two low reliefs that decorate its central part (Fig. 3.1).[7] Sculpted in 1677 by François Temporiti[8] these bas-reliefs are reminiscent of those on the walls of the Louvre. Among the symbols of greatness belonging to the astronomical world are two pendulums, one on each side. Both are beautiful, elongated instruments, one with a square box, the other rhomboidal. This sculpture on the walls of the "*Maison d'Uranie*,"[9] with very strong symbolic significance, helps us to reconstitute the exterior aspect of the objects featured in this chapter. One can compare the illustrations of Huygens' *Horologium oscillatorium* with the representations on the facade (Fig. 3.2): they are indeed the same type of object.

The pendulum clock, as much as the telescope, the other symbol of the Observatory's techniques and practices, was also depicted on the frontispieces of certain publications of the Academy. To the best of my knowledge (and unlike many *lunettes*, *verres* or other instruments used in the practice of the Academy of Sciences such as barometers and thermometers), no pendulum used at the Observatory mentioned in letters and manuscripts has survived to the present day. Thus, the morphology and characteristics of the objects in question are an issue in themselves.[10] Before considering the letters describing these pendulums, let us take a look at the state of French pendulums during the century of Louis XIV.

The pendulums that have come down to us, according to Pierre Kjellberg and Jean-Dominique Augarde, were luxury objects built and used essentially for decorative purposes, in great demand since the seventeenth century.[11] They were characterised by allegorical representations in their decoration throughout the reign of Louis XIV.[12] According to Kjellberg's study, the *Grand Siècle* pendulums essentially have three

[7] On material culture and objects the literature is huge. See Roche (1997), Findlen (2013), Gerritsen and Riello (2015), and the several publications by Bernasconi with special attention to clocks (2016, 2016, 2015). On scientific instruments see Daumas (1953), Morrison-Law, Schechner, Brenni (2017), and the recent special issue of *Artefact*, edited by Lamy (2023). See also the online dictionary on instrument-makers and craftsmen Beaudouin, Brenni and Turner (2018).

[8] Savoie (2015).

[9] Huygens (1673).

[10] In the presentation of an abridged version of this text at the workshop held at the University of Neuchâtel (December 2021), the question raised by Paolo Brenni was pertinent: what does posterity hold for these objects? By ruining pendulums with use or weather conditions (particularly the wooden parts), the best parts in the most stable objects were surely recycled, because they were obviously invaluable. Which explains at least in part why very few pendulum specimens from the Observatory, especially from this period, have come down to us. We must add that Wolf (1902, pp. 139–140) writes about a clock made by Thuret, consisting of a silk thread and a pendulum with a movable slider to regulate it.

[11] Augarde (1996), Kjellberg (1997).

[12] Augarde (1996, p. 126). As the book is organised by theme, several parts are dedicated to the Grand Siècle. See in particular "L'état des pendules à la mort de Louis XIV," starting at p. 247. See also Kjellberg (1997), the section "Le siècle de Louis XIV," starting at p. 32.

Fig. 3.1 Inconnu, "Observatoire de Paris: bas-reliefs de la façade sud par Temporiti," Bibliothèque numérique-observatoire de Paris, https://bibnum.obspm.fr/ark:/11287/3R7h4

forms: one called "*religieuse*" (the front of the object, where the dial was located, resembled the façade of a church), the more curvilinear "doll's head" pendulums, and the taller, more austere objects called "*pendules de parquet*".[13] According to Augarde, the pendulums that could beat the accuracy to the second were essentially the *parquet* ones, the most imposing objects, with a curved shape.[14] A specific length for pendulums beating to the second is also confirmed by the letters in our possession, as we will see shortly.

Halfway between the category of luxury decorative pendulums and those used for astronomical observations was the category of decorated astronomical pendulums, which provided information on one or more aspects of the sky or the calendar.[15] There are today several exemplars of various objects of this type for this era and the famous Fardoil pendulum, maintained in precise working order today at the *Observatory*, must be mentioned in this context. Pierre Fardoil (died around 1725), clockmaker to the king and author of numerous machines for the Academy, was the founder of a dynasty of watchmakers of the same name working in the French

[13] Kjellberg (1997, pp. 36–65).

[14] Augarde (1996, p. 124).

[15] Many examples of this kind of clock can be found in the two mentioned books.

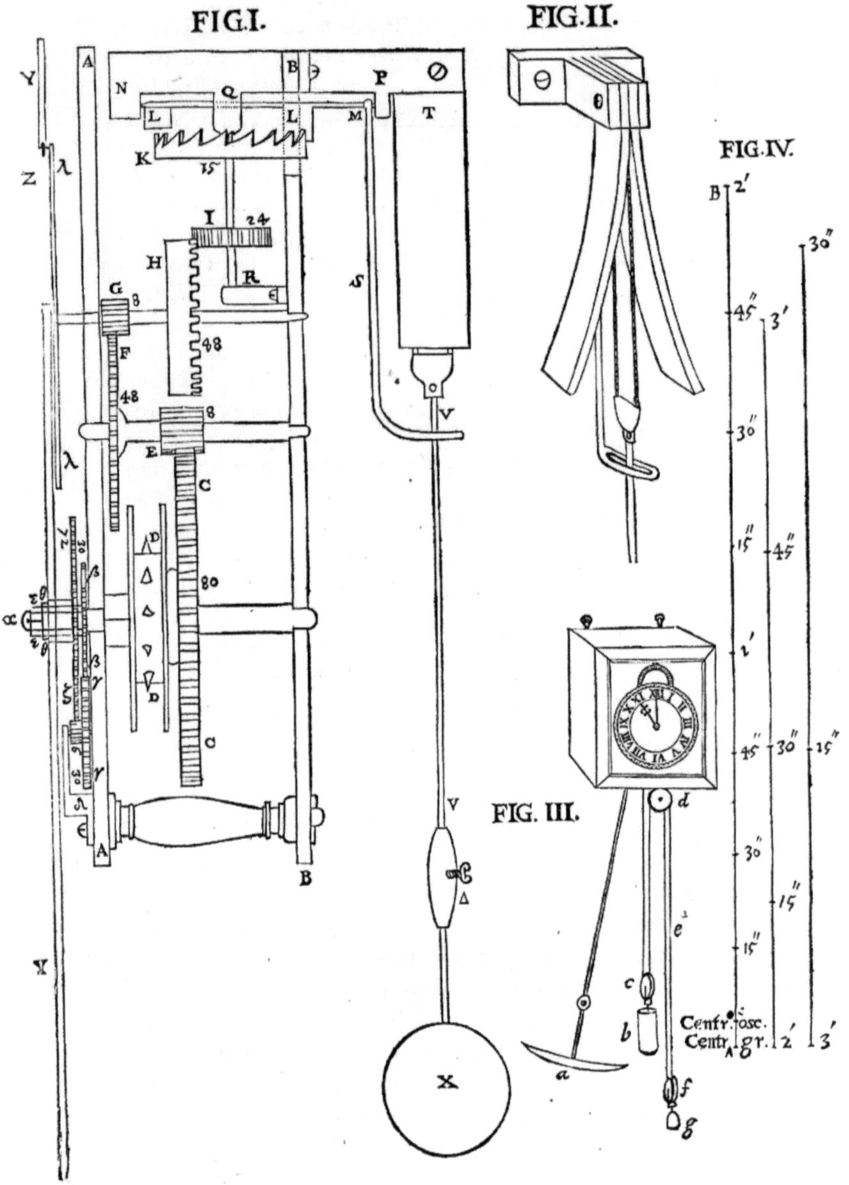

Fig. 3.2 Huygens, C. (1673). *Horologium Oscillatorium*. Paris: Muguet, p. 4. Gallica-Observatoire de Paris, ark:/12148/bpt6k3360z

capital. According to Jean Feray (author of the only existing contribution on this object), the astronomical clock, built between 1730 and 1739 by Pierre Fardoil,[16] could record a great deal of information on the state of the heavens—season, date of the month and respective lengths of day and night depending on the date, was similar in its characteristics to that of salon clocks. Feray speculates that the clock may originally have been a gift from Louis XIV to the Observatory itself.[17] Fardoil's clock, the parts of which are described in detail in the article, remains today one of the rare surviving witnesses to the role of time and space measurement under Louis XIV at the Observatory—and to its role approaching that of a museum during the *Grand Siècle*. Although this precious object was brought to the Observatory after the *Grand Siècle*, it embodied certain aspects of the elite that visited the building during that period.

Returning to letters and manuscripts, these give us some information on what the Observatory's clocks were like. The *Comptes des batiments du roi*, detailing the Academy and Observatory expenditures throughout the period we are interested in, also point to other shapes, probably less utilised but much closer to the François Temporiti ones and the ones used at the Academy and at the Observatory.[18] In response to Manfredi's letter quoted at the beginning of this chapter, Bianchini, in March 1702, wrote about "a minute and second clock, whose characteristic long pendulum makes it resemble French and English ones,"[19] providing some details about the instruments made in Rome by a mathematics teacher.

The beating of the second gave a significant length to these objects which, at the time, seemed to be characteristic of French clocks. In one of his most important publications, the *Istituzioni astronomiche*[20] on the state of astronomy at the time, Manfredi confirmed this in later years, stressing that long pendulums were the best for astronomical observations:

> For the exact determination of the timing of two observations taken at the same elevation, it is necessary to use a long pendulum clock, such as those, whose simple vibrations indicate the second, described by Huygens in his treatise on the Oscillatory Clock, and later refined by a few other later devices. It is not necessary for this clock to display the time [...], it is sufficient for its operation to be stable and regular.[21]

[16] On this clock, see the only existing article in the literature: Feray (1971, pp. 97–101). I would like to thank Suzanne Débarbat, Virginie Barbet and Robert Zeganadin of the Observatory library for their patience and the help provided for the research on the Fardoil clock.

[17] Feray (1971, p. 99).

[18] Among the clocks made by Martinot in 1685, one clock "shaped as an antique one, the other as a vessel (*cassolette*)," Geoffrey, *Comptes des bâtiments du roi*, vol. 2, p. 584.

[19] "The instruments made in Rome by Francesco Neri, professor of mathematics, including a clock with minutes and seconds that, with its long pendulum, resembles French and English ones." Letter from Francesco Bianchini to Eustachio Manfredi (Rome, 4 March 1702). Ms U 17, 20v. Vallicelliana Library, Rome.

[20] Manfredi (1749).

[21] Manfredi (1749, p. 41). Manfredi is referring here to Huygens's *Horologium oscillatorium*. A clock is also depicted in another publication by Manfredi (1736, p. 95), in a figure illustrating an angel geometer verifying the law of falling bodies through the clock.

The question of who designed the clocks for the Observatory is a complex one. Before attempting to answer it, it is necessary to make a premise. When we speak about clockmaking, we are referring to the construction of the device that gave movement to the pendulum (the mechanism was usually found inside the boxes). As each pendulum of this period was a complex machine, with various parts and materials, its construction required the artistry of various specialists working on the same object.[22] Kjellberg mentions three French families working during the *Grand Siècle* among the many clockmakers in France: the Thurets, the Martinots and the Hanets.[23] The *Comptes des batiments du Roy* testify that the Academy's pendulum clocks came in this period mostly from the workshops of Martinot and Thuret,[24] in particular Jacques Thuret.[25] Martinot and Thuret were the first clockmakers at the time to use the *balancier* as conceived by Huygens. Isaac II Thuret approached the Academy of Sciences, eventually becoming one of this institution's official clockmakers, thanks to a dense network of contacts that allowed him to get close to Huygens and have connections at the royal court.[26]

French clocks could also be characterized by a ringing striking device, at least if we go by the *Comptes des batiments du Roy*,[27] as the handwritten archives (such as the letters we have noted above or the Academy archives) make no mention of them. The striking device was very useful for astronomers, as they could count down the time without looking at the clock, and thus reduced the number of people present at an observation. Although astronomical observations (always made with various instruments and objects) generally involved several people working together (at least one at the *lunette*, the other at the clock). Astronomy, at the epoch is very often a matter of collaboration,[28] this device theoretically allowed a single person to observe at the lunette while keeping time on certain types of observations, such as the duration of passages. However, this type of device was very difficult to integrate into clock construction in the seventeenth and eighteenth centuries, as the beating interfered with the oscillations.

What made Observatory pendulums superior to others? As we are starting to see, the first part of the answer lies in their manufacture (including the artisans involved), which provided the main guarantee of quality These objects were expected first of all not to lag behind due to changes in physical or meteorological conditions. The second part of the answer concerns the calibration and usage techniques developed by the astronomers. Let us see how.

---

[22] The enamellers and the cabinetmakers, who looked after the box and its dial. In fact, the very term "clockmaker" can indicate various realities, see Augarde (1996, p. 9, pp. 13–14).

[23] Kjellberg (1997, p. 36). See also Augarde (1996, pp. 72-73), and the last part of the book, with a biographic encyclopaedia of the clockmakers working at the time in France.

[24] Guiffrey (vol. 1, pp. 475, 647, 723; vol. 2, p.584); Cardinal and al. (2008, p. 31 ff.).

[25] Letters by the astronomers of the Observatory confirm this (Deias 2020a and 2020b).

[26] For information about Martinot and Thuret, see Augarde (1996, pp. 72–73).

[27] Guiffrey (vol. 1, p. 647).

[28] Ruellet (2016).

For pendulums in this period one aspect of precision was the capacity of time-measuring objects to measure the smallest duration.[29] However, although important, measuring the smallest duration was not the main feature of good clocks at the time, as confirmed by the various manuals of the eighteenth century, including the *Essai d'horlogerie*, published in 1763 by the Swiss manufacturer Ferdinand Berthoud.[30] If we rely on Berthoud's lexicon, which indeed uses this term, it is not only the smallest part of the time that the clock signals, but also its stability, due to the fact—the result of a great deal of experimentation – that its manufacturing materials hardly expand or contract in response to temperature changes. As we shall also see in Abbé Beauchamp's observatory in Baghdad, the essential factor is stability or accuracy, that is the non-accumulation of delays or their ability not to jam. This was the main shortcoming of clocks carried on ships or exposed to strong winds or high temperatures. Too sensitive to humidity, temperature fluctuations, and the physical conditions of observation, they often accumulated significant delays.

Calibrating the clock was the astronomer's first indispensable act for any subsequent astronomical observation. At the Observatory, the pendulums were adjusted daily to the rhythms of the sky using a meridian line. During the seventeenth century, the presence of a meridian line is in this sense fundamental[31]: indicating the time of solar noon, it was possible to tell whether a clock was on time or had fallen behind, using the Equation of time from true noon to mean noon.[32]

For the observations at the Observatory, once calibrated within the walls of this place, the simplest use of pendulum clocks was for chronometric phenomena and determining the duration of passages (usually for stars, constellations or planets).[33] Thanks to the number of its oscillations, a pendulum could be used to deduce the duration of an astronomical event or experiment (such as the free fall of a body along

---

[29] This stability is clearly visible if we follow the sources over time. In 1668, the astronomer and academician Jean Picard experimented with clocks to get them to beat to the second. Later, academicians at the Louvre Library combined the use of large quarter-circles with two clocks, one for seconds, the other for half-seconds. See Wolf (1902, p. 136).

[30] See Berthoud (1763, for instance vol. 2, starting at the p. 221). Berthoud worked in Paris under Louis XV and was the first to receive the title of *Horloger Mécanicien du Roi et de la Marine* (Clockmaker to the King and the Navy).

[31] Wolf (1902, pp. 141–141).

[32] Although the clocks we use mark hours of perfectly identical duration (for obvious civil requirements), the Sun's movement does not correspond to such exact quantities. The Sun moves more slowly in summer than in winter, for example, and no two days have nights of exactly the same duration. The Sun's highest point on the horizon (noon) is therefore not at the time given by our clocks every day, but it is around 12 o'clock every day, at a different civil time. By the late seventeenth century, astronomers had already established an equation of time that allowed to find the time of "true" noon (solar time) in relation to civil noon (mean or mechanical time). Explanations of this and the mathematical expression of the equation can be found in any positional astronomy textbook or treatise of the time. I dealt with this subject in chapter 3 of my PhD dissertation (Deias 2020a). See also Heilbron (1999), as well as the Paris Observatory meridian website, in the section "*L'image du Soleil*," following "*Midi, à midi?*" (http://vo.imcce.fr/meridienne/equation.php, accessed 7 August 2023).

[33] See for instance among other sources the letter from Giovanni Domenico Cassini to Jean Picard (1672). Archives of the Paris Observatory, Ms B 4/9.

a tunnel, or the duration of an eclipse). From the second half of the seventeenth century, in observational practice, clocks were commonly associated with quarter-circles (*quarts-de-cercle*) of various dimensions.[34] Celestial coordinates of stars and celestial bodies (in particular for equatorial coordinates, known as right ascension and declination, independent of the observer's position on Earth)[35] were also the object of observation thanks to these clocks. We will see how this relates to the rhythm and organisation of collaborative work.

When observing eclipses, such as that of Jupiter's satellites, the mastery of timing, the knowledge of instrument reactions and the practical use of instruments were also fundamental, as each phenomenon presented its own difficulties and specificities during observations, and the astronomer must anticipate everything.[36] The relationship between the instrument and the observer, which Giovanni Domenico Cassini and his colleagues from the Observatory tried to teach and regulate through their texts, was also important, and it became common practice in observatories to create instructions that observed and mastered the astronomer himself, even during travel.[37]

Instructions for Astronomical Observations in Paris and Abroad

Pendulums were among the essential objects to be carried on journeys.[38] Indispensable to beat the pace of work at the Observatory within the walls of the building (and using the walls and the floor themselves for this), the pendulum clocks also had another vital function: they figuratively permitted to move the Observatory out of the

[34] The astronomer Gallet, in a letter to Cassini in 1682, mentioned the promise of a clock with a quarter-circle that had been made to him by Abbé Gallois from the Academy. See the letter dated 27 September 1682. Archives of the Paris Observatory, Ms B 4/10, piece no. 70. In his seminal work on the history of the Paris Observatory, Charles Wolf underlines the same use of a clock by Thuret: Wolf (1902, pp. 139–140).

[35] In Bianchini's *De Nummo Gnomone Clementino* (1703), certainly made by the clockmaker Thuret. On September 6, 1702, Bianchini enthusiastically informed Manfredi that the clock Cassini had ordered for him, similar to those Cassini used at the Observatory, was ready. This was undoubtedly the instrument Bianchini used the following October 30 in Rome to measure the difference between the horizontal celestial coordinates of the Sun and the star Sirius, the brightest star in the night sky. In addition to Cassini's advice from Paris on the construction of the Rome meridian, Bianchini was able to count on the Italian's mediation in Paris for the circulation of high-precision clocks. As we can see from one of Bianchini's figures in *De Nummo Gnomone Clementino*, a telescope positioned on the same axis as the meridian can be used to observe the passage of a star and the Sun at noon during the day.

[36] See for instance the *Instruction générale pour les observations géographiques à faire dans les voyages*, in PVARS, December 1681, 9bis, fol. 146v–147r.

[37] See Schaffer (1988) on the framing of observational practices by astronomers at Greenwich for their students from the eighteenth century onwards, and the role of observatory techniques in this process. See also Deias (2020a), for the case of the Paris Observatory.

[38] See the several publications of Dew in the reference list on the journeys undertaken by the Academy, in particular linked with these *Instructions*. See also Deias (2020a).

building, i.e. to recreate the observation conditions (almost as inside the Observatory) even in distant locations, for example in the open countryside or on a mountain whose geographical situation one wants to know about. This was true outside France. Pendulums were certainly a fundamental tool for orientation in foreign lands, particularly on cartographic voyages where astronomical measurements had to serve the practical purposes.

In fact, in whatever place people were observing, the pendulum was the most important element in deducing for that place the structure of the sky, a fundamental step in any observation. Very generally, the Observatory-trained astronomer on voyages must totally master the knowledge of instruments among which surely were, as Jean Deshayes wrote to Cassini, to "*regler et connoistre les horloges* (setting and knowing clocks)."[39] An emblematic example in this regard are the voyages to Asia and Africa organized in conjunction with the Academy by the Parisian astronomers from 1681. In these voyages, they also had to deal with collaboration and the work group.

A series of memoirs and instructions were drafted by the academic astronomers (one can find them among Cassini's manuscripts, but certainly other royal astronomers and Cassini's colleagues such as Jean Picard or Philippe de la Hire, helped to write them) so that the transport of the entire Observatory observation conditions along with the pendulums was done as efficiently, conscientiously and accurately as possible, minimizing the subjectivity of the astronomers and the impact of new observing conditions.[40]

The *Instruction générale à faire dans les voyages* (*General Instruction to Be Made During Voyages*), presented to the Academy in 1681, make the role of pendulums particularly visible, especially through a series of techniques conceived at the Observatory. Proceeding chronologically through the journey, the first task was to transport the best pendulums *en route* after synchronizing them with those on the Observatory's sundial, whose alignment was maintained using a plumb line. These had to thereafter be arranged at the appointed time and at the place it was deemed convenient to observe. Outside the Observatory, especially in foreign places far from the Observatory and France, the main role of this instrument was to help the astronomers use the celestial time to determine their spatial position. Thus, the first use was to reconstruct the path and positions of the Sun (first and foremost, at Noon). These techniques (described in the *Instruction générale*) relied almost exclusively on the

[39] In a letter dated January 29, 1682, from Cassini I's astronomer-pupil Jean Deshayes, the astronomer diligently repeats to his master the instructions the Italian gave him before leaving, stressing the importance of the clocks. See the letter from Deshayes to Cassini (Dieppe, 29 January, 1682), n. p. University Library, Pisa, Ms 423, fasc.20, pièce num.4, and Deias (2020a, annexe R).

[40] I am referring here in particular to: *Instruction générale pour les observations géographiques à faire dans les voyages*. Archives of the Paris Academy of Sciences, Procès-verbaux de l'Académie Royale des Sciences de Paris (PVARS), December 1681, vol. 9bis, fol. 141r–147v; *Instruction particulière pour le voyage des Canaries*, PVARS, 1681, vol. 9, fol. 135r–137r; *Instruction pour les observations qu'il faudra faire dans l'Isle de S.t Thomé*, PVARS, 1681, vol. 9bis, fol. 150v–154r. Archives of the Paris Observatory, *Instruction à ceux qui travailleront à l'observatoire par Cassini* I, Ms D 1/13, item no.8, fol. 1r–9v. All these documents have been also transcribed in Deias (2020a).

use of pendulums. This included taking symmetrical measurements of the Sun's elevation to calculate its position, a task that would have taken much longer without a clock.

With the exact calculation of this position, it was finally possible to set up a sundial, as was done at the Observatory. This enabled the study of the solar image, the various key positions along the Sun's path, of the Moon, and the position of the fixed stars relative to the Sun.

The interesting aspect of this method lies precisely in the expression "*a une minute prez* (down to the minute),"[41] used by the savants: clocks beating to the second made it possible to know the position of Noon to the nearest minute, a measurement to which other observations, and therefore the coordinates of the stars, can be related. In the countryside, for example, or in foreign lands, this is a good accuracy for knowing the motions of the sky.

Importantly, the use of the pendulum often preceded and enabled observation through the telescope. By observing the elevation and right ascension of the Sun and the fixed stars, astronomers could deduce the time difference for the meridian passage of the latter. This was fundamental information, as we will also see in the next section, because the meridian passage of stars at different latitudes was crucial for improving star atlases and for verifying the phenomenon of refraction.

The study of solar motion and sky times was indispensable to the visualization and reconstruction of the space (celestial and extraterrestrial) surrounding the observers: *en route*, it was also a matter of deducing the elevation of the celestial Poles, as well as the spatial morphology of the surrounding environments, particularly at sea.[42]

A series of cross-observations of the Sun's edges and elevations with telescopes and pendulums, made several times a day and over several days, also made it possible to ascertain whether one or more pendulums delayed or advanced the hour of Noon (or the rhythm of the sky in general).[43]

Although there were various methods (direct or indirect, to be used by day or night) for calculating the celestial coordinates of the stars, astronomical publications, as well as many contemporary letters, agree that the most accurate is that which Eustachio Manfredi calls the "*orologio esatto* (exact clock)" method in his *Istituzioni astronomiche*.[44] This is simply the published version, appearing several years later, of the same technique outlined in the *Instruction générale* at the Observatory for tracing the movements of the Sun. Normally used with a pendulum beating the second, this method operated on the same principle as that employed by traveling academicians, who used it to calculate the Sun's right ascension and declination.

Just like the clockmaker who first made it and set it in motion, the observing astronomer had first to know the technical object. Given the sensitivity of most pendulums to any touch of the hand, blow of the wind, change of temperature,

---

[41] *Instruction générale pour les observations géographiques à faire dans les* voyages, fol. 142v.

[42] See the *Instruction générale*, fol. 141r–147v.

[43] See the *Instruction générale*, fol. 141r–147v.

[44] Manfredi (1749, p. 147).

knowing how to verify and on occasion arrange for the repair of the pendulum was indispensable to its use.[45]

Of course, even while traveling, the pendulums themselves were also the subject of study, of improvement. The *Instruction* referred to the different behaviour of pendulum oscillations at Cayenne, that is, at the equator, observed in 1672 on a journey organized by the Academy.[46]

Synchronising clock noon with the noon indicated by the meridian line was the preferred method during the seventeenth century. However, other methods were also used, in particular those based on the observation of fixed stars (which would become the most widely used method in the following century). Another method involved using distant sights to set the clock based on the known duration of a celestial event passing through them (for example, the eclipse of the planet Mercury as observed from the Luxembourg Tower).

The regulation of clocks on fixed stars was developed at the Observatory before 1681, in particular by Philippe de la Hire, according to the minutes (*procès verbaux*).[47] This technique was first only experimented with at the Observatory, then improved until it became prevalent, as one of the synchronisation methods that flank and correct on the sundial. What does this method consist of? Using the measurement for the diurnal acceleration of fixed stars found by Jean Picard at the Observatory, an astronomer had to observe for two days following the instant of the passage of a star on the meridian, or with a telescope or simply observing the disappearance of the star behind any building. If on the second day a shift for this star of less than 3′56″ was found, the astronomer could be sure that the watch was well adjusted.[48]

Berthoud took up the idea that to check the accuracy of an astronomical clock. As a matter of fact, the method based on the movements of the fixed stars was indisputably the best in his day:

> The most accurate method had to be used, since it meant not neglecting half a second in 24 hours; yet the passage of the sun through a dial or an ordinary meridian could not give such great accuracy. The most convenient and accurate method is to use the passage of the fixed stars through the meridian, or through a certain point in the sky.[49]

In general, pendulums also contributed to one of the fundamental objectives of the Observatory's sciences and techniques: the management of errors in measurements and instruments, both within and beyond its walls.

The *Instructions* referred first and foremost to simple carelessness on the part of the observer and inevitable instrument failures. These errors could easily be corrected

---

[45] See the letter that Deshayes sent to Cassini from Dieppe in 1682.

[46] See the *Instruction générale, fol. nom. 144ve.*

[47] "Monsieur de la Hire proposed that of the two clocks to be made in the Observatory by order of Monsignor Colbert, one should be set to the fixed stars." PVARS, 13 December 1681, 9bis, fol. 134v.

[48] Lalande described this method in detail. See Lalande (1764, p. 1520).

[49] Berthoud (1763, vol. 1, pp. 122–128, quot. p. 122).

by cross-checking and repeated observation of the same phenomenon.[50] With rare exceptions, the most reliable value was the one that appeared most frequently, consistently found across repeated measurements, and the least likely to be affected by phenomena such as parallax or refraction, which were known to astronomers but only approximately accounted for. Networking observations also increased the reliability of certain measurements, especially when they could be compared with those produced by the Observatory, which were reputed to be more accurate.[51] As well as repeating measurements, astronomers sought to identify systematic errors due to instrument faults.

In the original idea of the Observatory according to Cassini, one of the main objectives of this building (and in general of a building made for observing the sky) was to be able to function as a large material instrument, with the purpose, first of all, of reducing measurement errors, which in all instruments, especially small ones, were inevitable.[52] The observations made in the network made this (among other things) also possible.

Beauchamp's Covered Terrace in Baghdad, and His Pendulums (1786–1789)

An example of the application of the Observatory's techniques outside its walls (and the uses of pendulum clocks) was the covered terrace built in Baghdad by Joseph de Beauchamp, abbot, vicar-general of Babylon, vice-consul of Baghdad and corresponding member of the Paris Academy.[53]

Sent to Iraq in 1782, his assignment was mostly scientific. Beauchamp was trained in Paris at the Observatory by Joseph Jérôme Lalande,[54] learning to apply all the techniques and tricks seen in this chapter, many of which went back to Giovanni Domenico Cassini (Cassini I) and other academicians such Picard or de la Hire. Beauchamp's scientific mission focused on observing the stars from a different latitude, improving the mapping of Iraq, gaining a better understanding of its maritime areas and lakes, and, more generally, reporting to France all the distinctive features of the Iraqi sky and landscape. The observatory operated for three years, from 1786

[50] See for instance the *Instruction générale*. See also the letter from Cassini to Picard (Paris, 14 January 1672). Archives of the Paris Observatory, Ms B 4/9, item no. 101.

[51] Among the publications on the network of observations (linked to the Observatory culture) see the several works by David Aubin. The most recent is Aubin (2015).

[52] Deias (2020a). See also the correspondence between Cassini and Manfredi.

[53] See on this subject the literature on Beauchamp in the reference list, especially Bigourdain (1930), Potts (2018) and Pierre (1990).

[54] See the referenced publications by Lalande in the reference list. For the contents of astronomy taught to Beauchamp, see *L'Astronomie* (1764). For Beauchamp's mission, see the *Bibliographie astronomique* (1803); for his life and instruments the *Notice sur la vie et les travaux du C. de Beauchamp* (1802).

until 1789 and French institutions covered all the fees for instruments, equipment and repairs for this site.[55]

In addition to being an understudied research subject, the significance of Beauchamp's observatory lies in its confirmation of the techniques used at the Paris Observatory and in Parisian astronomy more broadly, particularly the fundamental role of pendulums in both the physical construction of the observatory and in the observations themselves.

Beauchamp's observatory began with instruments transported from France. Some were entrusted to him directly (with others arriving later) by Lalande, Jacques Cassini (Cassini II) or Antoine-Laurent Lavoisier. These included pendulums crafted in France by the royal clockmaker Jean-André Lepaute.

Let us focus on the pendulum clocks at the Baghdad Observatory, because Beauchamp's scientific mission, particularly the establishment of his observatory would take us too far away from the issue of this article.[56] In his observation manuscript journal,[57] from 1786 to 1789, Beauchamp mentioned two pendulums, one to seconds (called "the big pendulum"[58]) and one to half seconds, both made by Lepaute, in addition to a few portable pendulums, probably smaller than the others, about which we have very little information. In full accordance with the tradition of the Paris Observatory, pendulums served three primary functions: tracing the observatory's sundial, measuring the duration of celestial passages, and supporting all astronomical tasks linked to timekeeping. The two pendulums were indispensable for celestial observations, as they were used alongside telescopes, especially the meridian telescope (*lunette méridienne*), which was typically paired with the large pendulum.

With the meridian *lunette* and the large pendulum beating the second, and using the methods previously discussed, Beauchamp mapped the structure of the sky above his terrace and materially organised all the elements of his observatory. Exactly above the meridian *lunette*, Beauchamp arranged a lamp to make the clock visible from the telescope.[59] Thanks to this lamp, the observation of the meridian telescope and the measurement of the passage times were theoretically possible by only one person.

---

[55] Lalande (1787, p. 301).

[56] On this subject, see Bret and Deias (2024).

[57] Bibliothèque de l'Institut de France, Paris, Ms 2196, Pierre-Joseph de Beauchamp, *Journal d'observations astronomiques, meteorologiques, physiques,* etc. *(1781–1785)* (hereafter, Beauchamp, Journal, vol. 1); Bibliothèque nationale de France, Paris, Manuscrits, Nouvelles acquisitions françaises, Ms 10157, Pierre-Joseph de Beauchamp, *Second volume d'observations astronomiques (1786–1787)* (hereafter, Beauchamp, Journal, vol. 2); Bibliothèque de l'Institut de France, Paris. Ms 2197, Pierre-Joseph de Beauchamp, *3e volume d'observations astronomiques faites à Bagdad en Asie par Pierre Joseph de Beauchamp, Vicaire g[é]n[ér]al de l'évéché de Babilone, Correspondant de l'Academie royale des sciences de Paris, 1788* (1788–1789, 1795) (hereafter, Beauchamp, *Journal*, vol. 3).

[58] "I prefer to take the passage of the planets with the big clock that I have set on the first mobile because the escapement is stronger and more sensitive. The passage will be more stable." Beauchamp, Journal, vol. 2, 3 October 1788, fol. 127r.

[59] A small cardboard reflector allowed not to disturb the observations (Beauchamp, *Journal*, vol. 2, 25 January 1787, fol. 55v).

Since 1786, this arrangement had enabled him to make very accurate observations at the meridian of the planet Mercury—which is very difficult to observe at Parisian latitudes and which Lalande particularly recommended to him—to observe various constellations, as well as all the stars of sixth magnitude (brightness and apparent diameter) of John Flamsteed's *Atlas*.[60]

Although the half-second pendulum was also useful for observations, Beauchamp preferred the seconds pendulum, which was larger but less sensitive to wind or positional shifts. By this point, Beauchamp had fully adopted the technique of synchronizing clocks using fixed stars: a method developed at the Paris Observatory. The main clock was synchronized accordingly. Beauchamp's second pendulum was adjusted, as needed, either to the stars or to noon, with the aid of the sundial he had drawn on his terrace.

The techniques and objects mentioned by Manfredi in 1702 circulated as far as Beauchamp's personal terrace. Early revolutionary events forced Beauchamp to abandon his observations in Baghdad, forgo the instruments promised to him by France, and leave.

Conclusion

In this chapter, I deliberately cross-referenced major eighteenth century publications on astronomy and horology with various manuscript texts dating from the second half of the seventeenth century to the eighteenth century. This approach allowed me to highlight both the status of pendulums at the Paris Observatory and the significant circulation of knowledge associated with these objects. The Observatory and its instruments developed their reputations in tandem, each reinforcing the other.

The Observatory was one of the earliest sites of celestial observation to leave behind a substantial corpus of texts and manuscripts detailing observation techniques and outlining a structured astronomical observation program. These documents attest to the fact that pendulum clocks were considered fundamental and indispensable to astronomy at the time, especially when used alongside *lunettes* and *quarts de cercle*. I focused on the pendulums of the Observatory, objects for which no direct material traces remain but which nevertheless bear witness to an important observational tradition. More than any other object, pendulums enabled synchronisation between the site and the instruments with the rhythm of the sky. The observational techniques described in this chapter show that pendulum clocks made it possible to measure the duration of phenomena, and the cadence of observations. They also played a critical role in addressing the essential issue of defining and minimizing observational errors.

We have also seen that in the Observatory's program placed great emphasis on missions to other continents beginning in 1671 and even more from 1681 onward. These missions focused on key locations around the globe in terms longitude, such as Asia. Through its expanding network, the Observatory was able to function as

[60] Fortin (1776).

though its operations extended well beyond the French borders. Pendulums, more so than telescopes, enabled astronomers to conduct precise observations in distant locations. In particular, the mapping of the Earth and research on longitude became possible through on-site observations using the Observatory's techniques and instruments (which were more accurate, more reliable, and less prone to error). Joseph de Beauchamp's observatory on the terrace in Baghdad represents the latest example of this circulation of knowledge regarding the pendulum and the Paris Observatory's observational tradition. Thanks to Parisian instruments, particularly its pendulums, Beauchamp was able to establish a functioning, self-contained observatory under the Iraqi sky.

During the eighteenth century the use of pendulums was replaced by that of marine chronometers. However, the few specimens of pendulums that have come down to us do testify—paradoxically—to the importance of these objects.

References

Manuscripts

Ms 10157. Pierre-Joseph de Beauchamp, *Journal d'observations astronomiques (1786–87)*, vol. 2. Bibliothèque nationale de France, Paris.

Ms 2196. Pierre-Joseph de Beauchamp, *Journal d'observations astronomiques, meteorologiques, physiques, etc. (1781–85)*, vol. 1. Bibliothèque de l'Institut de France, Paris.

Ms 2197. Pierre-Joseph de Beauchamp, *Observations astronomiques faites à Bagdad en Asie par Pierre Joseph de Beauchamp, Vicaire g[é]n[ér]al de l'évéché de Babilone, Correspondant de l'Academie royale des sciences de Paris (1788–1789)*, vol. 3. Bibliothèque de l'Institut de France, Paris.

Ms 423. University Library, Pisa.

Ms U 17. Vallicelliana Library, Rome.

Ms B 4/9. Archives de l'Observatoire de Paris.

Ms D 1/13. Archives de l'Observatoire de Paris.

Procès-verbaux de l'Académie Royale des Sciences de Paris (PVARS), vol. 1–23. Archives de l'Académie des Sciences, Paris.

Printed Texts

Augarde, J.-D. (1996). *Les ouvriers du temps. La pendule à Paris de Louis XIV à Napoléon 1er.* Antiquorum.

Andrewes, W. J. H. (Ed.). (1998). *The Quest for Longitude: The Proceedings of the Longitude Symposium.* Collection of historical scientific instruments, Harvard University.

Andryesse, C. D. (1998). *Christian Huygens. Biographie.* Michel.

Aubin, D. (2015). L'observatoire. Régimes de spatialité et délocalisation du savoir. In D. Pestre, S. van Damme, K. Raj & O. Sibum (Eds.), *Histoire des sciences et des savoirs. De la Renaissance aux Lumières,* vol. 2, pp. 55–72. Seuil.

Beaudouin, B., Brenni, P., Turner, A. (2018). *A Bio-bibliographical Dictionary of Precision Instrument-Makers and Related Craftsmen in France and Switzerland 1430–1960.* Paris

Sciences & Lettres (PSL). Retrieved 2 Aug 2023 from https://bibnum.explore.psl.eu/s/dictionarypim/page/prolegomena.

Bernasconi, G. (2016a). L'objet comme document: Culture matérielle et cultures techniques. *Artefact, 4*, 31–47.

Bernasconi, G. (2015). *Objets portatifs au Siècle des Lumières*. CTHS.

Bernasconi, G. (2016). Luxe, précision et mesure du temps: une montre de John Arnold comme document matériel. In L. Hilaire-Pérez, F. Simon & M. Thébaud-Sorger (Eds.), *L'Europe des sciences et des techniques, XVIe–XVIIIe siècles* (pp. 77–80). Presses universitaires de Rennes.

Berthoud, F. (1763). *Essai d'horlogerie*, 2 vols. Jombert, Musier and Pancoucke.

Berthoud, F. (1802). *Histoire de la mesure du Temps par les horloges*. 2 vols. Imprimerie de la République.

Bianchini, F. (1703). *De nummo et gnomone clementino*. Aloisius.

Bigourdan, G. (1930). Les instruments et les observations de P. -J. de Beauchamp. *Comptes Rendus De L'académie des Sciences, 190*, 697–700.

Blamont, J. (2001). La mesure du temps et de l'espace au XVIIe siècle. *Dix-Septième Siècle, 213*, 579–611.

Boistel, G. (2006). De quelle précision a-t-on réellement besoin en mer? Quelques aspects de la diffusion des méthodes de détermination astronomique et chronométrique des longitudes en mer en France, de Lacaille à Mouchez (1750–1880). *Histoire & Mesure, 21*(2), 121–156.

Bret, P., & Deias, D. (2024). The astronomical observatory of the Abbot de Beauchamp in Baghdad: Instruments, observations, and visual sources. *Nuncius, 40*(1), 134–167.

Cardinal, C., & Arizzoli-Clementel, P. (2008). *Ô temps! Suspends ton vol: Catalogue des pendules et horloges du Musée des arts décoratifs de Lyon*. EMCC.

Cassini, G. D. (1668). *Ephemerides bononienses mediceorum syderum*. Tinassi.

Cavazza, M. (2002). The institute of science of Bologna and the royal society. *Notes and Records of the Royal Society of London, 56*, 105–123.

Cavazza, M. (1990). *Settecento inquieto. Alle origini dell'Istituto delle scienze di Bologna*. Il Mulino.

Costabel, P. (1978). Isochronisme et acceleration, 1638–1687. *Archives Internationales D'histoire des Sciences, 28*(102), 3–20.

Daumas, M. (1953). *Les instruments scientifiques aux XVIIe et XVIIIe siècle*. Presses universitaires de France.

Débarbat, S. (2006). La méridienne di Cassini à l'Observatoire de Paris. *Giornale di Astronomia, 32*, 54–58.

Débarbat, S., & Dumont, S. (1999). Les académiciens astronomes, voyageurs au XVIIIe siècle. *Comptes Rendus De L'académie des Sciences, 327*, 415–429.

Débarbat, S., & Dumont, S. (1997). *Les débuts de la cartographie scientifique: l'apport des astronomes français*. Académie Royale de Belgique.

Débarbat, S., & Wilson, C. (1989). The Galilean Satellites of Jupiter from Galileo to Cassini, Römer, and Bradley. In R. Taton & C. Wilson (Eds.), *The General History of Astronomy, Vol. IIA: Planetary Astronomy from the Renaissance to the Rise of Astrophysics* (pp. 144–157). Cambridge University Press.

Deias, D. (2020a). *Inventer l'Observatoire. Sciences et politique au temps de Giovanni Domenico Cassini (1625–1712)*. PhD dissertation, Ecole des hautes études en sciences sociales, Paris.

Deias, D. (2020b). The Paris Observatory in the Construction of the Early Modern Ecosystem of Knowledge (1669–1712). In M. Feingold & G. Giannini (Eds.), *The Institutionalization of Science in Early Modern Europe* (pp. 138–173). Brill.

Dew, N. (2010a). Scientific travels in the Atlantic world: the French expedition to Gorée and the Antilles, 1681–1683. *British Society for the History of Science, 43*(1), 1–17.

Dew, N. (2011). *Orientalism in Louis XIV's France*. Oxford University Press.

Dew, N. (Ed.). (2008). *Science and Empire in the Atlantic World*. Routledge.

Dew, N. (2010). Vers la ligne. In N. Dew (Ed.), *Science and Empire in the Atlantic World* (pp. 52–72). Routledge.

Dew, N. (2015). Un Colbertisme scientifique? In D. Pestre, S. van Damme, K. Raj & O. Sibum (Eds.), *Histoire des sciences des savoirs. De la Renaissance aux Lumières*, vol. 1, pp. 431–445. Seuil.

Eyraud, C. H. (2004). *Horloges astronomiques au tournant du XVIIIe siècle: de l'à-peu-près à la précision.* PhD dissertation, Université Lumière Lyon 2.

Feray, J. (1971). La pendule astronomique de Fardoil de l'Observatoire de Paris. *Les Monuments Historiques De France, 4*, 97–101.

Findlen, P. (Ed.). (2013). *Early modern things: Objects and their histories, 1500–1800.* Routledge.

Fortin, J. B. (1776). *Atlas céleste de Flamsteed approuvé par l'Académie royale des sciences et publié sous le privilège de cette compagnie.* F. G. Deschamps.

Galilei, G. (1610). *Sidereus Nuncius.* Baglionum.

Gerritsen, A., & Riello, G. (Eds.). (2015). *Writing Material Culture History.* Bloomsbury.

Giuntini, S. (2001). Il carteggio fra i Cassini e Eustachio Manfredi (1699–1733). *Bollettino di Storia Delle Scienze Matematiche, 21*, 7–180.

Grillot, S. (1976). Le problème des longitudes sur Terre. *Vistas in Astronomy, 20*, 81–84.

Guiffrey, J. (Ed.). (1881–1901). *Comptes des Batiments du roi sous le regne de Louis XIV*, vol. 5. Imprimerie Nationale.

Heilbron, J. L. (1999). *The Sun in the Church.* Harvard University Press.

Huygens, C. (1673). *Horologium oscillatorium sive de motu pendulorum ad orologia aptato demonstrations geometricae.* Muguet.

Kjellberg, P. (1997). *Encyclopédie de la pendule française. Du Moyen-âge au XXe siècle.* Les éditions de l'Amateur.

Lalande, J. (1764). *L'Astronomie*, vol. 2. Desaint et Saillant.

Lalande, J. (1787). Nouvelles littéraires. *Journal des savants* (May), p. 301.

Lalande, J. (1802). Notice sur la vie et les travaux du C. de Beauchamp. *Magasin encyclopédique*, pp. 386–391.

Lalande, J. (1803). *Bibliographie astronomique.* Imprimerie de la République.

Lamy, J. (Ed.). (2023). Le renouveau de l'histoire des instruments scientifiques. *Artefact* 17.

Lamy, J. (2007). *L'Observatoire de Toulouse aux XVIIIe et XIXe siècles. Archéologie d'un espace savant.* Presses universitaires de Rennes.

Landes, D. (1987). *L'heure qu'il est. Les horloges, la mesure du temps et la formation du monde moderne.* Gallimard.

Manfredi, E. (1736). *De gnomone meridiano Bononiensi ad divi Petronii deque observationibus astronomicis eo instrumento ab ejus constructione ad hoc tempus peractis auctore Eustachio Manfredi.* Lelio della Volpe.

Manfredi, E. (1749). *Instituzioni astronomiche opera postuma del dottore Eustachio Manfredi.* Dalla Volpe.

Mascheretti, P. *La misura del tempo: prospettiva storica.* PhD dissertation, Dipartimento di Fisica "A. Volta," Università di Pavia.

Morrison-Low, A. D., Schechner, S. J. & Brenni, P. (2017). *How Scientific Instruments Have Changed Hands.* Brill.

Pantin, I. (Ed.). (1992). *Galileo Galilei, Sidereus nuncius. Le Messager céleste (1610).* Introduction, édition critique, traduction et commentaire par Isabelle Pantin. Les Belles Lettres.

Pastoureau, M. (1979). *La cartographie des origines au XVIIIe siècle.* Société pour la science.

Pelletier, M. (2002). *Les Cartes Cassini, la science au service de l'état et des régions.* CTHS.

Pestre, D., van Damme, S., Raj, K., & Sibum, O. (Eds.). (2015). *Histoire des sciences et savoirs. De la Renaissance aux Lumières*, vol. 3. Seuil.

Pierre, M. L. (1990). *Pierre-Joseph Beauchamp (1752–1801), missionnaire, astronome, géographe, diplomate.* Lycée Claude-Mathy.

Potts, D. T. (2018). "*Observatorium in Bagdad constructum*". the observatory of "Babylon" from Ctesias to Louis XVI. *Canadian Society for Mesopotamian Studies Journal, 13*, 5–14.

Raj, K. (2007). *Relocating Modern Science: Circulation and the Construction of Knowledge in South Asia and Europe, 1650–1900.* Palgrave Macmillan.

Raj, K., Schaffer, S., Roberts L., Delbourgo J. (Eds.). (2009). *The Brokered World, Go-betweens and Global Itelligence, 1770–1820*. Science History Publications.
Roche, D. (1997). *Histoire des choses banales: naissance de la consommation dans les sociétés traditionnelles (XVIIe–XIXe siècle)*. Fayard.
Ruellet, A. (2016). *La maison de Salomon: histoire du patronage scientifique et technique en France et en Angleterre au XVIIe siècle*. Presses universitaires de Rennes.
Savoie, D. (2015). Les bas-reliefs de Temporiti à l'Observatoire de Paris. *Cadran Info, 31*, 85–96.
Schaffer, S. (1988). Astronomers mark time: discipline and the personal equation. *Science in Context, 2*(1), 115–114.
Shapin, S., & Schaffer, S. (1985). *Leviathan and the Air-pump: Hobbes, Boyle, and the Experimental Life*. Princeton University Press.
Sobel, D. (1998). *Longitude: l'histoire vraie du génie solitaire qui résolut le plus grand problème scientifique de son temps*. Seuil.
Table ronde du CNRS, Paris, 27–29 March 1979: Huygens et la France. (1982). Vrin.
Van Helden, A. (1998). Longitude and the Satellites of Jupiter. In W.J. H. Andrewes (Ed.), *The Quest for Longitude: The Proceedings of the Longitude Symposium* (pp. 86–100). Collection of Historical Scientific Instruments, Harvard University.
Wolf, C. E. (1902). *Histoire de l'Observatoire de Paris de sa fondation à 1793*. Gauthier-Villars.

Dalia Deias holds a PhD in the history of science from the Centre Alexandre Koyré/École des Hautes Études en Sciences Sociales, Paris. She is currently a researcher in the history of science at the SPH, Bordeaux. Her research topics, which have led to several publications in French, Italian and English, include the circulation of knowledge in early modern correspondence, Catholic learned astronomers (especially Jesuits and Capuchins), sites and practices of astronomical observation from the sixteenth to the eighteenth centuries.

Chapter 4
The Chronometry Collection of the Palermo Astronomical Observatory

Ileana Chinnici

Abstract The Palermo Astronomical Observatory was founded in 1790 and was equipped with the best astronomical clocks made in England and in France. Its equipment changed over the years, of course, including excellent 19th- and 20th-century pendulums and chronographs. Today these instruments, after restoration, are on display at the Museo della Specola and form a unique collection, showing the evolution of the typical standard timekeeping equipment of an astronomical observatory.

Introductory Notes on the Palermo Astronomical Observatory

The Palermo Astronomical Observatory was founded in 1790 by Giuseppe Piazzi (1746–1826), a Theatine priest from Northern Italy, who was called to hold the chair of astronomy at Palermo and to build an astronomical observatory in the Sicilian town.[1] He made it one of the most important observatories of his time, with a sound scientific programme and cutting-edge equipment. Established under the Kingdom of the Two Sicilies, the Observatory was the expression of a cultural project that was rooted in the Enlightenment and managed by Freemasons. The location of the Observatory at the top of a tower of the Royal Palace somehow symbolized the alliance between Science and Power that was typical of the Enlightenment. Its advanced scientific programme and its modern equipment were built thanks to international Freemason connections.[2]

The Palermo Observatory was one of the most Southern observatories in Europe and adopted the general programme of star cataloguing that was ongoing in most

[1] Foderà Serio (1993).

[2] Coniglio and Chinnici (2022); Coniglio et al. (2020).

I. Chinnici (✉)
INAF—Istituto Nazionale di Astrofisica, Osservatorio Astronomico di Palermo, Palermo, Italy
e-mail: ileana.chinnici@inaf.it

G. Bernasconi et al. (eds.), *Astronomical Observatories and Chronometry, 18th-20th Century*, History of Physics, https://doi.org/10.1007/978-3-031-95874-8_4

European observatories in those years. This research activity had produced important results in the eighteenth century, such as the discovery of light aberration and the nutation of the Earth's axis by James Bradley (1693–1762) at the Royal Observatory of Greenwich. It was expected to improve the previously published star catalogues with more accurate measurements and revisions, especially for the austral celestial hemisphere, partially visible from Palermo.

During his directorship, Piazzi provided the Observatory with first-class instruments, mostly acquired during a stay in London, in the years 1788–1789, that was aimed at securing the best equipment available at that time. From the well-known instrument maker Jesse Ramsden (1735–1800), he obtained the construction of a vertical circle, the Ramsden Circle,[3] which is considered a masterpiece. Thanks to this instrument, especially conceived for astrometric measurements, Piazzi would obtain remarkable scientific achievements: in 1803 he published a star catalogue containing the celestial coordinates of more than 6,000 stars, and a second edition in 1814, extended to more than 7,000 stars; moreover, during the observations for the star catalogue, he discovered the first minor (today dwarf) planet, Ceres, in 1801.

In the second half of the nineteenth century, the Palermo Observatory was one of the most active centres of solar physics in Europe and beyond. Pietro Tacchini (1838–1905), astronomer at Palermo, was a pioneer in solar spectroscopy and established the first spectroscopic society in 1871. The *Memoirs of the Italian Spectroscopic Society*, edited by Tacchini and published in Palermo in the last quarter of the nineteenth century, are today considered the first journal of astrophysics ever published.[4] These results show that, at that time, astronomical research at Palermo Observatory was very advanced and up-to-date.

Later, at the beginning of the twentieth century, the Observatory went into a long decline, due to lack of human and financial resources, obsolete equipment and increasing light pollution in the sky of Palermo, which altered the quality of the observations. Things changed in the Seventies, when Giuseppe Salvatore Vaiana (1935–1991) was appointed director. He had pioneered the development of X-ray astronomy in the United States and, once back at Palermo, maintained international collaborations and made the Palermo Observatory a leader in this field. He revived the Observatory by attracting and training young researchers and trying to develop all its potentialities, including the creation of a laboratory for X-ray astronomy (XACT), in operation since the early 1990s.

Nowadays, the Observatory is still active and is incorporated in the network of the National Institute for Astrophysics (INAF): research, laboratory and outreach activities are still ongoing. It has kept its original site on the tower of the Royal Palace, as the astronomical work is mainly conducted through satellite observations. As we shall see below, it has preserved its precious heritage, thanks to a prolonged and huge effort: the most important instruments are on display in its museum, its library is rich in ancient books and its archive contains documents dating back to the epoch of the Observatory's establishment. These historical collections of instruments, books

[3] Chinnici et al. (2001).

[4] Chinnici (2008).

and documents prove the tight connection between astronomy and chronometry at the Observatory from the end of the eighteenth century to the beginning of the twentieth century.

The Astronomical Museum

In 1990 Vaiana strongly encouraged the difficult process of retrieving the material heritage of the Observatory, which lay in very bad condition: rusty scraps of instruments were scattered in various rooms, together with mouldy and ruined books. The work necessary to recover, restore and exhibit this important heritage was challenging. He entrusted this huge task to Giorgia Foderà, an associate from the University of Palermo. She called in the most reputable specialists in the field and started a fruitful collaboration with Paolo Brenni for restoring the main instruments. In about ten years, Foderà and her collaborators catalogued, restored and reinstalled the ancient instruments in their original sites and cases.[5] In March 2000 the Museo della Specola was opened to the public in the original rooms of the Observatory.[6] Visitors can admire the original furniture, paintings, instruments as much as possible as they were in the past, and have an immersive experience of the working environment of the astronomers of those times.

This small but rich museum is not conceived as a mere container of objects, but as a reconstruction of the past. Behind every object, there is an accurate philological work about its position, its functioning, its use, etc. For this reason, the collections are not physically assembled together, but virtually collected in the printed and digital catalogue of the museum.[7]

The historical collections "tell" of the research work that was carried out at the Observatory in the past two centuries and that can be deduced from the type and quality of the equipment.[8] The museum hosts about a hundred instruments dating back to the eighteenth century until the beginning of the twentieth century and related to astronomy, chronometry, meteorology, seismology, physics, spectroscopy, geomagnetism, but also globes, furniture, artworks, historical calculators and typewriters: all elements connecting the visitors with the extraordinary as well as daily past activity of the Observatory.

It is worth mentioning that similar work in recovering and restoring has also been done for the library and the archive: many ancient books and documents have been restored and are today preserved and available to scholars. Moreover, adopting a LAM approach (library, archive and museum viewed as interconnected units forming

---

[5] Foderà Serio and Chinnici (1997).

[6] Chinnici (2000).

[7] Foderà Serio and Chinnici (1997); https://www.astropa.inaf.it/museodellaspecola/ (accessed on 15 October 2024).

[8] Chinnici (2006).

Table 4.1 Synoptic table detailing the chronometry collection of the Palermo observatory

	Eighteenth century	*Nineteenth century*	*Twentieth century*
Pendulums	Cumming & Grant (UK) Mudge & Dutton (UK) Vulliamy (UK) Janvier (F) Danè (I)	Frodsham (UK) Dent (UK)	Frodsham (UK) Riefler (D)
Chronographs		Hipp (CH) Cavignato (I)	Wetzer (D)
Marine chronometers		McGregor (UK) Whiffin (UK)	Nardin (CH)

a whole), some projects of preventive conservation have been recently activated[9] to reduce the effects of degradation processes on the historical collections kept in the Observatory museum, library and archive.[10]

The Chronometry Collection

The small chronometry collection is one of the most important at the Observatory. Today it includes fifteen items (some instruments we know of, through historical inventories or reports, disappeared in the years 1950–1970, probably thrown away or stolen). As in the case of the other collections, it consists of decommissioned instruments that had been in use at the Observatory. They are listed below, identified by the name of the maker and their national origin (abbreviated) (Table 4.1).

From this list it is easy to recognize that most of the clocks of the Palermo Observatory were made in the United Kingdom. This is not surprising, as Sicily kept important political and commercial relationships with this country during the eighteenth and nineteenth centuries[11]: under the Bourbon government it was a bastion of the English naval and commercial fleets, while in the nineteenth century, several English families established their residence on the island and traded in wine, pistachios, almonds and other typical local products. We see from the list that the distribution in time is quite constant (5–6 instruments per century); this means that the supply was regular: over time, the Observatory provided for the purchase and renewal of the chronometric instruments for its scientific activities. In the list above, the instruments are grouped into three classes, which represent the typical chronometric equipment of an astronomical observatory. Astronomical pendulums were the usual timekeepers in the observatories and provided the sidereal time and the mean time; in the nineteenth century they were mostly used as regulators, sending an electric impulse to

[9] Chinnici et al. (2021).

[10] Carotenuto et al., (2021, 2022).

[11] D'Andrea (2006).

other clocks for their synchronization. The chronographs, generally connected to a regulator, provided a graphic track on paper ribbons, to record the time measurement (depending on the personal equation). The marine chronometers were versatile and quite reliable portable instruments that could be transported and used during scientific expeditions and whenever the time measurements taken at the Observatory were not immediately available.

Below we give a short description of each instrument, grouped by type.

Astronomical Pendulums

- *Astronomical Pendulum by Cumming & Grant (London, 1790)*

This pendulum was part of the original equipment of the Observatory. It was bought by Piazzi during his stay in London in the years 1788–1789 to provide the sidereal time measurements for the astronomical observations made with the Ramsden Circle. At Piazzi's request, the clock was tested by Count Hans Moritz von Brühl (1732–1809) in his private observatory at Harefield, in England.[12]

The pendulum is signed "Cumming & Grant/London"; it was made by clockmakers Alexander Cumming (1733–1814) and John Grant (1781–1810 active), Cumming's nephew, who took over the business.

The pendulum has an escapement "à palette et détente", quite rare, invented by Cumming for compensating gear errors. The system includes an additional independent escapement on the same wheel, for correcting the deviations due to cyclical gear errors.[13] Although ingenious, this system was less effective than expected and was soon abandoned.

Piazzi gave a detailed description of this pendulum in his report on the establishment of the Observatory[14]; he complained about its lack of precision and reports that it was repaired by a local clockmaker. It provided the sidereal time for the Ramsden Circle, but was soon replaced (see below).

Today this is the only extant pendulum with Cumming's escapement system: it is hence the most important and rare item of the chronometric collection of the Palermo Observatory.

- *Astronomical Pendulum by Mudge & Dutton (London, 1790)*

This clock was also part of the original Observatory equipment; it belonged to Count von Brühl and was purchased by Piazzi in 1791.

The instrument is signed "Tho[ma]s Mudge Will[ia]m Dutton London"; it was made in London by Thomas Mudge (1715–1794) and William Dutton (1720–1794), two of the finest English clockmakers of the time. Initially set to mean time, from

---

[12] Piazzi (1792, p. 53).

[13] Cumming (1766).

[14] Piazzi (1792, pp. 51–54); Azzarello and Foderà Serio (1992).

1794 it replaced Cumming's pendulum to provide the sidereal time for the Ramsden Circle during the observations for the second edition of Piazzi's star catalogue (1814).

Its mechanisms are admirably perfected: the pendulum has a grid compensation with steel and brass pipes, having different coefficients of thermal expansion; the lever free escapement (invented by Mudge in 1769) has a steel wheel and transparent corundum pallet, quite rare, whose hardness ensures long-lasting operation without deterioration.

The clock is complete in all its original parts and can be considered a masterpiece of eighteenth century English clockmaking.

- *Astronomical pendulum by Janvier (Paris, 1790)*

This clock was purchased by Piazzi in 1791 from Parisian clockmaker Antide Janvier (1751–1835), through an intermediary, astronomer Jérôme de Lalande (1732–1807), whom he considered his master, mentor and adviser. It is signed "Janvier/aux menus plaisirs du Roi". Piazzi remarked that it was not as accurate as the two English-made pendulums, but its price was more than two-thirds lower, and its error did not exceed 10 s within a year, which he considered quite acceptable.

It was set to mean time in the gallery and, at the beginning of the twentieth century, placed in the meridian room and adapted to serve as a regulator for Hipp's chronograph (see below).

- *Counter by Vulliamy (London, 1790)*

This counter for minutes and seconds was purchased by Piazzi from Count von Brühl at the same time as the Mudge & Dutton clock in 1791. It is signed "Vulliamy London" and was built in London by Benjamin Vulliamy (1747–1811), known for the construction of the Regulator Clock, the main timekeeper of the King's Observatory at Kew and official regulator of time in London.

The counter was installed in the meridian room to serve the transit instrument: the observations of the transit of a star on the local meridian was, in fact, crucial to the determination of the instant of the transit, which was obtained, at that time, by interpolating the time beaten by the ticking of the clock (eye and ear method). It was restored in the 1990s.

- *Counter by Dané (Palermo, ca. 1803)*

Of this clock, a counter for minutes and seconds, only a few original parts remain: the mechanism, the dial—signed "Giacomo Dané/Palermo"—and the hands. Case, lens, weights and clock chain were replaced in the twentieth century. The instrument was to serve a Troughton equatorial telescope, purchased by Piazzi in 1803, and no longer extant.

- *Chronographic Pendulum by Frodsham (London, 1865)*

This English-made clock was purchased in 1865 by Gaetano Cacciatore (1814–1889), director of the Palermo Observatory. The instrument is signed "CHARLES FRODSHAM/N° 84, STRAND, LONDON/Clock Maker/To the Queen", serial number 1039. The clock was tested at the Greenwich Observatory by the Astronomer

Royal, George Biddell Airy (1801–1892) and, with Hipp's chronograph (see below), was to serve the large Merz equatorial telescope, installed at the Observatory in the very same year.[15] The pendulum, having mercury compensation, is described by Cacciatore[16] as provided with a complete circuit electrical device for registering the seconds on a chronograph. It was to serve as a regulator for the Hipp chronograph, but from 1883 it was connected to the Cavignato chronograph (see below). The firm of Charles Frodsham (1810–1871), who was Vulliamy's successor, is still in business in London today.

- *Regulator pendulum by Frodsham (London, 1901)*

This instrument was purchased by director Filippo Angelitti (1856–1931) with university funds; price: 1,755.80 Italian liras.[17] It is signed "Cha[rle]s: Frodsham/ Clockmaker to the Queen/115 New Bond St[reet] London", serial number 2069. It was used as a regulator for a Wanschaff's zenith refractor, which had been purchased the year before (this telescope is now on display at Cagliari Observatory on permanent loan). The clock has a mercury compensation mechanism, typical of Frodsham clocks. The instrument is in very good condition and still working. Frodsham regulators were high quality, very reliable clocks, and were part of the usual observatory equipment between the end of the nineteenth century and the beginning of the twentieth century.

- *Regulator pendulum by Dent (London, 1907)*

This instrument was also purchased by Angelitti in 1907. It is signed "DENT/ CLOCKMAKER TO THE KING/61. STRAND &/4 ROYAL EXCHANGE/ LONDON", serial number 55226. It is a very common, inexpensive regulator often found in observatories in the early twentieth century. It was to serve the Salmoiraghi transit instrument that Angelitti purchased at the end of the nineteenth century.

- *Regulator pendulum by Riefler (Nesselwang near Munich, 1928)*

This clock was bought by Angelitti with funds obtained by the City Council of Palermo. The Observatory had been downgraded in 1923 as a university facility for astronomy and the Director was obliged to seek subsidies at the local level.

It is signed "Clemens Riefler/München/1928", serial number 472. The clock was purchased through the watchseller O. Stönner Zollikofer who had his shop in Palermo; price: 24,039 Italian liras.

It is a free escapement clock under a pneumatic bell. To alleviate friction, the pendulum swung in a vacuum cylinder where air was pumped out. It sent electrical impulses to the Wetzer chronograph (see below). In the archives of the observatory there is a printed paper dating from the time, with the description and instructions for use. The system was patented by Clemens Riefler (1820–1876) and his successor, his son Sigmund (1847–1912); their precision instrument industry, still in

---

[15] Chinnici and Brenni (2015).

[16] Cacciatore (1865).

[17] Foderà and Chinnici (1997, pp. 94).

existence, was renowned worldwide. Riefler regulators were excellent instruments, widely distributed in astronomical observatories at the beginning of the twentieth century.

Chronographs

- *Cylinder chronograph by Hipp (Neuchâtel, ca. 1864)*

In the second half of the nineteenth century, the use of electromagnetic chronographs reduced the reaction time of the observer (the so-called personal equation) to determine the instant of the transit of a star (namely, its passage at the meridian). This chronograph was made by the Swiss manufacturer Matthaus Hipp (1813–1893).[18] It was purchased around 1869 and electrically connected to the Mudge & Dutton clock to serve the new Pistor & Martins meridian circle. It recorded on paper around the cylinder the electrical signals coming from an impersonal micrometer and, at the same time, the impulses of the pendulum.

The instrument is signed "M. HIPP NEUCHATEL SUISSE", serial number 2926. This type of instrument was used in determining the difference in longitude between Geneva and Neuchâtel and a very detailed description is contained in the report concerning these operations[19]; at the time, it was considered a standard for this type of measurement.

In 1872 Hipp's chronograph was used to determine the difference in longitude between the observatories of Palermo and Naples (and previously, in 1870, between the observatories of Rome and Naples), where another Hipp chronograph is also preserved (as well as in Rome Observatory). The one in Palermo was repaired at the beginning of the twentieth century by technicians of the Naples Observatory. The chronograph is still installed in its original wooden cabinet with neo-Gothic decorations similar to those in the meridian hall. It is complete in its parts except for the scribing points; two spare cylinders are still preserved and, in the archives, a catalogue of Hipp's telegraph instruments.

- *Dry chronograph by Cavignato (Padua, 1883)*

This electromagnetic chronograph was purchased by Cacciatore in 1883 to be connected to the Frodsham 1039 clock; price: 49 Italian liras. It was built by Giuseppe Cavignato, technician at the Padua Observatory, according to the system adopted by Rudolf Fuess (1838–1917) in his "Bandchronograf". The instrument is signed "CAVIGNATO GIUSEPPE/MECCANICO/DEL R[EGIO] OSS[ERVATORI]O DI PADOVA". It recorded the electromagnetic signals from the pendulum and the micrometer on strips of paper. Signal transmission was telegraphic; the chronograph is dry-type, because it did not use ink nibs, but points which perforated the paper,

---

[18] Peyer et al. (1902).

[19] Plantamour and Hirsch (1864).

activated by electromechanical arms. The instrument is still installed on its table and protected by a wooden case with a glass front opening.

- *Four-points chronograph by Wetzer (Pfronten [Germany], 1929)*

This electromagnetic chronograph was purchased in 1929 together with the Riefler clock; price: 7619.64 Italian liras. It is a telegraph chronograph with bands having four ink nibs on arms operated by electromagnets. It could therefore record four different signals. It is signed "H. WETZER/PFRONTEN BAYERN", serial number 466. It also has two additional reels for winding the paper strips.

Marine Chronometers

- *Marine chronometer by McGregor (Glasgow, ca. 1868)*

Over the second half of the nineteenth century, the Palermo Observatory purchased some marine chronometers. These precision and easily transportable clocks were widely used for scientific expeditions. The oldest marine chronometer in the Palermo collection dates from around 1868. It was built by Duncan McGregor (1828–1893) who had started his watchmaking business in Greenock in 1836 and then opened a shop in Glasgow in 1844, with branches in Liverpool in 1879 and London in 1886.

The Palermo instrument is signed "D. McGREGOR & Co[mpany]/Makers to the Admiralty/GLASGOW & GREENOCK", serial number 186. It was set to sidereal time, but it was not considered reliable enough. It is still kept in its double wooden case, usual for this type of instrument, the second being padded and used for transport. It is quite well preserved.

- *Marine chronometer by Whiffin (London, ca. 1870)*

This instrument was purchased on the occasion of the Italian scientific expedition to the South-East of Sicily to observe the total solar eclipse of 1870[20]; price: 40 sterling.[21] Its acquisition was suggested by Airy, who tested the instrument at the Greenwich Observatory. Later, it was used during the Italian expedition to India to observe the transit of Venus on the sun in 1874. At the beginning of the twentieth century it was set to mean time and used for the noon signal. It is signed "Swann Whiffin/London", serial number 342. In the mahogany case there is an ivory plaque reading "S. Whiffin/N° 342". It is very well preserved.

- *Marine chronographic chronometer by Nardin (Locle [Switzerland], 1913)*

From the test certificate, sent by the manufacturer and kept in the archives, we can deduce that this chronometer was purchased by Angelitti in 1913. It is a chronographic chronometer, which can send an electrical signal to a chronograph. It is signed

[20] Chinnici (2023).

[21] Foderà and Chinnici (1997, pp. 98).

"ULYSSE NARDIN/LOCLE SUISSE"; we find engraved on the dial the inscription "GRAND PRIX/PARIS" and the serial number 531. In the red mahogany case there is a plaque reading "LOCLE—ULYSSE NARDIN—SUISSE" and another with the serial number 531; on the edge of the lid there is another plaque with the name and address of the seller: "O. Stönner Zollikofer/PALERMO/Corso Vittorio Emanuele 142". The certificate was issued by the Neuchâtel Observatory and it contains an essential description of the chronometer and the errors in its operation; the document is signed by the assistant astronomer Hermann Stroele (1880–1921).

Final Remarks

This text can be considered but a starting point for a more in-depth study of the collection of clocks at the Palermo Observatory and an invitation to experts to further examine and study them. Studies on their technical details could provide interesting information about their construction and further archival research could reveal other important details or connections with other observatories and astronomers.

For the number and types of instruments which have just been briefly illustrated here, we can affirm that this collection is entirely representative of the typical equipment of astronomical observatories from the end of the eighteenth century to the beginning of the twentieth century.

It shows the evolution of the technical details of clocks used in astronomy as well as the leadership of different manufacturers (and countries) in the chronometry industry.

Preserved here are quite rare clocks of English manufacture, dating from the eighteenth century; they prove the relationships established by the astronomers of Palermo with their foreign colleagues[22] and their exchanges regarding the purchase and testing of these instruments. It also shows the role of certain observatories, such as those of Greenwich and Neuchâtel, in the dissemination of chronometry instruments in the international network of astronomical observatories.

This important heritage must be preserved as well as it can be, in ours and other observatories where chronometric instruments are still kept. For this reason, workshops on conservation, functioning and maintenance of astronomical clocks, as well as collaboration with experts in this field are to be promoted and encouraged. Time determination being one of the main observatory tasks, astronomical clocks are a significant part of the material heritage of these institutions. In Italy, for instance, more than one hudred items including sidereal and mean time pendulums, chronometers and chronographs are preserved in the astronomical observatories of the National Institute for Astrophysics (INAF). A simple search in the web portal of the INAF astronomical heritage "Polvere di Stelle" (https://www.beniculturali.inaf.it/opac/musei [accessed on 15 October 2024]) will confirm this statement. It is of great importance that research institutions do not disperse their historical collections

[22] Devoy and Chinnici (2023).

and make efforts to find appropriate solutions to catalogue, safeguard and display them as much as possible in their original context: making the scientists of today aware of the past is the best strategy to preserve this heritage for the future. A recent workshop on "Timekeepers in Historical Observatories" held in Palermo in March 2025, has increased the awareness of the importance of this heritage and established a network of experts, astronomers and restorers exchanging information about conservation, restoration and maintenance of historical chronometric instruments in the observatories.

References

Barbera Azzarello, M., & Foderà Serio, G. (1992). *Orologi e orologiai a Palermo.* Sellerio Editore.

Cacciatore, G. (1865). Annunzi. *Bullettino Meteorologico del R. Osservatorio di Palermo* 3, p. 11.

Carotenuto, M. R., et al. (2021). Astronomical collections in Observatories: promoting preventive conservation. In R. M. Ros, B. Garcia & S. R. Gullberg, et al. (Eds.), *Education and Heritage in the era of Big Data in Astronomy, Proceedings IAU Symposium No. 367, 2020* (pp. 505–507). Cambridge University Press.

Carotenuto, M. R., et al. (2022). Preventive conservation projects on LAM materials at the INAF-astronomical observatory of Palermo "G.S. Vaiana." *Bulletin of the American Astronomical Society* 54(2).

Chinnici, I. (2000). Il Museo dell'Osservatorio Astronomico di Palermo G.S. Vaiana. *Giornale di Astronomia, 26*(2), 49–51.

Chinnici, I. (2006). Un percorso interdisciplinare tra storia e scienza: l'Osservatorio di Palermo e le sue collezioni. *Quaderni Del Museo Geologico Gemmellaro, 8*, 69–74.

Chinnici, I. (2008). The Società degli Spettroscopisti Italiani: birth and evolution. *Annals of Science, 65*(3), 393–438.

Chinnici, I., & Brenni, P. (2015). The Palermo merz equatorial telescope: an instrument, a manuscript, some drawings. *Nuncius, 30*, 228–279.

Chinnici, I., et al. (2001). The Ramsden circle at the Palermo astronomical observatory. *Bulletin of the Scientific Instrument Society, 71*, 2–10.

Chinnici, I. (2021). Il Museo della Specola: Progetti per la conservazione, la fruizione, la comunicazione (2018–2020). *Rapporto Tecnico INAF*, 2021.

Chinnici, I. (2023). A traveling chronometer. *Science Museum Group Journal* 20.

Coniglio, M., & Chinnici, I. (2022). L'Osservatorio Astronomico di Palermo e l'Archivio Pascoli. *Giornale di Astronomia, 48*(3), 14–19.

Coniglio, M., et al. (2020). *Urania Ferdinandea: la fondazione dell'Osservatorio Astronomico di Palermo.* INAF-Osservatorio Astronomico di Palermo.

Cumming, A. (1766). *The Elements of Clock and Watch Work adapted to Practice.* A. Millar & D. Wilson.

D'Andrea, D. (2006). Great Britain and the Mediterranean islands in the Napoleonic Wars—the "insular strategy" of Gould Francis Leckie. *Journal of Mediterranean Studies, 16*(1/2), 79–90.

Devoy, L., & Chinnici, I. (2023). Observatories across the miles: how astronomers at Palermo and Greenwich worked together through the 19th century to share knowledge, data and instruments. Poster presented at the XLII SIC Symposium, Palermo, 18–22 September 2023.

Favager, A. (1924). *L'électricité et ses applications à la chronométrie.* Editions du Journal Suisse d'Horlogerie et Bijouterie.

Foderà Serio, G., & Chinnici, I. (1997). *L'Osservatorio Astronomico di Palermo. La storia e gli strumenti.* Flaccovio Editore.

Foderà Serio, G. (1993). On The History Of The Palermo Astronomical Observatory. In J. Linsky & S. Serio (Eds.), *Physics of Solar and Stellar Coronae* (pp. 21–33). Kluwer Academic Publishers.
Hipp, M. (1869). *Prix-courant illustré de la Fabrique de Télégraphes et appareils électriques à Neuchâtel.* Imprimerie J. Attinger.
Peyer, Favarger and Cie. (1902). *Prix courant de la fabrique de télégraphes et appareils électriques fondée par M. Hipp en 1860. Catalogue B: Appareils scientifiques, instruments de mesure et de contrôle pour les sciences et l'industrie: Instruments de mesures électriques, instruments scientifiques pour la mesure du temps, appareils divers, enregistreurs météorologiques.* Wolfrath and Sperlé.
Piazzi, G. (1792). *Della Specola Astronomica de' Regj Studj. Libri IV*. Stamperia Reale.
Plantamour, E., & Hirsch, A. (1864). Détermination télégraphique de la différence de longitude entre les observatoires de Genève et de Neuchâtel. *Mémoires De La Société De Physique Et D'histoire Naturelle De Genève, 17*, 288–436.

Ileana Chinnici is an astronomer at the INAF Osservatorio Astronomico di Palermo, where she works mainly in the field of history of nineteenth-century astronomy. She has been curator of the historical collections of the Observatory for many years and is currently scientific advisor for these collections, with a unified approach to instruments, books, and archives. Co-author of the catalogue of the old scientific instruments of the Observatory and author of many publications in specialist journals, she was awarded the 2021 Osterbrock Book Prize. Recently involved in projects for the preventive conservation of the scientific heritage of the historical observatories, in 2024 she was elected President of the Scientific Instrument Commission for the triennium 2025–2028.

Part II
Time at the Observatory in the Longue Durée: The Quest for Precision After 1800

Chapter 5
"Hour by Hour, Day and Night": Splitting Time Between the Astronomical and Physical Observatories

Ion Mihailescu

Abstract This paper argues for a clearer distinction between the regimes of observation established in astronomical and physical observatories in the mid-19th century. Although both observatories often shared the same spaces, directors, and techniques—such as the mechanization of observations—they studied phenomena characterized by fundamentally different temporal structures. Astronomers primarily carried out observations of phenomena with highly predictable patterns, requiring fixed and regular times of observation, whereas meteorological and magnetic observations focused on irregular and unpredictable phenomena, demanding constant, hourly observations day and night.

> The watch-towers of science now cover the whole earth, and the sentinels never sleep.

Mitchel (1850, p. 21).

> ...the eye of science remains, so to speak, constantly open to everything that happens on the surface of our planet.

Quetelet (1853, pp. 701–702).

In a landmark article, Simon Schaffer argued that the managers of the great observatories developed a new chronometric regime of discipline and surveillance to tackle the problem of the personal equation. Their solution relied on a division of labour within the observatory, a network of observing sites and the mechanization of observations by self-registering instruments such as the chronograph which was introduced by George Biddell Airy at Greenwich in 1854: "Airy's new transit circle and his electric barrel-chronograph dictated a new regime."[1]

Many historians have come to understand Schaffer's paper as showing that self-registering instruments (referred to as a general class of instruments) were adopted in

I. Mihailescu (✉)
Institut d'histoire, Université de Neuchâtel, Neuchâtel, Switzerland
e-mail: ion-gabriel.mihailescu@unine.ch

[1] Schaffer (1988, p. 120).

G. Bernasconi et al. (eds.), *Astronomical Observatories and Chronometry, 18th-20th Century*, History of Physics, https://doi.org/10.1007/978-3-031-95874-8_5

response to a particular astronomical problem within astronomy (the personal equation). In a paper on self-recording instruments in physiology, Soraya de Chadarevian briefly summarized the key point of Schaffer's article as: "Managers of astronomical laboratories, notably Airy in Greenwich (1854), introduced *automatic recording systems* in reaction to the problem of the personal equation of the observers."[2] Jimena Canales, employing the same generalization, was led to erroneously revert the chronological order in which self-registering instruments were actually introduced in observatories.

One of the main motivations for developing self-registering apparatuses arose from the desire to automatically track the rotation of the Earth against the stars in order to record time continuously. Instruments for recording temperature and other meteorological data were *later* employed at weather stations and observatories for the same purpose.[3]

In fact, a self-registering wind-gauge was already in operation at Greenwich in 1841, and photographic self-registration for the magnetometer, thermometer, barometer and hygrometer were introduced in 1847; the chronograph was only added in 1854.[4]

The identification of self-registering instruments in the observatory with the chronograph has become so powerful that Kevin Donnelly portrayed Adolphe Quetelet's assistant, Jacques Crahay, as if he had conducted astronomical observations by the "eye-and-ear" method which was then made obsolete by the chronograph.[5] Crahay actually took meteorological observations "hour upon hour" for many years until "a more precise type of observation had been substituted which left the pencil to record the data itself."[6] This was the system of meteorological "observations registered mechanically (*observations enregistrées mécaniquement*)" introduced in 1847; the first chronograph seems to have been acquired by the Brussels observatory only after Quetelet's death in 1874.[7]

This is not a simple problem of chronology, or of setting the record straight. Recent studies on the history of meteorology, tides, or geomagnetism have often remarked on the development or use of self-registering instruments in the 1830s and 1840s.[8]

---

[2] De Chadarevian (1993, p. 276).

[3] Canales (2016, p. 507).

[4] Airy (1843, pp. 82–83); Airy (1849, pp. 83–90). For the "chronographic clock and barrel" see Airy (1856, pp. 6–7).

[5] Donnelly (2014, p. 487): "Observations at the beginning of the century were conducted using the 'eye-and-ear' method, where an observer would look for a star to cross his vision while listening for a clock to sound the time. In the Brussels Observatory, Quetelet noted that his assistant Jacques Crahay conducted the 'painful obligation of observing by oneself, hour upon hour'. Yet the move to a chronograph in the 1830s hardly lessened the demands or increased the amount of individual participation. Observers still had to patiently watch the skies, only now the time was recorded automatically." Crahay's observations, which were only meteorological, are accurately described in Donnelly (2016, p. 82).

[6] Quetelet (1855, p. 127); republished, with small modifications in Quetelet (1866, p. 253).

[7] Quetelet (1847, p. 9); *Annuaire de l'Observatoire royal de Bruxelles* (1890, p. 247).

[8] MacDonald (2018); Reidy (2008); Anderson (2005).

However, the fundamental differences between the practices of observation with self-registering instruments in astronomy and the physical sciences have remained unacknowledged. The case of the introduction of the chronograph at Greenwich Observatory, as presented by Schaffer, still forms an essential reference for how historians conceptualize the mechanization of observation and the role of self-registering instruments.[9]

My goal is to disentangle the early history of the mechanization of observation by distinguishing more carefully between astronomical and physical (i.e., magnetic and meteorological) regimes of observation. Such distinctions can be subtle and have often been obscured by the fact that throughout the middle of the nineteenth century astronomical and physical observations could be carried out by the same observers, in the same places and under the auspices of the same institution, overseen by the same director. Thus, it is not surprising that historians have mostly focused on explaining what brought the astronomical and physical sciences together: at a scientific level, one hoped to find a correlation between astronomical and physical phenomena; at a practical level, both astronomical and physical observations shared many of the same "observatory techniques" of manipulating instruments, handling data, or coordinating networks of observers.[10]

Paradoxically, it was this very close identification of physical and astronomical observatories which also led to tensions between their regimes of observation. The new system of hourly or bi-hourly observations, introduced in 1840 for the Magnetic Crusade and designed for naval and military observers, engendered much hardship for the astronomical staff. A close analysis of Airy's implementation of this system of observation at Greenwich will allow a reassessment of certain historiographic clichés, such as the organization of the observatory as a factory, or the conflation of sources concerning astronomical and physical observations.[11] This will provide a context for reconceptualizing the different role and status of astronomical, and magnetic and meteorological self-registering instruments.

The Physical Observatory

Because his paper was centred on the personal equation, Schaffer did not always carefully distinguish between the mechanization of astronomical and physical observations. The blending of quotes about both types of observations, but without preserving

[9] Daston and Galison (1992, p. 83): "As the photograph promised to replace the meddling, weary artist, so the self-recording instrument promised to replace the meddling, weary observer. It was not simply that these devices saved the labor of human observers; they surpassed human observers in the laboring virtues: they produced not just more observations, but better observations.".

[10] Aubin et al. (2010).

[11] Before Schaffer (1988), the case for Airy introducing at Greenwich a "factory mentality" was made by Chapman (1985). A cogent criticism of this analogy has been made in Smith (1991), who prefers to see Airy's observatory as an "accountant's office", and more recently in Daston (2022, p. 109).

the distinction between them, has led some historians to believe that the discussion of self-registering instruments must apply, foremost, to the chronograph.[12] For example, Schaffer draws a parallel between the remarks about astronomical observations of Ormsby Mitchel, an American astronomer responsible for the development of the chronograph in the late 1840s, and those of John Herschel about self-registering meteorological and magnetic instruments:

The demarcation between the superintendent and the observers was often stressed. In his popular lectures aimed at raising funds for new American observatories, Mitchel argued that this demarcation was necessary because "the observer himself is but an imperfect and variable machine, utterly incapable of marking the exact moments required" (Mitchel, 1860, 176). John Herschel agreed. "The abridgement of the merely mechanical work of such observatories by self-registering apparatus is a subject which cannot be too strongly insisted on," he told his audience at the British Association. "The merely mechanical part of the observer's duty" must be minimized in order to liberate the astronomer from error and degradation.[13]

These quotes from Herschel's presidential address to the British Association for the Advancement of Science from 1845 were not referring to astronomy and astronomers, but rather to "physical observatories" which were supposed to keep a "perfectly unbroken and normal registry of all the meteorological and magnetic elements and of tidal fluctuations where the locality admits."[14] These "physical observatories" were to be run continuously and supported by governments in a fashion that until then had been confined exclusively to astronomy:

Terrestrial Physics, therefore, form a subject every way worthy to be associated with Astronomy as a matter of universal interest and public support, and one which cannot be adequately studied except in the way in which Astronomy itself has been by permanent establishments keeping up an unbroken series of observations.[15]

Self-registering instruments were needed in "such [i.e. physical] observatories" not to liberate the observer "from error and degradation", but rather to "alleviate" his duties so that he can "apply himself to the theory of his subject" and to "the most

---

[12] For such readings of Schaffer's article see, for example, De Chadarevian (1993), Donnelly (2014), Canales (2016).

[13] Schaffer (1988, p. 126). It should be remarked that Mitchel (1860, p. 180) did not draw a stark contrast between "imperfect" human observers and machines. Only a couple of pages later he discussed "*instrumental* imperfections" and the "imperfect machine" which was the astronomical clock: "No clock has ever been made which can keep *perfect* time; and the great object of the observer is to learn the peculiarities of his clock, to determine its deviations from absolute accuracy, and to be able to mark these deviations, if possible, from minute to minute". One mastered personal and instrumental imperfections through very similar means. A similar conclusion is reached by Hoffmann (2007, pp. 364–365) that Bessel's research on errors and the personal equation "does not testify to any distrust in the human observer or allow the conclusion that the observer was framed in wholly different terms to his instruments."

[14] J. Herschel (1846, p. 34).

[15] J. Herschel (1846, pp. 33–34). The concept of a "physical observatory" was first introduced in Herschel (1830, pp. 213–214), but the name was coined soon after by David Brewster. On Herschel's concept see MacDonald (2018, esp. pp. 24–48).

important of all his duties, the systematic deduction from the registered observations of the mean values and local coefficients."[16]

Though encouragements for the development of self-registering instruments were common throughout the 1830s, they usually had a different character: they mainly targeted particular types of instruments, such as wind-and tide-gauges, that had to be observed almost constantly throughout the day and night to obtain a reliable estimation of daily fluctuations and averages.[17] These self-registering instruments were deemed necessary because such continuous registers of observations were unavailable and unlikely to be obtained, for long periods of time, by direct observation. However, when Herschel gave his address in 1845, a large network of British and foreign observatories had been carrying hourly or bi-hourly magnetic and meteorological observations for more than four years, and had amassed millions of values that still needed to be reduced, published and, to justify such an effort, transformed into laws or theories.[18]

This project, generally referred to as the Magnetic Crusade, came about in the late 1830s when several prominent scientists lobbied the British government to establish a series of magnetic observatories across the British Empire to match those already in operation in Continental Europe.[19] The campaign was led by Edward Sabine and Herschel who in the end secured the funding for an Antarctic expedition and a system of fixed observatories: Greenwich (under the direction of Airy), Dublin, Toronto, St Helena, the Cape of Good Hope, and Van Diemen's Land; several other observatories at Shimla, Madras, Singapore and Bombay were supported by the British East India Company. In addition to these eleven British observatories (including the "movable" observatory of the Antarctic expedition), twenty-two foreign observatories agreed to participate in this system of magnetic observations.[20]

The inspiration for the Magnetic Crusade came from Carl Friedrich Gauss's work on magnetism and the network of over twenty observatories he established across Germany and Russia in 1834, which he coordinated from Göttingen with the help of Wilhelm Weber.[21] This union of observatories (*Magnetische Verein*) followed a system, first proposed by Humboldt, of simultaneous magnetic observations on a few set days, *Termine* or "term days". In Humboldt's system the readings were supposed to be carried out hourly for forty-four hours, on seven different occasions, but Gauss shortened the distance between readings to three minutes because he found it necessary "to follow the phenomena in much closer time intervals".[22] The system proved to be too demanding for some of the other participants, so the interval was fixed to five minutes for the sake of uniformity. For the same reason, the number of

[16] J. Herschel (1846, pp. 34–35).

[17] On self-registering tide-gauges see Reidy (2008, pp. 141–145). For wind gauges Forbes (1830), Harris (1845).

[18] On the handling of this data see Goodman (2018a).

[19] Cawood (1979); Carter (2009).

[20] Lloyd (1842, pp. 5–6).

[21] O'Hara (1983).

[22] Humboldt (1830, pp. 238–239); Gauss and Weber (1837, p. 8).

hours was reduced to twenty-four, as was the number of term days: first to six (four *Haupttermine* and two *Nebentermine*), and then to only four by 1839.

The British adopted Gauss's system, but expanded the number of term days to twelve, one for each month. On the rest of the days, with the exception of Sundays, simultaneous magnetic and meteorological observations were taken every two hours at agreed upon times (the even hours in Göttingen time).[23] Neither Gauss nor most of the German observatories joined the British in this intense system of observations.[24] The foreign members consisted of ten Russian observatories (which were also part of the *Magnetische Verein*), and a series of small observatories: Prague and Milan (Austria), Munich (Bavaria), Brussels (Belgium), Cairo (Egypt), Algiers (France), Lucknow (Oudh), Breslau (Prussia), Cadiz (Spain), Trivandrum (Travancore), Philadelphia and Cambridge (USA).[25]

What could have been the reason behind the British proposition for hourly observations, and why was it accepted (though not necessarily followed) by so many observatories? Though some dedicated observers were known for their long-term series of hourly observations (mostly taken during the day, though it was not uncommon for short-term observations to also be taken throughout the night), these were isolated cases. As shown by the case of Gauss's *Magnetische Verein*, increasing the size of a network of observatories often required reducing the number of observations, if simultaneity and uniformity were to be preserved. The case of the Magnetic Crusade was different for several reasons. First, it was initially conceived by Edward Sabine, James Clark Ross, and Humphrey Lloyd as an Arctic expedition. Ships already employed a continuous, twenty-four-hour watchkeeping system, which made them ideally suited for continuous observations and could thus be transformed into "itinerant observatories", as Herschel called them in his 1845 address.[26] Second, the new British land observatories, which some historians have argued were added to the initial plan to gain Herschel's support, were staffed by army and naval non-commissioned officers.[27] Third, the foreign observatories were either able to secure cheap labour for the observations, or they considered the effort worthwhile because it gave them a chance of firmly establishing their observatories on the scientific map.[28]

The new British system was also probably accepted because it was only temporary. The funding for the land observatories had initially been approved for three

[23] "Report on Meteorology" (1840).

[24] For Gauss's objection to the British system of observations see Cawood (1979, p. 512).

[25] Lloyd (1842, pp. 5–6).

[26] J. Herschel (1846, p. 35). On the various British attempts at transforming the naval watchkeeping system and log books into a system of meteorological observations see Naylor (2015a); Naylor (2015b). On a French expedition in which the naval watch was used for meteorological observations, and then extended to a similar regime of observations on land see Locher (2007).

[27] See Carter (2009).

[28] For the participation in the magnetic crusade of the newly established Trivandrum observatory (of the Indian princely state of Travancore) see Ratcliff (2016); for the participation of the Brussels observatory see Locher (2007).

years (1840–1842), but was readily renewed for another three as shipping the instruments and setting them up had taken almost a full year.[29] In 1845, in parallel with the meeting of the British Association where Herschel gave his presidential address, a magnetic conference was held to discuss the continuation of the system of observations. In a letter to Herschel published in a report, Airy pointed out that the observations for the previous six years had barely been reduced and published (Greenwich having reached only as far as 1842), and that these tasks could not be hastened if the observations were to continue. For Airy, "the object of these [physical] observatories is totally different from that of astronomical observatories;" their goal was not "the accurate determination of the present state of certain elements," which, indeed, would have required an ever-increasing number of observations, but the discovery of laws. However, if the "continuous and simultaneous observations" of the past five years were not sufficient for finding such laws, then he could "hardly conceive that any number of years would be found sufficient."[30]

While Herschel supported the continuation of observations, he was also receptive to Airy's criticism that "as long as the observations shall be continued, so long *at least* will those discussions be delayed, and so long *at least* will the real intellectual progress of the science be put off."[31] Herschel's solution of replacing direct observation by self-registering instruments was purposefully asserted to placate Airy. When it was decided that the magnetic observatory at Greenwich "be permanently continued, upon the most extensive and efficient scale," Airy got the following proposition to also be adopted:

That it is highly desirable to encourage by specific pecuniary reward the improvement of self-recording magnetical and meteorological apparatus: and that the President of the British Association and the President of the Royal Society be requested to solicit the favourable consideration of Her Majesty's Government to this subject.[32]

Greenwich Revisited

In 1835, George Airy, then director of the Cambridge University observatory, was appointed Astronomer Royal at Greenwich. After some unexpected delays, a magnetic observatory was established in 1839 which followed Gauss's system of observation: on four term days the magnetic declension was measured every five minutes for twenty-four hours. For meteorological observations, Airy followed a system proposed by Herschel in 1835, which recommended four daily measurements, except on the equinoxes and solstices when readings were to be taken on the

[29] For the renewal process see Carter (2009); for the shipping and installation of instruments see Naylor and Goodman (2020).

[30] "Proceedings Magnetical Conference" (1846, pp. 52–53).

[31] Ibid, p. 53.

[32] Recommendations (1846, p. xvi); Airy (1896, pp. 170–171).

hour.[33] While some had urged Airy to employ a dedicated observer for the magnetic observatory, "the [magnetic] observations were distributed among the six assistants and myself, each person taking the observations of two hours at a time."[34] Airy's preference for this arrangement is somewhat surprising given that historians have often touted the "rigid hierarchy of staff" and "tight division of labor" at Greenwich.[35] It might very well be that Airy saw these observations as a useful training opportunity for his assistants because, as Gauss wrote to Airy, "there is no better preparatory school for practical astronomy, than observing at the magnetometer."[36]

Lee Macdonald has persuasively argued that Airy was forced to change this arrangement when he learned about a plan, probably proposed by Edward Sabine, to establish a new magnetic and meteorological observatory near London.[37] To preserve the magnetic pre-eminence of his observatory, Airy offered to match the new system of observations which required readings to be taken every two hours, besides the usual term days. For this purpose, a whole new magnetic and meteorological department was created with a superintendent and two assistants. James Glaisher, who had been Airy's first assistant at the Cambridge observatory and had followed him to Greenwich, was transferred from the "the Circle Department" and put in charge. Glaisher occupied this position for almost forty years while pursuing a diverse range of activities which brought him fame and various honours.[38] Two calculators, Edwin Dunkin and John Russell Hind, were promoted as his assistants.[39]

The working conditions and wages of assistants and calculators at Greenwich under Airy has attracted a considerable amount of attention.[40] In a pioneering article, Allan Chapman associated the "factory mentality" introduced by Airy with (among other factors) "a rigid hegemony of staff, in which all personnel knew their places in a strict chain of command":

> Beneath the graduate assistants came the observers, computers and "obedient drudges", who were generally appointed as very young men, and trained exclusively in the performance of one or other specialised task, such as the calculation of lunar places or the operation of a particular instrument.[41]

This picture of Airy's disciplinary regime, echoed by Schaffer and others, has been mostly based on the account of Walter Maunder, one of Airy's last assistants in the 1870s.[42] However, in later accounts, Chapman has come to consider that

---

33 J. F. W. Herschel (1835, pp. 4–6).

34 Airy (1840, p. lix). On the dedicated observer see Macdonald (2018).

35 Smith (2003, p. 159).

36 Gauss quoted in Macdonald (2018, p. 20).

37 Macdonald (2018, p. 27).

38 Hunt (1996), Chapman (2012), Anderson (2005, pp. 95–99).

39 Airy (1842, p. 2).

40 For some of the more recent studies see Croarken (2009), Chapman (2012), Aubin (2017), Belteki (2023).

41 Chapman (1985, p. 322).

42 The most often quoted remarks have been: "It was impossible for a man to be at his best for long under the old regime, and from forty-six to forty-seven has been an ordinary age for an assistant

"Maunder was not the ideal man to write of the assistant's lot at Greenwich," and that Edwin Dunkin's autobiography "depicts Airy differently and with much more understanding."[43] This is not necessarily surprising since Dunkin had a success story: he started work as a calculator at the age of seventeen, but eventually, with Airy's help, would rise to become deputy astronomer royal, fellow of the Royal Society and president of the Royal Astronomical Society.[44]

After a close analysis of archival materials, Daniel Belteki has concluded that Dunkin was the exception.[45] Because most of the recent debate has concerned calculators, it is worth considering these issues from the perspective of the mag. and met. department. The second assistant, John Russell Hind, the son of a lace manufacturer from Nottingham, was also aged seventeen when hired as assistant; four years later he was recommended by Airy as astronomical observer to a private observatory in London where he searched for comets.[46] A third assistant, James Paul (aged sixteen), was hired in 1840 as a temporary replacement of Glaisher in the Circle Department, but then moved to mag. and met. in 1843, but resigned by the autumn of 1844.[47] Paul's replacement, Hugh Breen Jr, was another young man who had started as a calculator in 1838 (aged sixteen); after less than three years, he was moved to the astronomical department and raised to be fourth assistant.[48] The next three assistants had less luck because the number of positions was reduced from three to one after the method of photographic self-registration was introduced in 1848. Dilkes Lovelace/ Loveless, aged sixteen, was hired in 1844; George Humphreys, aged twenty-four, in 1846; Thomas Downs, hired as a calculator at the age of seventeen in 1839, was promoted as assistant in 1845. The only lucky one was John Charles Henderson, aged twenty-two, who was hired in 1848 as assistant (the only one at this point) and promoted in 1852 as astronomical assistant.[49]

Between 1840 and 1850, out of eight mag. and met. assistants, three were promoted internally as astronomical assistants, while a fourth joined a private observatory. Three of the assistants had been promoted from the ranks of calculators, while the other five were externally hired. The shift from calculator to assistant brought a substantial salary increase: a mag. and met. assistant had a salary of £120 per

to break down under the strain"; or, "For thirty-five years of his administration the salaries of his assistants remained discreditably low, and his treatment of the supernumerary members of his staff would now probably be characterized as 'remorseless sweating'": Maunder (1900, pp. 117–140).

43 Chapman (1992, p. 108; 2012).

44 Chapman (1992), Chapman (2004). For the different trajectory of his brother, who also worked as a calculator at Greenwich, see Belteki (2023).

45 "Dunkin's story, that of an extremely rare computer promoted within the observatory ranks, is very much the exception and not the rule": Belteki (2023, p. 73).

46 Hutchins (2004).

47 Airy (1846, p. 1).

48 Afterwards, Breen left the observatory to establish a private school in Ireland: see Chapman (2012, pp. 10–11). Hugh Breen Sr., who became superintendent of the calculators, is discussed in Belteki (2023, pp. 75–76), though not the later career of the son.

49 All the appointments were announced in in the annual reports of the observatory; for a summary see Dolan (n.d.-a).

year, only slightly less than the £130 p.a. earned by the lowest ranking astronomical assistant, and roughly double the salary of a calculator.[50] The mag. and met. position came with another advantage: assistants had the opportunity, which few calculators had, of carrying out astronomical observations with the meridian instruments.[51] This might have helped considerably because all the new astronomical assistants, who had been hired between 1836 and 1852 (Dunkin, Breen and Henderson), had first served for several years as mag. and met. assistants.

The work of the mag. and met. assistants was "probably the most monotonous, trying, and distasteful of any that the observatory had to show." recalled Walter Maunder, who joined the observatory in 1873 when these duties were still "within the memory of members of the staff still living."[52] Dunkin had a similar recollection that the two-hourly observations were "always a harassing work for the observers."[53] Though a rotation system insured that night duties were repeated every three days, observers must have been bored and frustrated waiting for the next reading. An anonymous article, attributed to James David Forbes, explained what made the mag. and met. observations so much more frustrating than the astronomical ones: "The observations being perfectly regular and unintermitted day or night, independently of the state of the weather, would seem to be more harassing in that particular."[54] Airy seemed to have recognized the special nature of this regime of observation and, as he later recalled in his "Autobiography", "refused to commence this [system of observations] until I had secured a 'Watchman's Clock' for mechanical verification of the regular attendance of the Assistants."[55] The annual reports provided a detailed description of this arrangement that was introduced "in order to give reasonable security to myself and to the superintendent, that the assistants have really been present at the time at which their observations profess to have been made."[56]

The watchman's clock which, as also remarked by Airy, was "frequently used in large manufactories," has been invoked by historians as proof of the transformation of the observatory into a factory.[57] Chapman considered that "Airy really 'industrialised' the Royal Observatory when he instituted clocking-on for the assistants, to make

[50] The salaries of calculators could vary greatly based on their productivity: see Dolan (n.d.-b); Belteki (2023). The salaries of the assistants, which increased with rank, "were almost twice those paid by most universities or private observatories and three times those paid to assistants at the Nautical Almanac Office": Croarken (2009, p. 388).

[51] The astronomical observations of the mag. and met assistants are mentioned in the annual reports of the observatory; see for example Airy (1844, p. 31), Airy (1845, pp. 29–32), Airy (1850, p. 2). For the fact that calculators were often denied permission to use the astronomical instruments, see Belteki (2023).

[52] Maunder (1900, pp. 241–242).

[53] Dunkin (1892, p. 81).

[54] Forbes (1850, p. 353).

[55] Airy (1896, p. 141).

[56] Airy (1843, pp. 84–85).

[57] Airy (1843, pp. 84–85).

sure that they did not cheat on their timings."[58] Similarly for Schaffer, the "factory discipline" was represented by "clocking on":

The coordination of self-registration, prompt and accurate reduction of observations, and a rigid timetable for assistants' work, including the practice of clocking on introduced by Airy for all his subordinates, promoted a severe disciplinary regime at Greenwich (Airy, 1896, 203).[59]

However, both Chapman and Schaffer forget to mention that the watchman clock was only introduced and employed for the mag. and met. assistants engaged in bi-hourly observations; we even know that the watchman clock was located in the ante-room of the magnetic observatory.[60] Its use was immediately discontinued when the bi-hourly system of observations ceased in 1848: "this irksome system of control is now happily superseded by Mr. Brooke's invention of the photographic registration of the magnets."[61]

Conclusions

Even historians who have described the naval regime of meteorological observations, such as Simon Naylor, have been prone to be too hasty in identifying it with the regime of observation in astronomical observatories such as Greenwich.

Naval seamen were meant to be the meteorological equivalents of the "obedient drudges" who helped to operate astronomical observatories and to treat their work and their ship in just the way that astronomical technicians were expected to operate in the observatory.[62]

However, as the above analysis has shown, such conclusions are mostly based on a confusion of the activities of the astronomical, and magnetic and meteorological observatories. In fact, the astronomical staff, though commonly described by historians as "obedient drudges", found the system of bi-hourly observations to be particularly harrowing. Airy installed his factory-like system of surveillance, the watchman's clock, precisely because the two regimes of observations were not equivalent. It is an irony of the historiography that this difference has been erased and the features historians have attributed to the astronomical observatory were actually drawn from descriptions of the magnetic and meteorological observatories.

---

[58] Chapman (1985, p. 327).

[59] Schaffer (1988, p. 120).

[60] G. B. Airy (1843, p. 1): "the 'check-clock', or 'watchman's clock', is in the ante-room, affixed to the dividing wall, nearer to the bifilar-magnet than to the balance-magnet." For a plan of the magnetic observatory with the location of the watchman's clock see Glaisher (1844). This use of the watchman's clock by the mag. and met. department was clearly stated in the earlier literature on Greenwich, see Meadows (1975, p. 10).

[61] Forbes (1850, pp. 353–354). On Brooke's photographic method of registration see MacDonald (2018, pp. 75–76), Ronalds (2016, Chapter 15).

[62] Naylor (2015a, p. 774); see also Naylor (2015b).

During the Magnetic Crusade, the observatories staffed by officers excelled at this new regime of observation in comparison to those staffed by astronomers. Some, out of a spirit of competition, switched to hourly observations and boasted that their "observatory will yield to none, and … be second to none, in the mass & value of the data they will collect."[63] Meanwhile, the astronomical observatories found even the bi-hourly observations too taxing, and discontinued them at the first opportunity.[64] If the officers took pride in their heroic regime of observation, the astronomers felt that, in the words of Karl Kreil (assistant and future director at the Prague observatory), hourly observations had "made the observer a slave of his instrument and almost an automaton himself."[65] Besides, while the colonial observatories staffed by officers sent their readings to Edward Sabine's magnetic department at Woolwich Arsenal for processing and publication, the astronomical observatories took this responsibility upon themselves. When faced with a choice between continuing the hourly observations or publishing the accumulated results, astronomers opted for the latter.

It is in this context that the first self-registering *systems* of observation (rather than mere isolated instruments such as Osler's wind-gauge) were introduced, initially, in the astronomical observatories which took part in the magnetic crusade. Johann von Lamont, at the Munich observatory, and Karl Kreil at the Prague observatory, developed their own systems of mechanical registration, while Airy adopted a photographic system for Greenwich. At the Brussels observatory, Quetelet substituted the bi-hourly system of direct observations for Kreil's autographs. These early instruments never surpassed the precision of direct observations, but this did not matter that much. Airy insisted that the value of precision in positional astronomy should not be blindly extended to meteorology. Though Sabine, whom Airy dismissed as a mere collector of data, was committed to improving the precision of observations, he also agreed that the precision of isolated measurements was not relevant. One needed a global network of observatories to increase the coverage of the phenomena being studied. While both astronomical and physical observations gained from the spatial dispersal of observatories (though not necessarily to the same extent, as pointed out by Herschel), the temporal dimension affected these two disciplines differently. Besides the physical measurements which had to be quantified over time (as daily averages), many believed that the best bet for a scientific breakthrough was the observation of magnetic, electric and meteorological disturbances. This required the simultaneous reading of multiple instruments not only in one place, but across all the observatories; and not only during term-days, but day and night, across the year.

---

[63] Joseph Henry Kay quoted in Goodman (2018b, p. 202). Kay was an officer in the Royal Navy and the director of the observatory on Van Diemen's Land, known today as Tasmania. The Toronto observatory, under the command of Lieutenant John Henry Lefroy, also switched to hourly observations from 1842 throughout 1848: Toronto Observatory (1853, p. 22).

[64] For example, Johann von Lamont, the director of the Munich observatory, announced in 1845 at the magnetic conference that he would "give up the present system of two-hourly observations, and will make only three or four observations a-day": "Proceedings Magnetical Conference" (1846, p. 22).

[65] Kreil (1843, p. 263).

The temporal distinction between astronomical and physical observatories is particularly subtle because not only their histories overlap, but also their rhetorical devices. As shown in the opening quotes, the idea of a continuous register of observations could be similarly depicted by same metaphor of an unbroken vigil. However, as this paper has argued, one must be cautious about such cultivated analogies. In fact, there are two different temporal structures of continuity. In the case of astronomy, the watchtower metaphor implied a *successive continuity*: "the sentinel upon the watchtower is relieved from duty, but another takes his place, and the vigil is unbroken."[66] The new regime of observation of the physical sciences, however, required a *simultaneous continuity*: the observers' eyes had to be open at the same time, at all locations. While the astronomer could follow the temporal course of the phenomena, the physical observers required a temporal grid of observations, besides the spatial network of observatories, to ensure "that no natural phenomenon of any importance can manifest itself without having been seen and observed."[67]

References

Airy, G. B. (1840). *Astronomical and magnetical and meteorological observations made at the Royal Observatory Greenwich, in the year 1839*. Barker and Murray.

Airy, G. B. (1842). *Astronomical and magnetical and meteorological observations made at the Royal Observatory Greenwich, in the year 1841*. Barker and Murray.

Airy, G. B. (1843). *Magnetical and meteorological observations made at the Royal Observatory Greenwich, in the years 1840 and 1841*. Palmer and Clayton.

Airy, G. B. (1844). *Astronomical observations made at the Royal Observatory Greenwich, in the year 1842*. Palmer and Clayton.

Airy, G. B. (1845). *Astronomical and magnetical and meteorological observations made at the Royal Observatory Greenwich, in the year 1843*. Palmer and Clayton.

Airy, G. B. (1846). *Astronomical and magnetical and meteorological observations made at the Royal Observatory Greenwich, in the year 1844*. Palmer and Clayton.

Airy, G. B. (1849). *Astronomical and magnetical and meteorological observations made at the Royal Observatory Greenwich, in the year 1847*. Palmer and Clayton.

Airy, G. B. (1850). *Astronomical and magnetical and meteorological observations made at the Royal Observatory Greenwich, in the year 1848*. Barker and Murray.

Airy, G. B. (1856). *Astronomical and magnetical and meteorological observations made at the Royal Observatory Greenwich: in the year 1854*. Eyre and Spottiswoode.

Airy, G. B. (1896). *Autobiography*. Cambridge University Press.

Anderson, K. (2005). *Predicting the weather: Victorians and the science of meteorology*. University of Chicago Press.

Annuaire de l'Observatoire royal de Bruxelles. (1890). Hayez.

Aubin, D. (2017). On the epistemic and social foundations of mathematics as tool and instrument in observatories, 1793–1846. In J. Lenhard & M. Carrier (Eds.) *Mathematics as a Tool: Tracing New Roles of Mathematics in the Sciences* (pp. 177–196). Springer International Publishing.

Aubin, D., Bigg, C., & Sibum, H.O. (2010). Introduction: Observatory techniques in nineteenth-century science and society. In D. Aubin, C. Bigg, & H.O. Sibum (Eds.) *The Heavens on Earth:*

---

[66] Mitchel (1850, p. 5). The quote was used and commented on by Schaffer (1988, p. 127).

[67] Quetelet (1853, pp. 701–702): "qu'aucun phénomène naturel de quelque importance ne puisse se manifester sans avoir été vu et observé.".

Observatories and Astronomy in Nineteenth-Century Science and Culture (pp. 1–32). Duke University Press.
Belteki, D. (2023). The winter of raw computers: The history of the lunar and planetary reductions of the Royal Observatory, Greenwich. *The British Journal for the History of Science, 56*(1), 65–81.
Canales, J. (2016). Recording Devices. In B.V. Lightman (Ed.) *A Companion to the History of Science* (pp. 500–514). Wiley.
Carter, C. (2009). Magnetic fever: Global imperialism and empiricism in the nineteenth century. *Transactions of the American Philosophical Society* 99(4):16, 1–168.
Cawood, J. (1979). The magnetic crusade: Science and politics in early Victorian Britain. *Isis, 70*(4), 493–518.
Chapman, A. (1985). Sir George Airy (1801–1892) and the concept of international standards in science, timekeeping and navigation. *Vistas in Astronomy, 28*, 321–328.
Chapman, A. (1992). George Biddell Airy, FRS (1801–1892): A centenary commemoration. *Notes and Records of the Royal Society of London, 46*(1), 103–110.
Chapman, A. (2004). Edwin Dunkin (1821–1898), astronomer. In *Oxford Dictionary of National Biography*. Oxford University Press.
Chapman, A. (2012). Airy's Greenwich staff. *The Antiquarian Astronomer, 6*, 4–18.
Croarken, M. (2009). Human computers in eighteenth- and nineteenth-century Britain. In E. Robson & J. A. Stedall (Eds.) *The Oxford Handbook of the History of Mathematics* (pp. 376–403). Oxford University Press.
Daston, L. (2022). *Rules: A Short History of What We Live By*. Princeton University Press.
Daston, L., & Galison, P. (1992). The image of objectivity. *Representations, 40*, 81–128.
De Chadarevian, S. (1993). Graphical method and discipline: Self-Recording instruments in nineteenth-century physiology. *Studies in History and Philosophy of Science Part A, 24*(2), 267–291.
Dolan, G. (n.d.-a). The Assistant Grades. Retrieved May 5 2023 from http://www.royalobservatorygreenwich.org/articles.php?article=999.
Dolan, G. (n.d.-b). Pay 1836–1871. Retrieved from http://www.royalobservatorygreenwich.org/articles.php?article=945.
Donnelly, K. (2014). On the boredom of science: Positional astronomy in the nineteenth century. *The British Journal for the History of Science, 47*(3), 479–503.
Donnelly, K. (2016). *Adolphe Quetelet, Social Physics and the Average Men of Science, 1796–1874*. Routledge.
Dunkin, E. (1892). Sir George Biddell Airy, K.C.B. *The Observatory, 185*, 74–94.
Forbes, J. D. (1830). Description of a new anemometer. *The Edinburgh Journal of Science, 2*, 31–42.
Forbes, J. D. (1850). National observatories—Greenwich. *The Edinburgh Review, 184*, 299–356.
Gauss, C. F., & Weber, W. (1837). *Resultate aus den Beobachtungen des Magnetischen Vereins im Jahre 1836*. im Verlage der Dieterischen Buchhandlung.
Glaisher, J. (1844). The magnetic and meteorological Royal Observatory, just completed at Greenwich. *Illustrated London News*, 16 March 1844, 163–164.
Goodman, M. (2018a). Follow the data: Administering science at Edward Sabine's magnetic department, Woolwich, 1841–57. *Notes and Records: The Royal Society Journal of the History of Science, 73*(2), 187–202.
Goodman, M. (2018b). From "Magnetic Fever" to "Magnetical Insanity": Historical Geographies of British Terrestrial Magnetic Research, 1833–1857 (PhD). University of Glasgow.
Harris, W. S. (1845). Report on the Working of Whewell's and Osler's Anemometers at Plymouth during the years 1841, 1842, 1843. In *Report of the British Association for the Advancement of Science: 1844*. (pp. 241–266). Murray.
Herschel, J. (1830). *A Preliminary Discourse on the Study of Natural Philosophy*. Longman & Taylor. J.
Herschel, J. (1846). Address. In *Report of the British Association for the Advancement of Science: 1845*. (pp. 27–44). Murray.

Herschel, J. F. W. (1835). *Instructions for Making and Registering Meteorological Observations in Southern Africa, and Other Countries in the South Seas, and also at Sea.* Reprinted for private distribution.

Hoffmann, C. (2007). Constant differences: Friedrich Wilhelm Bessel, the concept of the observer in early nineteenth-century practical astronomy and the history of the personal equation. *The British Journal for the History of Science, 40*(3), 333–365.

von Humboldt, A. (1830). De l'inclinaison de l'aiguille aimantée dans le nord de l'Asie, et des observations correspondantes des variations horaires faites en différentes parties de la terre. *Annales De Chimie Et De Physique, 44,* 231–243.

Hunt, J. L. (1996). James Glaisher FRS (1809–1903) Astronomer, Meteorologist and Pioneer of Weather Forecasting: "A Venturesome Victorian." *Quarterly Journal of the Royal Astronomical Society, 37,* 315.

Hutchins, R. (2004). Hind, John Russell (1823–1895), Astronomer. In *Oxford Dictionary of National Biography.* Oxford University Press.

Kreil, C. (1843). *Astronomisch-meteorologisches Jahrbuch für Prag,* vol. 2. Haase.

Lloyd, H. (1842). *Account of the Magnetical Observatory of Dublin and of the Instruments and Methods of Observation Employed There.* Graisberry and Gill.

Locher, F. (2007). The Observatory, the land-based ship and the crusades: Earth sciences in European context, 1830–50. *The British Journal for the History of Science, 40*(4), 491–504.

MacDonald, L. (2018). *Kew Observatory and the Evolution of Victorian Science, 1840–1910.* University of Pittsburgh Press.

Macdonald, L. T. (2018). The origins and early years of the magnetic and meteorological department at Greenwich Observatory, 1834–1848. *Annals of Science, 75*(3), 201–233.

Maunder, E.W. (1900). *The Royal Observatory, Greenwich: A glance at its history and work.* The Religious Tract Society.

Meadows, A. J. (1975). *Greenwich Observatory,* vol. 2. Taylor & Francis.

Mitchel, O. (1850). *The Planetary and Stellar Worlds: An Exposition of the Discoveries and Theories of Modern Astronomy.* Thomas Nelson.

Mitchel, O. (1860). *Popular Astronomy.* Routledge.

Naylor, S. (2015a). Log books and the law of storms: Maritime meteorology and the British admiralty in the nineteenth century. *Isis, 106*(4), 771–797.

Naylor, S. (2015b). Weather Instruments all at Sea: Meteorology and the Royal Navy in the Nineteenth Century. In F. MacDonald & C.W.J. Withers (Eds.) *Geography, Technology and Instruments of Exploration* (pp. 77–96). Ashgate.

Naylor, S., & Goodman, M. (2020) Atmospheric Empire: Historical Geographies of Meteorology at the Colonial Observatories. In M. Mahony & S. Randalls (Eds.) *Weather, Climate, and the geographical Geographical Imagination: Placing Atmospheric Knowledges* (pp. 25–42). University of Pittsburgh Press.

O'Hara, J. G. (1983). Gauss and the Royal Society: The reception of his ideas on magnetism in Britain (1832–1842). *Notes and Records of the Royal Society of London, 38*(1), 17–78.

Proceedings Magnetical Conference. (1846). Proceedings connected with the magnetical and meteorological conference, held at Cambridge in June 1845. In *Report of the British Association for the Advancement of Science: 1845* (pp. 1–82). Murray.

Quetelet, A. (1847). *Rapport adressé à M. Le Ministre de l'interieur sur l'etat et les travaux de l'observatoire royal pendant l'année 1847.* Imprimerie Hayez.

Quetelet, A. (1853). Quelques remarques sur l'influence des Académies, des Congrès et des Conférences scientifiques. In *Bulletins de l'Académie royale des sciences, des lettres et des beaux-arts de Belgique* (pp. 694–704). Hayez.

Quetelet, A. (1855). Notice sur Jacques-Guillaume Crahay. *Annuaire De L'académie Royale des Sciences, des Lettres Et des Beaux-Arts De Belgique, 27,* 117–130.

Quetelet, A. (1866). *Sciences mathématiques et physiques chez les Belges au commencement du 19e siècle.* H. Thiry-van Buggenhoudt.

Ratcliff, J. (2016). Travancore's magnetic crusade: Geomagnetism and the geography of scientific production in a princely state. *The British Journal for the History of Science, 49*(3), 325–352.
Recommendations. (1846). Recommendations Adopted by the General Committee at the Cambridge Meeting in June 1845. In *Report of the British Association for the Advancement of Science: 1845* (pp. xv–xvi). Murray.
Reidy, M. S. (2008). *Tides of History: Ocean Science and Her Majesty's Navy*. University of Chicago Press.
Report on Meteorology. (1840). *Report of the Committee of Physics and Meteorology of the Royal Society Relative to the Observations to be Made in the Antarctic Expedition and in the Magnetic Observatories*. Richard & John E. Taylor.
Ronalds, B. F. (2016). *Sir Francis Ronalds: Father of the Electric Telegraph*. Imperial College Press.
Schaffer, S. (1988). Astronomers mark time: Discipline and the personal equation. *Science in Context, 2*(1), 115–145.
Smith, R. W. (1991). A national observatory transformed: Greenwich in the nineteenth century. *Journal for the History of Astronomy, 22*(1), 5–20.
Smith, R. W. (2003). Remaking Astronomy: Instruments and Practice in the Nineteenth and Twentieth Centuries. In C. Lindberg, R.L. Numbers, & R. Porter (Eds.) *The Cambridge History of Science*, vol. 5. pp. 154–173. Cambridge University Press.
Toronto Observatory. (1853). *Observations Made at the Magnetical and Meteorological Observatory at Toronto in Canada. 1853*, vol. 2. Longman.

Ion Mihailescu is a researcher at the Institut d'histoire at the Université de Neuchâtel and a former postdoctoral researcher in the Laboratory for the History of Science and Technology at the Swiss Federal Institute of Technology in Lausanne (EPFL). His work focuses on the use of graphical methods in the physical sciences and the development of self-registering instruments. He is currently investigating the development of atomic clocks at Neuchâtel, and its role in the 1967 redefinition of the second.

Chapter 6
Timekeeping and Time Balls at the Hamburg Observatory

Gudrun Wolfschmidt

Abstract In the nineteenth century, timekeeping was central to navigation, urban coordination and maritime activities in Hamburg, a major port city and Germany's gateway to the world. This paper examines the historical development of timekeeping at the Hamburg Observatory focusing on its technological advances, influential figures and contributions to maritime navigation and public life. A central element of this study is the role of time balls in the dissemination of accurate time. These time balls, synchronised by telegraphic signals with observatory clocks, played a crucial role in ensuring navigational accuracy by calibrating ship's chronometers. The analysis extends beyond Hamburg to provide a global perspective on the evolution and eventual obsolescence of time balls, which were replaced by wireless radio signals in the early twentieth century.

Introduction

In a port city like Hamburg, which served as Germany's gateway to the world, timekeeping and time services have always played a crucial role, especially in the nineteenth century—the age of railways, telegraphy, and the flourishing pendulum clock and chronometer making industry. Accurate clocks were essential for navigation, the precise determination of longitude at sea, and the regulation of public clocks. The Hamburg Observatory, originally located in Millerntor (*1825) and since 1912 in Bergedorf, played a central role in timekeeping. Time was determined there by observing the meridian transits of stars with transit instruments or meridian circles. This time was then transmitted to the harbour to trigger the time ball on the roof of Kaispeicher No. 1, which was in operation from 1876 to 1934.

This paper delves into the complex history of timekeeping at the Hamburg Observatory, examining its technological innovations, the contributions of key figures, and the impact of its precise time distribution on maritime navigation and urban life in Hamburg. It then explores the role of time balls in the dissemination of time in

G. Wolfschmidt (✉)
GNT, Hamburg Observatory, University of Hamburg, Hamburg, Germany
e-mail: gudrun.wolfschmidt@uni-hamburg.de

G. Bernasconi et al. (eds.), *Astronomical Observatories and Chronometry, 18th-20th Century*, History of Physics, https://doi.org/10.1007/978-3-031-95874-8_6

Hamburg, which was crucial to the city's bustling maritime activities. My analysis of time balls not only focuses on those present in Hamburg and Germany but also extends to a global perspective.

Timekeeping at Hamburg Observatory

Johann Georg Repsold (1770–1830), founder of the Repsold Instrument Workshop in 1799, supplied instruments to observatories worldwide and built his private observatory at Stintfang on the Albertus Bastion of the Hamburg Wall in 1802. This observatory, along with his next observatory Millerntor (1825) on the Henricus Bastion, and the Danish observatory in Altona (1823), played an important role in surveying, time measurement, and navigation in Hamburg.[1] The Millerntor Observatory (east dome) was combined with the School of Navigation (west dome), and the observatory became the Hamburg State Institute in 1833.

Inspired by the *Rota Meridiana*, an idea of Ole Christensen Rømer (1644–1710), the first modern meridian circle was introduced by Johann Georg Repsold in 1802–1803. At the Hamburg Observatory, time was determined by observing the meridian transits of stars using Repsold's instruments (see Fig. 6.1). The equipment at Millerntor was gradually improved with a meridian circle (1802/1818), a transit instrument (1829), and a new meridian circle (1836) made by Adolf & Georg Repsold, featuring two circles each with a diameter of 103 cm. Later at the new Hamburg Observatory in Bergedorf, a transit instrument in a pavilion was added (see Fig. 6.2), and a sophisticated meridian circle (1909) was installed, made by Adolf Repsold & Söhne. The transit clock used to determine the time was always set to sidereal time. The pendulum clocks at the Millerntor Observatory were made by Kessels and Knoblich, both from Altona. The master clock was placed in the meridian hall and connected to the slave clocks via electrical contacts. Concerning the observation of star transits, one can recognize a development from subjective (visually and manual time registration) to objective methods:

- Eye-ear method (Metronome)—Counting the beats of the metronome.
- Eye-hand method (Hipp Chronograph)—Visually and moving a thread with automatic registration, used in Millerntor Observatory since 1868.
- Impersonal (self-registering) Micrometer, A. Repsold & Sons (1870)
- Modern objective methods: Photographic (1920s) and Photoelectric (1950s) with computer registration (1967).

George Rümker (1832–1900), second director of Hamburg's Millerntor Observatory and Navigation School (1857/67–1900), was particularly interested in timekeeping and navigation, and was also chairman of the examination board for

[1] Wolfschmidt (2007a, pp. 111–115).

Fig. 6.1 Measuring the transit with a meridian circle, in Hamburg-Bergedorf (credit: Archive Hamburg Observatory)

Fig. 6.2 Transit instrument pavilion (1912), in Hamburg-Bergedorf (credit: Archive Hamburg Observatory)

helmsmen and captains. Rümker proposed to the Hamburg Senate and the Imperial Admiralty that a *Chronometer Testing Institute*[2] be established at the Hamburg Observatory, following the example of England. This institute was founded in 1876 and given its own building in the garden of the observatory.

The Hamburg Observatory for the new Hamburg Observatory, built in Bergedorf (1906–1912), a new transit instrument and a meridian circle were acquired, made by A. Repsold & Sons, in addition a large collection of 22 pendulum clocks, seven electric second slave clocks, eleven marine and pocket chronometers, as well as accessories, especially for telegraphy and radio communication. The main clock room (Fig. 6.3), the centre for timekeeping at the Hamburg Observatory, was located in the basement of the main building, a room with a perfectly constant temperature. Due to the presence of a large park in Bergedorf Observatory, each telescope and instrument was housed in a separate dome or pavilion and was equipped with two clocks, regulated by sidereal and solar time.

Of the numerous pendulum clocks and chronometers used at the Hamburg Observatory, only five still exist in Bergedorf: the Riefler pendulum clocks no. 10 and no.

[2] Schramm (2010, p. 125).

Fig. 6.3 Clock room, centre for timekeeping in Hamburg observatory, around 1912 (right: contact clockwork made by J. & A. Unger of Strasbourg), (credit: Archive Hamburg Observatory)

11 by W. Bröcking of Hamburg (1910), no. 14 by Adolf Pohl of Hamburg (1913), the marine chronometers no. 30 by Kittel of Altona (1914) and no. 70 by John Poole (1818–1867) of London (1919). The remaining clocks were sent in 1972 to the Staatliche Uhrmacherschule Hamburg, formerly in Altona, now in Farmsen, to the Astronomisch-Physikalisches Kabinett, Museumslandschaft Hessen-Kassel, and to the Hamburg Planetarium. However, some clocks and chronometers were taken out of service because they could no longer be repaired or were lost during an eclipse expedition to Russia in 1914 at the beginning of the First World War.

Time Balls for Time Distribution Services

The "Prehistory": Distribution of Time by Noon Guns

Today, in the age of satellite navigation, it is hard to imagine that for centuries it was impossible to determine the exact position of a ship on the high seas. The invention of the Harrison chronometer (1759) provided a solution, but the chronometers had to be compared as regularly as possible with the time determined daily by astronomers through transit observations, and indicated by highly accurate pendulum clocks. In this way, the accuracy of each chronometer could be recorded by "rating" the chronometers, i.e. by determining their exact error, since the deviation of the chronometers had to be taken into account when calculating the ship's position, the longitude.

Until the nineteenth century, captains of sailing ships had to take their chronometers to an observatory to compare them with the exact time of the pendulum clocks there. In Marseille, for example, this was done until the beginning of the twentieth century, when the captains had to walk from the port to the Bourse building, where there was a main reference clock connected to the observatory, which from the 1860s was far away, on the Plateau Longchamp. However, not every port had the possibility of comparing chronometers with pendulum clocks, even though the number of ships equipped with chronometers increased steadily from the beginning of the nineteenth century. It was for this reason that the idea of displaying the exact time to ships in ports was born.

The midday gun as a historical time signal, such as at the Cape Observatory, Cape Town, South Africa (1806) or at Edinburgh Castle (1861), was not accurate enough to synchronise the ships' chronometers. The speed of sound and the distance of the time gun must indeed be taken into account when determining the exact time.

The First Time Balls as Visual Time Signals

As an alternative to gunfire, the first time ball was invented by British naval officer and future admiral Robert Wauchope (1788–1862) in 1818 while stationed on St Helena. In 1824 he submitted to the Admiralty his plan for checking the accuracy of ship's chronometers by means of optical signals. The plan was to raise a large hollow metal body, five feet in diameter, which Wauchope called a "time ball," to the top of a mast, and then use a special mechanism to drop it every day at a fixed time, usually one o'clock. The time ball had to be large, spherical, painted black and visible from all directions. Construction was not easy, because during the first experiments the time ball broke when it fell. It took five years before the first successful construction was tested in 1829 at Portsmouth Dockyard, Royal Navy College, in the English Channel.

The first time balls were built at the Royal Greenwich Observatory (1833) by John Pond (1767–1836), made by Maudslay, Sons and Field, for ships on the Thames, then at Ladder Hill Observatory on St Helena (1834), Cape Town (1836), Liverpool (1844), Edinburgh (1853), Glasgow (1858), and later in Dublin (1874–1915) and Brighton (1887).

The Hamburg Time Ball (1876–1934)

In Hamburg, it was George Rümker who first decided to erect a time ball (1876) in a prominent place in the harbour, called Kaiserhöft. He also proposed to the

Fig. 6.4 Timeball in the port of Hamburg, museum for Hamburg history (Photo by Gudrun Wolfschmidt)

Hamburg Senate and the Imperial Admiralty the establishment of a Chronometer Testing Institute at the Observatory, as at Greenwich; it was founded in the same year and was housed in a separate building in the Observatory's garden. The building was a timber-framed structure walled in with clinker bricks.[3] The time ball was placed on top of the 40-m-high tower above its west façade (Fig. 6.4). The statics of the Kaispeicher were not designed to take much extra weight, so the construction had to be light but strong enough to withstand the strong wind and ensure perfect operation. Four legs with additional springs were combined in a ring to which the mast with the time ball was attached. The whole construction, invented and built by Carl Bamberg of Berlin, could be tilted up to 12° without damaging the structure and was protected by a lightning conductor.[4]

The black signal ball was 1.5 m in diameter and hung 53 m above the Elbe at high tide (Fig. 6.5). Ten minutes before the exact time was announced, the ball was released, pulled halfway up, and three minutes before that it was pulled to the top of the mast—enough time for the navy staff to be ready to observe the moment of release of the time ball. It was then lowered 3 m and a spring cushioned the fall. A

[3] Maass (1990).

[4] Wolfschmidt (2012). Cf. Company publications (*Firmenschriften*) from the *Carl Bamberg-Werk*, Archives of the Deutsches Museum in Munich and Deutsches Technikmuseum Berlin.

signal was triggered to indicate to the operators in the observatory that the time had been correctly displayed. In the event of a malfunction, e.g. due to a storm or ice on the mast in winter, a 40 cm red ball with a hemp rope (not a metal cord to protect against thunderstorms) was raised 5 min later to indicate that there had been a time error.[5] The time ball was dropped for the first time on 16 September 1876. The time ball was controlled by the contact clockwork of the Hamburg Observatory, built by J. & A. Unger of Strassburg (1899). This system controlled the timeballs in Hamburg, Bremerhaven and Cuxhaven, the light signal systems, the telegraphic overseas time connection as well as the telephone time signal.

The importance of time ball systems at this time is reflected in the fact that the Report on the Scientific Instruments at the Berlin Trade Exhibition in 1879 (*Bericht über die Wissenschaftlichen Instrumente auf der Berliner Gewerbeausstellung im Jahre 1879*) devoted ten pages to describing the transmission of time signals by time balls.[6] The Hamburg time ball operated for several years without any major problems until it was destroyed by fire in its warehouse on 2 April 1892. Reconstruction work was quickly completed and the rebuilt time ball was put into operation on 1 February 1893.

The time ball was dropped at 12 h 0 m 0.s0 Greenwich Mean Time which is 12 o'clock 39 min and 53.7 s Hamburg local time.[7] Greenwich Mean Time (GMT) was introduced as Universal Time (UT) in 1884, adopted during the International Meridian Conference in Washington D. C. by 25 countries. By Imperial Act (*Reichsgesetz*) of 12 March 1893, the mean solar time at the 15th meridian east of Greenwich was finally adopted as the legal time throughout the German Empire—Central European Time (CET).[8] The transmission to the time ball was initially done manually by the astronomer pressing a button on the switchboard of the clock system, which was connected to an underground cable running from Hamburg's Millerntor Observatory to the time ball. In 1899, it was triggered electrically by a contact clockwork (*Kontakt-Uhrwerk*), made by J. and A. Unger of Strasburg (1899) (Fig. 6.6).[9] Public clocks were also erected in 1876, one at the Chamber of Commerce and Stock Exchange (*1841), one at the Millerntor Observatory (later the *Museum für Hamburgische Geschichte*), and from 1907 also in the flood gauge tower (*Pegelturm, Flutmesserturm*), the St. Pauli Landing Bridges, and the shipyard at Reiherdamm in Steinwerder, all controlled by the observatory. The distribution of time to the public was important for the city's commercial interests, but also for the timetables of coaches, trains, and ships.

[5] This was a rare event (1–2%) as shown in the journals written by the astronomers and published in the annual reports of the Hamburg Scientific Institutions (*Jahrbücher der Hamburgischen Wissenschaftlichen Anstalten*), 1891-1900, and of Hamburg Observatory (*Jahresberichte, Hamburg-Bergedorf*), 1903–1934.

[6] Loewenherz (1880)

[7] Amtliche Bekanntmachung vom 13.9.1876, Staatsarchiv Hamburg, Bestand Sternwarte 364–9/F13c.

[8] Reichs-Gesetzblatt (1893)

[9] Schramm (2010, p. 210).

Fig. 6.5 Functioning model of the Hamburg time ball (Scale 1: 10), Design: Heinz Otto Kult, 2010 (Hamburg Observatory)

Fig. 6.6 Unger contact clockwork (1899) of the Hamburg observatory in Bergedorf (Archives of Hamburg Observatory)

Time Balls in the German Empire

In the German Empire, time balls were erected in cities along the coasts of the North and Baltic Seas. The first time ball was put into operation in 1874—almost 40 years after the Greenwich time ball and two years before the Hamburg time ball—at the Imperial Naval Observatory (*Marineobservatorium*) in Wilhelmshaven (1874–1905) and then at the port of Hamburg in Cuxhaven (1875–1934). The head of the Water Engineering Inspection (*Wasserbauinspektion*) in Cuxhaven, Hugo Leutz, designed a device that was mounted on a 16-m column and put into operation on 16 October 1875. The ball was triggered electrically by a standard clock in the telegraph office, which was transmitted via a separate telegraph line from the clock at the Hamburg Observatory.[10] In Bremerhaven a time ball column was erected in 1876 (1876–1928), then a time ball on the harbour office (*Hafenamt*) (1895), also in Emden in the harbour (1911). Today there is still a time ball on the tower of the old Port Authority office in Dortmund harbour.[11]

In the cities bordering the Baltic Sea, time balls were installed in Kiel (observatory 1884, Imperial Wharf (Kaiserwerft) 1886, artillery depot 1908), in some (German, now Polish) ports (Swinemünde, now Śwínoujście, 1879), in the port of Danzig (now Gdańsk, 1876–1894; Stettin, now Szczecin, 1908) and in Riga, Latvia (1885), and finally in the German colony in China, Tsingtau (now Qingdao, 1909).[12]

Time Balls Around the World

Time balls were quite common before the advent of radio time signals. About 243 time balls, 14 time flaps, 7 light signals and 49 time cannons were erected worldwide.[13] They were typically located at observatories, navigation schools, in harbours, on lighthouses, semaphore signals, or on port authority buildings, port buildings, port offices, warehouses, custom houses, hydrographic offices, ferry buildings or telegraph buildings. Even inland, in cities such as Berlin, Vienna, Zurich, Madrid or New York, there were such installations. The publishing house of the *Nürnberger Zeitung* on the Fleischbrücke had a time ball.

[10] Stadtarchiv Cuxhaven (2010).

[11] Ellerbrock (1999).

[12] From 1898 to 1914, the Chinese city of Tsingtau was a colony of the German Empire. The Tsingtau Observatory (1898) was used for astronomy, but also for timekeeping, meteorology, typhoon warning, geomagnetism and seismology. For a complete list of time ball locations within the German Empire, see Hulse (2023).

[13] Hülse (2023).

After Great Britain (about 65 time balls), where they became very popular, and South Africa,[14] in North America the first time balls were erected in the United States (the oldest was installed at the United States Naval Observatory in Washington, D. C. in 1845),[15] and in Canada (Quebec, 1852), then in Australia, at the Sydney Observatory (1858), and others in Wellington, New Zealand, starting from 1864.[16]

In Africa, it is worth mentioning the presence of time balls in South Africa and Angola (NOTA 32), while in Asia, prominent time balls can be found in Shanghai (1907) and in the Hong Kong Observatory (1885), in some cities of Japan, whose time balls date back to the beginning of the twentieth century; however, the oldest time balls in this continent are located in South East Asia: India, mostly set up by the East India Company, later (after 1858) by the Indian Office,[17] Burma/Myanmar, Ceylon/Sri Lanka and Batavia/Indonesia. Only a few time balls have been installed in South America, in Argentina (Buenos Aires, Cordoba Observatory) Brazil (Rio de Janeiro Observatório) and Chile (Valparaíso Escuela Naval).

There were many time balls in Europe, especially in Denmark, Norway, Sweden,[18] Finland, Northern France,[19] Spain, Portugal, Italy, Greece, Turkey, Ukraine, and time flaps only in Belgium and Netherlands.

Nowadays, only a few installations remain, and even fewer have been restored to working order. Working time balls can still be seen at the Royal Observatory Greenwich (1833), Edinburgh (1855), Gdańsk/Poland (1876), Navigationsskolan Gothenburg/Sweden (1862), Lorient/Bretagne (1928, 2021), Valparaíso/Chile (1894, 2016), Cape Town (1836), Washington D.C. (1845), Sydney (1858), Geelong, Australia (1862, 2006) and Port Chalmers/New Zealand (1867, 2020), Lyttelton/New Zealand (1876, 2018).

[14] St. Paul de Loanda Observatory in Angola had a time ball (1879), regulated by the observatory there, cf. Kinns (2021, pp. 328–330).

[15] Bartky and Dick (1981). San Francisco (1852), Cincinnati (1874), Boston (1878), St Louis (1881), New Orleans (1883), Aloha, Hawaii (1926).

[16] Wellington customs house in the harbour (1864, 1888), the time signals came from the Dominion Observatory, Port Chalmers, Otago province (1867–1970, restored to service in 2020), and Lyttelton port of Christchurch (1876-1934, rebuilt after earthquakes in 2011/12 and 2018).

[17] Time balls are recorded in Calcutta (1835), Bombay/Mumbai (1840), Madras/Chennai (1853) and Karachi 1874). See Kinns (2020).

[18] Time ball on navigational schools in Stockholm, Malmö, and Gothenburg, in addition in Lund and Uppsala, and on the military wharf in Karlskrona. Cf. Holmberg (2022).

[19] *Time flap* in Brest military port, école de navigation (1887) and Cherbourg port (1887), Dunkerque port (1902), and a *Boule horaire* in Lorient, Bretagne (1928, 2021), in the Atlantic Ocean, and a *Boule du temps* in Toulon military port, Observatoire de la Marine in the Mediterranean Sea (1880).

Modern Time Service

After the expansion of the port of Hamburg and the spread of steamships, the time ball was no longer visible everywhere. As more and more ships were equipped with a radio system, the transmission of the exact time by time balls was no longer necessary. Most time balls were decommissioned in the 1920s or replaced by improved light signalling systems (until 1939). The Cuxhaven and Bremerhaven time balls were decommissioned, and the Hamburg Observatory ceased to provide daily time signals at the end of 1928; from 1929, the time service was completely taken over by the Deutsche Seewarte. Only the Hamburg time ball was in use until 1934, when the Hamburg Observatory ceased to keep time. The time ball survived the Second World War, but in 1963 the remains and tower of the old warehouse were demolished to make way for a new warehouse.

Time balls were gradually replaced by electric light signal systems. In 1906 such a system was installed at Kaiser-Wilhelm-Port on Kuhwärder, and in 1909 at St. Pauli Landungsbrücken, with signals four times a day. An important innovation took place: On February 23, 1907, the Hamburg time signal was transmitted by radio technology[20] through the East Frisian coastal radio station (Küstenfunkstelle) Norddeich Radio Station, "KND" with four 65-m-high transmission masts.[21] In addition since 1908/11, telegraphic time signals were sent from Hamburg Observatory to Monrovia (Liberia), Horta (Azores), Vigo (Spain), Tenerife (Canary Islands), Pernambuco (Brazil) and to the German colony Lome (Togo, Africa). The radio time signals were also transmitted to the Großfunkstelle für drahtlose Telegraphie ("Large radio transmitter station for wireless telegraphy") in Nauen near Berlin, founded by Telefunken in 1906; the masts were first 100-m high, then in 1914, two masts of 260-m height, belonged to the highest buildings in Europe. Two years later, from 1909, the time signal was transmitted via the Eiffel Tower station in Paris. At the International Time Signal Conference in Paris in 1912, the main signal *ONOGO* was defined,[22] named after the Morse code transmitted—O (- - -), N (-.), G (- -.). The time service, which transmitted time signals by radio, enabled navigators on ships to synchronise their ship's chronometers regularly with their radio system. The time service of the Imperial Naval Observatory Wilhelmshaven (*1879) began in 1910.

[20] Concerning the invention by Heinrich Hertz and the development of radio communication, see Wolfschmidt 2007b and 2008. In the 1900s until 1912, there were two separate radio technology systems, the German *Telefunken Gesellschaft für drahtlose Telegraphie* (Company for Wireless Telegraphy) of Berlin (*1903), and the British *Marconi Marine Company* of London (*1900). A regular commercial service between *Marconi Towers* (4 wooden aerial towers, 200 feet (61-m) in height), Cape Breton Island, Nova Scotia, Canada, and Clifden, Ireland, opened officially on October 17, 1907, cf. Macleod (1992).

[21] Receiving radio station 1907–1924, Transmitting radio station 1907–1998, this was the world's largest maritime distress radio system on the 500 kHz frequency, which was finally replaced by the satellite-based GMDSS in 1999, cf. Hechtel (2005). Korth (2012), and Retrieved from https://www.norddeichradio.info/index.php.

[22] The precision pendulum clock Adolf Pohl, Hamburg (1913), clock inventory list (Wolfschmidt 2021), no. 13, was used for the signal *ONOGO*.

More generally, modern timekeeping in Hamburg was marked by the introduction of the telephonic time signal in 1907.[23] It soon became permanently retrievable under *H3 Alster 10.000* (1926), later *Hamburg 03* (1932)—at the request of many watchmakers. From 1911 to 1943, this service was available throughout the entire territory of the German Reich. In 1937, the telephone time signals were supplemented by the *speaking clock*[24] as a continuous time announcement (*Hamburg03* and *Hamburg05*). In addition, the time was supplied by the radio broadcasting apparatus.

During the First World War, in 1917, the Hamburg Observatory began transmitting the first radiotelegraphic time signals with automatic time signal transmitters. On 1 November 1919, however, the long and uninterrupted tradition of the observatory in controlling the time signals came to an end; the service was controlled by the *Deutsche Seewarte* (German Naval Observatory, founded in 1872) in Hamburg.[25] In 1920, the time service was moved to the new building of the Museum of Hamburg History, built on the site of the former Millerntor Observatory. Here the time signal was supplied by the contact clockwork and transmitted to the Hamburg standard clocks, the time balls, also in Cuxhaven and Bremerhaven, and the light signal systems. The accuracy was one tenth of a second per day.

Conclusion

Timekeeping with transit instruments was a typical task of observatories in the nineteenth century, and was especially important for navigation. This article has focused on the activities of the Hamburg Observatory at Millerntor and Bergedorf. The Observatory's precise pendulum clocks were connected by contact clockworks to send a signal to drop the time ball. In the same way, about 250 time balls around the world distributed the time before radio time signals were available. Accurate time was essential for controlling the ship's chronometers—essential in turn for determining the position of ships for safe navigation.

Telegraphically synchronised time balls were eventually replaced by wireless technologies, allowing chronometers to be calibrated at any time at sea. Accurate time signals were sent by radio from a station in Nova Scotia, Canada, by the Marconi Wireless Telegraph Company (1907), from Norddeich Radio (1907), Germany, using the Telefunken system, rival of Marconi, since 1911 DEBEG, and from the Eiffel Tower (1909), France, also using a Marconi system. Unfortunately, there are no time balls left in Germany today, but this type of heritage is still fascinating, as evidenced by the few examples that are still in operation, such as the time ball at Greenwich Observatory, which is very popular and has become an attraction for many visitors.

---

[23] Dohmen (1914).

[24] *Speaking clock*: Paris 1933, Great Britain 1936, Hamburg 1937.

[25] Deutsche Seewarte (1921, 1937), Wegner (2007, pp. 170–200), Schramm (2010, pp. 124–127).

References

Annalen der Hydrographie und Maritimen Meteorologie. (1910). *Zeitschrift für Seefahrt-und Meereskunde, Achtunddreißigster Jahrgang.* Ernst Siegfried Mittler und Sohn.

Bartky, I. R., & Dick, S. J. (1981). The first timeballs. *Journal for the History of Astronomy, 12*, 155–164.

Bundesgesetzblatt, vol. 1 (1978). Bundesanzeiger Verlagsges. m.b.H.

Deutsches Hydrographisches Institut. (1947). *Jahresbericht.*

Deutsche Seewarte. (1921, 1937). *Zweiundvierzigster und dreiundvierzigster Jahresbericht über die Tätigkeit der Deutschen Seewarte für die Jahre 1919–1920.* Köbner and Co. G.m.b.H. Zweiundsechzigster Jahresbericht über die Tätigkeit der Deutschen Seewarte für das Jahr 1936. H. W. Köbner & Co. G.m.b.H., Altona

Dohmen, K. (1914). Die Hamburger Zeitsignaleinrichtung für Fernsprechteilnehmer. *Telegraphen- und Fernsprech-Technik, 1*, 1.

Ellerbrock, K-P. (Ed.). (1999). *Dortmunds Tor zur Welt, einhundert Jahre Dortmunder Hafen.* Klartext Verlag, Essen.

Hechtel, D. (2005). *Zur Geschichte der Kommunikation auf See. Das Ende der Einsamkeit.* Verlag Convent, Hamburg.

Hülse, K. (2023). List of 243 timeballs, 14 time flaps and 49 time guns worldwide. https://www.leuchtturm-welt.net/#, 7.4.2023.

Holmberg, G. (2022). The production and distribution of synchronized time in Sweden, 1850–1914. In A. Ekström & S. Bergwik (Eds.) *Times of History, Times of Nature. Temporalization and the Limits of Modern Knowledge* (pp. 30–48). Berghahn Books.

Kinns, R. (2020). Time signals for mariners in India, Burma and Ceylon. *Journal of Astronomical History and Heritage, 23*(3), 523–552.

Kinns, R. (2021). Time signals for mariners in the Atlantic Islands and West Africa. *Journal of Astronomical History and Heritage, 24*(2), 315–336.

Korth, H.-G. (2012 [1900–1909]). Historisches rund um die Telegraphie Seefunktechnik der deutschen Handelsmarine. In *Jahrbuch der drahtlosen Telegraphie und Telephonie.* Johann A. Barth. Republished online (2012) http://www.seefunknetz.de/hist00.html

Loewenherz, L. (Ed.). (1880). *Bericht über die Wissenschaftlichen Instrumente auf der Berliner Gewerbeausstellung im Jahre 1879.* Julius Springer.

Maass, D. (1990). *Der Ausbau des Hamburger Hafens, 1840–1910.* Schiffahrts-Verlag HANSA.

Macleod, M. K. (1992). *Whisper in the Air—Marconi: The Canada Years.* Lancelot Books.

Oberkommando der Kriegsmarine. (1938). *Verzeichnis der Leuchtfeuer und Signalstellen aller Meere, Teil IIIA, Nordsee, Südlicher Teil.* E. S. Mittler and Sohn.

Stadtarchiv Cuxhaven. (2010). Schreiben vom 21.09.2010, Az. 107/10/K.

Reichsamt des Inneren. (1893). *Reichs-Gesetzblatt.* Reichsdruckerei.

Schramm, J. (2010). *Sterne über Hamburg. Die Geschichte der Astronomie in Hamburg.* Kultur-und Geschichtskontor.

Wegner, G. (2007). Aus der Geschichte der Deutschen Seewarte in Hamburg. In G. Wolfschmidt (Ed.) *Hamburgs Geschichte einmal anders. Entwicklung der Naturwissenschaften, Medizin und Technik,* vol. 1 (pp. 170–199). Books on Demand GmbH.

Wolfschmidt, G. (2007a). Sterne über Hamburg—Höhepunkte der Entwicklung der Astronomie. In G. Wolfschmidt (Ed.) *Hamburgs Geschichte einmal anders. Entwicklung der Naturwissenschaften, Medizin und Technik,* vol. 1 (pp. 103–137). Books on Demand GmbH.

Wolfschmidt, G. (Ed.). (2007b). *Von Hertz zum Handy—Entwicklung der Kommunikation. Ausstellung zum 150. Geburtstag von Heinrich Hertz (1857–1894).* Books on Demand.

Wolfschmidt, G. (Ed.). (2008). Heinrich Hertz (1857–1894) and the development of communication. In *Proceedings of the International Symposium in Hamburg, 8–12, 2007.* Books on Demand GmbH.

Wolfschmidt, G. (2012). Telescopes made in Berlin—from Carl Bamberg to Askania. In A.D. Morrison-Low, S. Dupré, S. Johnston, & G. Strano *From Earth-Bound to Satellite: Telescopes, Skills and Networks* (pp. 177–193). Brill.

Wolfschmidt, G. (2021). Anhang: Historisches Uhrenverzeichnis der Hamburger Sternwarte. In C. Prignitz, & G. Wolfschmidt (Ed.), *Zeit für Hamburg—Eine Uhr der Sternwarte und ihr historisches Umfeld* (pp. 84–94). tredition.

Gudrun Wolfschmidt is a German astronomer and historian of science. She served as a professor at the University of Hamburg's Centre for the History of Science and Technology (GNT) from 1997 to 2017. Wolfschmidt earned her doctorate with a dissertation titled *Analysis of Close Binary Systems* at Friedrich-Alexander University Erlangen-Nuremberg (FAU), with research conducted at the Dr. Karl Remeis Observatory in Bamberg. She also completed her first and second state examinations in physics and mathematics. From 1987 to 1997, she contributed to the Deutsches Museum in Munich, where she played a key role in the creation and curation of the museum's Astronomy exhibition, which opened in 1992 and remained on show until 2022. Wolfschmidt also worked as an assistant at the Research Institute for the History of Science and Technology, where she pursued teaching and her Habilitation in *Genesis of Astrophysics* at Ludwig Maximilian University of Munich (LMU). Since 1997, she has been a professor at the University of Hamburg, specialising in the history of science and technology.

Chapter 7
Determining Time, Eliminating the Personal Equation and Introducing Quartz Clocks: The Neuchâtel Observatory's Operating Chain in 1954

Julien Gressot and Romain Jeanneret

Abstract Until the redefinition of the "second" in 1967, the scientific measurement of time was a core aspect of the astronomers' expertise. The Neuchâtel Observatory (1858-2007) was particularly concerned with the problem of determining time. A structural lack of personnel led the institution to promote the automation of measurement procedures. At the mid-twentieth century, the Observatory acquired a photographic zenith tube (PZT). The PZT made it possible to automate the determination of time, while eliminating a certain number of technical errors in the meridian circles. Coordinated by quartz clocks, the PZT profoundly changed the functioning of the Observatory, especially with regard to the human factor. However, its success did not last long, as atomic time quickly became the new reference for determining time. This chapter examines the material and organisational developments at the Neuchâtel Observatory in order to assess their impact on the status of human beings at the institution, focusing on the operational chain for time determination in 1958.

In the twentieth century, time measurement was revolutionised by the introduction of new technologies using quartz and zenithal observations. These changes modified the practices and scientific instruments for determining time. At the Neuchâtel Observatory, since it was founded in 1858, measuring time has formed the epistemological

---

J. Gressot (✉) · R. Jeanneret
Institut d'histoire, University of Neuchâtel, Espace Tilo-Frey 1, 2000 Neuchâtel, Switzerland
e-mail: julien.gressot@unine.ch

R. Jeanneret
e-mail: romain.jeanneret@unine.ch

R. Jeanneret
Haute Ecole Arc Conservation-Restauration, HES-SO University of Applied Sciences and Arts Western Switzerland, Espace de l'Europe 11, 2000 Neuchâtel, Switzerland

G. Bernasconi et al. (eds.), *Astronomical Observatories and Chronometry, 18th-20th Century*, History of Physics, https://doi.org/10.1007/978-3-031-95874-8_7

basis for other scientific activities.[1] For more than a century, in order to remain at the forefront of time measurement, the institution has sought to integrate these various innovations into its "operating chains", a concept which describes the set of scientific instruments, calculations and techniques used by actors in a given space to produce accurate data—in this case, to accurately determine time.[2] This concept emphasises the need to analyse a series of material, mathematical and human factors, focusing not only on the main instruments or the main scientific operations, but also on the practical activities and the day-to-day running of the Observatory. To take full advantage of this concept, visualisations for representative periods (1861, 1913 and 1954) were carried out, in the context of broader research into the history of the culture of precision at the Neuchâtel Observatory[3] in order to understand the transformation of the daily activity of time measurement from the foundation of this institution to the dismissal of the astronomical practice in determining time around the middle of the twentieth century. The aim was to understand the relationship between the various operations and the evolution of the data at each step. In this paper we will focus on the operating chain of 1954. The Observatory had regularly carried out research and adaptations to maintain its operational chain at the best possible level of accuracy, according to the available resources. At the Observatory, astronomers were particularly concerned with the personal equation, the physiological variable of observation specific to each astronomer, and tried to eliminate it mathematically and mechanically in order to improve the accuracy of the data.[4] It was not until the mid-twentieth century that a new instrument made it possible to fully automate time measurement. This was the Photographic Zenith Tube (PZT), a scientific instrument for observing the zenith, which used photography to record the exact moment of a star's transit. This removed the human act of observation, and with it the effects of the personal equation on the data. To function, a PZT worked in coordination with quartz clocks,[5] which were gradually replacing mechanical clocks.

In this paper we want to illustrate how time was determined at the Neuchâtel Observatory in 1954, starting with the instrumental innovations represented by the PZT and the quartz clocks. We first examine the evolution of the technical system for determining time, in order to understand the developments that led to the operational chain of 1954. We then look at the way in which this operating chain was introduced, and the objectives pursued by the Observatory, before concluding with a discussion of the changes in practice brought about by the new instrumental configuration.

[1] Canales (2001); Schmidgen (2003); Gressot and Jeanneret (2022).

[2] Gressot and Jeanneret (2023a).

[3] Gressot (2024a).

[4] Schaffer (1988); Canales (2001); Schmidgen (2009); Gressot (2023).

[5] Katzir (2016).

One Century of Time Measurement at Neuchâtel

From the outset, the Neuchâtel Observatory specialised in astronomical time measurement, using the most advanced instruments and working methods to meet the needs of the watchmaking industry, which required very accurate time to certify the quality of its products.[6] A first operating chain set up to determine time consisted of an Ertel and Sohn meridian circle, an essential instrument in nineteenth- and twentieth-century astronomy.[7] It was complemented by a sidereal pendulum built by Joseph-Thaddeus Winnerl (1799–1886) and a printing chronograph crafted by Matthaüs Hipp (1813–1893). While the meridian circle provided the initial data by observing the transit of various reference stars, the chronograph was a central device because it allowed the data to be recorded on paper and compared with the seconds of the sidereal clock. This chronographic technique replaced the "eye and ear" method.[8] From then on, the astronomer closed an electrical contact each time he observed the bisection of the star with one of the vertical threads of the filar micrometer of the meridian circle. The observer's entire attention was then focused on the movement of the star, whereas before the astronomer also had to count the seconds of the pendulum, assess the moment of the bisection between two strokes in fractions of a second, and record, memorise or dictate these data to an assistant. The chronographic method, on the other hand, made it possible to multiply the number of threads of the reticle and thus the data recorded for each star, thus ensuring better statistical control and therefore greater accuracy. In addition, the chronograph and a chronoscope enabled Adolphe Hirsch (1830–1901), first director of the Neuchâtel Observatory, to carry out research on the personal equation, making the region an international research centre in this field in the 1860s.[9] He concluded that the personal equation varied not only between different people but also for the same observer, which led him to consider the body of the astronomer as part of the observing apparatus:

Such a variation, which considerably exceeds not only the errors of the averages of a daily series, but also the variation of the uncertainty of an isolated observation […] forces us to admit that physiological time for the same observer varies quite considerably from one day to the next, according to the disposition of his nervous system. In this respect, our body is like any other precision machine; its correction is variable, like that of a meridian circle or a pendulum.[10]

---

[6] Gressot (2024b).

[7] Belteki and Gressot (2023); Gressot and Jeanneret (2023b).

[8] See Slotten (1993). The eye and ear method involves the astronomer estimating the moment of the star's transit at each thread in relation to the beat of the seconds of the sidereal clock.

[9] Hirsch (1862 and 1863), Canales (2001), Schmidgen (2009).

[10] Hirsch (1862–1863, p. 365): "Une telle variation qui dépasse considérablement non-seulement les erreurs des moyennes de la série d'un jour, mais aussi la variation de l'incertitude d'une observation isolée […] force d'admettre que le temps physiologique pour le même observateur, varie assez considérablement d'un jour à l'autre, selon la disposition de son système nerveux. Notre corps est donc sous ce rapport tout-à-fait dans le cas des autres machines de précision; sa correction est variable, comme celle d'une lunette méridienne ou d'une pendule".

From then on, the Observatory, like other institutions, tried to include the observer's reaction time in the calculations for error reduction. As a result, the personal equation was often determined by observing an artificial star, a device occasionally installed on a sight placed 60 m from the meridian circle. In this way, the influence of the human factor could be quantified.

Despite the importance of compensating for the inaccuracies attributed to the human factor, from the last decades of the nineteenth century the emphasis was on automating astronomical observation, reducing the role of the astronomer. At the end of the nineteenth century, the Observatory decided to review its operating chain, in particular by changing its meridian circle. This change provided the opportunity to acquire an impersonal micrometer, a device designed by the company *Repsold and Söhne* at the end of the nineteenth century.[11] To acquire such a device, the Observatory turned to the *Société genevoise d'instruments de physique* (SIP), which produced a motorised impersonal micrometer using the "driven-wire method (*méthode du fil entraînée*)," which was activated by a motor and gears to follow the transits of the stars. Using a knob, the astronomer could vary this speed and thus synchronise the movement of the thread with that of a star. Contacts were placed at regular intervals on the micrometer diameter and sent an impulse to the chronograph every time the moving filament crossed a fixed one. From then on, all the observer had to do was to adjust the thread to the correct rhythm, whereas with the chronographic method, the astronomer had to press a button with the risk of anticipating or delaying each bisection. The impersonal micrometer therefore had the task of mechanically eliminating the personal equation. Despite this theoretical advantage, the device was not entirely satisfactory in practice, even though it was revised and adapted several times.

Automating Time Determination

The limited performance of the impersonal micrometer encouraged the Observatory to persist in finding solutions to improve the accuracy of time determination. This aspect was enhanced by advances in timepieces and the use of quartz as a resonator.[12] Edmond Guyot (1900–1963), third director of the Neuchâtel Observatory, launched a major project to reorganise the operating chain in order to take advantage of these new technologies. It called for the addition of two technical innovations: the acquisition of a PZT and the installation of quartz clocks.[13] This project was the first step towards a radical transformation of the way the Observatory operated. Whereas until then, astronomers had been the only scientific discipline represented, quantum physicists

[11] Brooks (1991, 2008).

[12] Katzir (2016, pp. 98–103), Lombardi (2011), Stephens (2008).

[13] "Rapport du Conseil d'État à l'appui d'un projet de décret concernant l'octroi d'un crédit pour l'agrandissement de l'observatoire cantonal et pour l'achat de divers instruments," session of March 15, 1948, published in the *Bulletin officiel du Grand Conseil* (BOGC), nr 113, pp. 702–740.

began to make their appearance and even gradually took over. The same happened with technicians and electronic engineers who gradually replaced the watchmakers who had collaborated with the Observatory.

The authorities of the Canton of Neuchâtel, which financed this public institution, had a technophile vision of progress and wanted to remain at the forefront. Despite the uncertainties surrounding the new equipment, the watchmakers did not want to take the risk of falling behind the other countries. Sydney de Coulon (1889–1976), an industrialist, politician and member of the Observatory Commission,[14] is an example of these ambitions. He believed that "the Neuchâtel Observatory must be the best equipped to tell the time with the greatest precision."[15] This desire was in line with the objective of keeping pace with technological progress. As the report of the Council of State explained, it is a question of looking to the future, because "progress continues and we must strive to move forward so as not to be left behind." The way to do this was to modernise and expand existing facilities as circumstances require." This argument cannot be understood without putting it in the context of the situation of the watchmaking industry in Neuchâtel in the aftermath of the Second World War, which was in a very favourable economic position and on which the canton's finances depended. However, it remained in competition, mainly with the Geneva watchmaking industry. From then on, the political authorities did everything in their power to favour the local watchmaking industry over its rivals. In this race for industrial development, the Observatory played an important role by certifying the quality of watchmaking. The watchmakers, aware of the economic stakes involved, contributed to the new operating chain, thus facilitating the implementation of an ambitious and costly project, given the financial resources of the Canton of Neuchâtel.[16] The project should therefore be seen as a coordinated initiative by the state, industry and science in the field of time measurement at the end of the Second World War.[17] It should be noted, however, that this involvement of the private sector was solely for its own interests, and the Observatory developed in the field of practical services, while scientific research in astronomy and other fields was gradually abandoned. This operating chain should be seen as a strengthening of the influence of industry and the State on the management of a scientific institution that was increasingly subordinated to economic and political needs.

---

[14] Barrelet (2008).

[15] "Séance de la Commission de l'Observatoire, le mardi 14 octobre 1947," *Procès-verbaux de la commission de l'Observatoire cantonal de Neuchâtel (PVCOC) 1939–1955*, Archives de l'État de Neuchâtel (AEN), 1EP-352, fol. 3.

[16] On the participation of the watchmaking associations, see: BOGC, 113, 1947–1948, p. 711.

[17] Bonneuil and Pestre (2015).

The 1948 Project: Constructing the PZT

The origins of zenith tubes date back to the seventeenth century, but it was not until the discovery of an irregularity in the oscillation of the Earth's axis, now known as the "Chandler wobble," towards the end of the nineteenth century, that the instrument aroused enthusiasm in the astronomical community.[18] In 1909, a grant was awarded by the International Association of Geodesy to install a "photographic reflex zenith tube" in Gaithersburg, an instrument that marked the birth of the PZT with the addition of photography.[19] This instrument was delivered to Gaithersburg in 1911 before being transferred to the United States Naval Observatory (USNO) in Washington in 1915.[20] Originally designed for the study of latitudinal variations, it was adapted for time measurement in 1930, expanding its functionality. In 1946, the potential of the PZT was raised at the International Astronomical Union (IAU), whose Committee 19 on Latitude Variations pointed out "the great interest there would be in establishing zenithal photographic telescopes at various points on the Earth, not only to improve knowledge of the movement of the pole, but also to collect indispensable data relating to the proper movements in declination, as well as data useful for determining the constants of the solar system."[21] Indeed, after the 1946 IAU meeting, Guyot proposed the acquisition of a PZT, pointing out that.

The success of the photographic zenithal telescope is recognised by many observatories which have decided to acquire such an instrument. [...] It is easy to foresee that, in the future, the new method of determining time will gradually supplant the transit method, so that meridian instruments will soon be used only for determining the positions of stars [...]. The Neuchâtel Observatory, which has always endeavoured to keep its time service up to date with modern requirements, cannot ignore this new advance in a field it holds dear.[22]

The PZT was considered an alternative to the meridian circle, which made it possible to improve the accuracy of timekeeping, thus justifying Guyot's decision to equip himself with one in order to remain at the forefront of timekeeping:

The zenithal telescope is currently the best way of determining the time, and the Neuchâtel Observatory cannot ignore it. This telescope has the advantage of completely eliminating the influence of the observer, since the moment of transit is obtained photographically. It also has the advantage of being used to determine latitude, which would be a very interesting field of research for our Observatory.[23]

---

[18] Carter and Carter (2000).

[19] Ross (1915, p. 7).

[20] Ross (1915, p. 7).

[21] Oort (1950, p. 64).

[22] Guyot (1946, pp. 333–334).

[23] "Séance de la Commission de l'Observatoire, le mercredi 12 mars 1947," *PVCOC 1939–1955*, AEN, 1EP-352, fol. 4. It should be noted here that this view of the status of the PZT is still shared and that it is considered, along with the Danjon astrolabe, to be the most advanced of its time. See Gurfil and Seidelmann (2016, pp. 19–20), Sutton (2012, pp. 46–47), Beutler (2005, p. 33).

Guyot mentions two advantages. Firstly, it would improve the accuracy of time measurement, and secondly, it would allow the region to participate in an international research programme to study latitudinal variations.

In August 1947, in order to obtain a PZT, the Neuchâtel Observatory contacted the USNO,[24] which informed them that the instrument developed was a prototype manufactured in its workshops and that no commercial version existed.[25] Guyot then contacted Grubb and Parsons of Newcastle-upon-Tyne, which had been developing a PZT since 1943 in close collaboration with the Greenwich Observatory on the basis of the USNO's work.[26] For the British institution, the aim was also to improve the quality of time measurement, as the Royal Astronomer Harry Spencer Jones (1890–1960) had pointed out:

The provision of a 10-inch photographic zenith telescope is required to increase the accuracy of time determinations. This is a necessary step in the attainment of a high precision in the radio time-signals distributed from the Royal Observatory. A great step forward has been made in this direction since the beginning of the war using quartz clocks for prediction purposes, the careful control of line lags, the elimination of mechanical relays etc. It is the errors inherent in time determinations by visual observations with a transit instrument that at present impose the limit on the accuracy that is attainable.[27]

These claims were challenged by various astronomers who did not share the enthusiasm for the PZT, such as "M. Moreau, of the Uccle Observatory, and Barros, a Spanish astronomer, who [...] criticised the photographic telescope for being too complicated."[28] As an illustration of these doubts about the PZT, several scientific institutions that were initially interested in the instrument delayed action, preferring to wait for the results of the British experiments.

On 20 September 1947, Sydney de Coulon urged the Observatory Commission to order the instrument quickly "if we do not want to be overtaken by other institutions, in particular the Geneva Observatory."[29] De Coulon's intervention was successful, as the Observatory stepped up its efforts to acquire the instrument. More than the

[24] Letter of the Département de l'Économie publique to Jean Humbert, Member of the Neuchâtel State Council, 12 august 1947, AEN, 1EP-364.

[25] Letter from Edmond Guyot to Jean Humbert, July 2, 1947, AEN, 1EP-364, lunette zénithale photographique 1947–1955. On 10 July, Humbert wrote to the Foreign Affairs Division to intervene via the Legation in Washington: Jean Humbert to the Département politique fédéral, Division des affaires étrangères, July 10, 1947, AEN, 1EP-364, lunette zénithale photographique 1947–1955.

[26] From 1943 the Greenwich Observatory was in discussion with Grubb & Parsons about the development of a PZT. Correspondence relating to these discussions can be found in the Cambridge Library, GBR/0180/RGO 9/110 and GBR/0180/RGO 9/111.

[27] Letter from Harry Spencer Jones to C.S. Wright, July 19, 1943, Cambridge Library, GBR/180/RGO-9–110, Correspondence on photographic zenith tubes, 1943–1954.

[28] "Séance de la Commission de l'Observatoire, le jeudi 3 mars 1949," *PVCOC 1939–1955*, AEN, 1EP-352, fol. 3.

[29] "Séance de la Commission de l'Observatoire, le mercredi 20 septembre 1950," *PVCOC 1939–1955*, AEN, 1EP-352, fol. 4.

desire to bring the Observatory's scientific instruments to the forefront of technological innovation, it was the fear of being overtaken by the Geneva watchmaking industry, considered at the time to be the main competitor in this economic sector, that prompted politicians and industrialists to push ahead with the installation of the new operating chain, despite the financial costs and technical doubts about the potential of the PZT.

On 13 May 1949, George Sisson, director of Grubb and Parsons, sent Edmond Guyot an estimate for two possible PZT models. The first cost £28,000 for a "Greenwich type" telescope and the second £22,000 for a simplified telescope.[30] These sums were beyond the means of the State, which turned to the watchmaking industry, which increased its financial contribution at Coulon's request.[31]

After some discussion, Grubb and Parsons submitted a new proposal in June 1950 with a Type 1A PZT; a compromise between Models I and II—planned for the Mount Stromlo Observatory in Australia—in response to criticism from astronomers at Greenwich that Model II removed too many essential functions.[32] Grubb and Parsons lamented the fact that it was necessary to develop three models of PZT, stating: "It is unfortunate that we should be faced with the prospect of making three different types of PZT, but of course astronomers are all men of rugged individuality where telescopes are concerned."[33] It was the very high accuracy targets set by the British Observatory that made it difficult to adapt PZTs to other local contexts. Finally, on 21 November 1950, the Neuchâtel establishment opted for a type 1A PZT, which was delivered on 3 April 1954.[34]

---

[30] Letter from George Sisson to Edmond Guyot, May 13, 1949, AEN, 1EP-364, lunette zénithale photographique. It should be noted, however, that at that time the models were still under development and there were no instruments with definite characteristics.

[31] Letter of the Chambre Suisse d'horlogerie to Sydney de Coulon, November 21, 1949, AEN, 1EP-364, lunette zénithale photographique, fol. 2.

[32] Letter from George Sisson to Edmond Guyot, June 30, 1950, AEN, 2IND-79, lunette zénithale photographique (PZT), achat auprès de la maison SIR HOWARD GRUBB PARSONS de Newcastle, correspondance, rapports, etc. 1949–1954. To see the differences between the models, see the various plans held at the Cambridge Library.: RGO 101/3/4/1, 101/3/4/2 et 101/3/4/3, as well as fund GBR/0180/RGO 103. The Tyne & Wear archives and MIH archives also hold various plans. We would like to thank Emma Saunders of the Cambridge Library for her invaluable help.

[33] Letter from George Sisson to D.S. Perfect, April 3, 1951, Cambridge Library, GBR/180/RGO 103, Photographic Zenith Tube: Records, c.1940s-1980s.

[34] For order information, see: Letter from Edmond Guyot, to George Sisson, November 21, 1950, AEN, 2IND-79, lunette zénithale photographique (PZT), achat auprès de la maison SIR HOWARD GRUBB PARSONS de Newcastle, correspondance, rapports, etc. 1949–1954. Letter from George Sisson to Edmond Guyot, November 15, 1950, AEN, 1EP-364, lunette zénithale photographique. The price finally reached £16,027 at the time of delivery. Letter from George Sisson to Edmond Guyot, December 1, 1953, AEN, 2IND-79, lunette zénithale photographique (PZT), achat auprès de la maison SIR HOWARD GRUBB PARSONS de Newcastle, correspondance, rapports, etc. 1949–1954.

Quartz Clocks

The use of quartz clocks is inseparable from that of the PZT and occupies an important place in the operating chain in 1954. Indeed, the degree of precision required by the PZT is such that mechanical clocks are not sufficient, as stated in the 1948 report: "The operation of a photographic zenithal telescope requires a quartz clock. This is the only type of clock capable of driving the photographic plate of the telescope with the required precision."[35] While the PZT justifies the need for a quartz clock, the reverse is also true. The greater accuracy achieved with quartz clocks requires an increase in the resources needed to control their operation.

The Observatory's first three quartz clocks were part of a series of six developed by Ernst Baumann (1909–1980), professor and head of the Institute of Low Current Engineering at the Swiss Federal Institute of Technology in Zurich (ETHZ), by his colleague Francis Berger, an engineer, and by Ébauches SA; the other three clocks were delivered respectively to the ETHZ, to the Swiss Post, Telegraph and Telephone Company (PTT) and to the Swiss Laboratory for Horological Research (LSRH).[36] The quality of the clocks had to be checked in collaboration with the Neuchâtel Observatory, as Baumann pointed out in the 1948 State Council report:

To test crystals, and to measure frequencies accurately, you need a frequency standard. The frequency standard is also a time standard. As the time base is the duration of the Earth's rotation, it is necessary to refer to it. In Switzerland, the most accurate time is given by the Neuchâtel Observatory, so it was natural to establish a link with it.[37]

The expertise of the Neuchâtel astronomers in the field of time measurement was recognised by the institutions involved in the development of quartz, making it the ideal place to experiment with this technology. The Observatory's first quartz clock

---

[35] "Le fonctionnement d'une lunette zénithale photographique nécessite une horloge à quartz. C'est la seule espèce d'horloge qui soit en mesure d'entraîner la plaque photographique de la lunette tout en assurant à cet entraînement la précision voulue." See "Rapport du Conseil d'État à l'appui d'un projet de décret concernant l'octroi d'un crédit pour l'agrandissement de l'observatoire cantonal et pour l'achat de divers instruments," session of March 15, 1948, BOGC, 113, 1947–1948, pp. 702–740, 708.

[36] Letter from Sydney de Coulon to the President of the Chambre Suisse de l'Horlogerie, the President of the Fédération Suisse des Associations de Fabricants d'Horlogerie (F.H.), the President of the Union des Branches Annexes de l'Horlogerie (UBAH), the President of the Association d'Industriels Suisse de la Montre Roskopf and the President of Ébauches SA, 5 August 1947, AEN, 1EP-364, lunette zénithale photographique 1947–1955. On the Neuchâtel network of institutions involved in microtechnical development, see: (Richter, Busca, Débely and Perret, 2000, especially pp. 105–112 and on the Observatory pp. 127–130). It should be noted that while three clocks are mentioned, only two are referred to in the various annual reports of the Observatory.

[37] "Pour essayer les cristaux, comme pour mesurer les fréquences avec précision, il est nécessaire de posséder un étalon de fréquence. L'étalon de fréquence est aussi un étalon de temps. Comme la base de temps est la durée de la rotation terrestre, il est nécessaire de s'y référer. Comme en Suisse, le temps le plus exact est donné par l'observatoire de Neuchâtel, il était naturel d'établir une liaison avec lui." See "Rapport du Conseil d'État à l'appui d'un projet de décret concernant l'octroi d'un crédit pour l'agrandissement de l'observatoire cantonal et pour l'achat de divers instruments," session of March 15, 1948, BOGC, 113, 1947–1948, pp. 702–740, 709.

(HQ1) arrived in July 1948[38] and the second on 21 July 1950 (HQ2).[39] These two clocks were the only ones of their kind until 1954. In 1950, the first observations raised doubts about the quality of this equipment, but Guyot tried to reassure the Commission about the performance of quartz: "[The] two clocks are built. The first, which was at the Observatory, is with Mr Berger; the second does not yet work. Mr Berger had some problems. It is not possible at the moment to give results, to judge the clock itself [...]. In any case, the quartz is excellent."[40] To demonstrate its doubts about the performance of quartz clocks, the Observatory acquired a second mechanical Leroy clock in 1949, and a third in 1952, when the company announced that it would cease production of mechanical clocks.[41] Quartz and mechanical technologies were developed in parallel within the operational chain, demonstrating that instrument calibration required time to adapt new technologies to the local context. Acquiring new equipment was not enough to improve accuracy. Many challenges had to be overcome in terms of installation, use of scientific instruments, organisation of work, training of staff, standardisation of observations and persuading the various political, scientific and industrial actors to keep the initially costly effort moving in the desired direction. As Guyot emphasised, he "believes in the future of quartz clocks; progress is very rapid at the moment, but we cannot yet guarantee the continuity [of the performance of quartz clocks over time, AN]".[42] This was confirmed in 1954, when the Committee mentioned the delay in the delivery of the third quartz clock by Oscilloquartz, a division of Ébauches SA. Jean Rossel (1918–2006), professor of physics at the University of Neuchâtel, stressed the positive aspects of this situation: "The Oscilloquatz [sic] department, after having overcome the major technical difficulties inherent in any new creation, has made great progress in terms of precision, which means that the Observatory will be in possession of well-developed equipment when the time comes".[43] Acquiring chronometric instruments while they are still under development entails the risk of being faced with dysfunctional prototypes that can compromise timekeeping and require long periods of calibration. At the Observatory, however, the debate over the performance of quartz and mechanical clocks soon died

[38] Rapport du directeur (RD) 1948, p. 6.

[39] RD 1950, p. 6.

[40] "[Les] deux horloges sont construites. La première, qui était à l'Observatoire est chez M. Berger; la seconde ne marche pas encore. M. Berger a eu quelques ennuis. Il n'est pas possible de donner des résultats actuellement, de juger la pendule elle-même [...]. Le quartz est en tous cas excellent." See "Séance de la Commission de l'Observatoire, le mercredi 27 septembre 1950," *PVCOC 1939–1955*, AEN, 1EP-352, fol. 6.

[41] RD 1949, p. 6, RD 1952, p. 4.

[42] "[Il] croit à l'avenir des horloges à quartz; les progrès sont très rapides actuellement, mais nous ne pouvons pas encore assurer la continuité." See "Séance de la Commission de l'Observatoire, du mercredi 12 mars 1952," *Séances de la Commission et Procès-verbaux 1949–1955*, AEN, 1EP-352, fol. 5.

[43] "Le département Oscilloquatz [sic] [après] avoir été en but à de grandes difficultés techniques inhérentes à toute création nouvelle, a obtenu de grands progrès dans la précision, ce qui fait que l'Observatoire se trouvera au moment voulu en possession d'appareils bien en point." See "Séance de la Commission de l'Observatoire, de mercredi 3 février 1954", *Séances de la Commission et Procès-verbaux 1949–1955*, AEN, 1EP-352, fol. 3.

down. By 1955, "mechanical clocks were no longer in use", because "the accuracy of quartz clocks is 10 to 20 times greater than that of mechanical clocks."[44] In just a few years, quartz clocks replaced mechanical clocks once and for all. The mechanical clocks were kept, in case the quartz clock's battery failed, before being given to museums in 1959 and 1961, thus moving from the status of scientific instruments to that of objects of cultural heritage.[45] In the meantime, the number of quartz clocks had increased and the first models were already considered obsolete, demonstrating a process of technological acceleration. The Q4 and Q5 oscillators, manufactured by Oscilloquartz, were installed at the beginning of 1954,[46] followed in 1956 by the Q6 and Q7 clocks.[47] Q9, manufactured in the United Kingdom, and Q10, designed in the United States, were introduced between the end of 1959 and the beginning of 1960.[48] The latter is considered to be particularly accurate. At the same time, the Q1, Q2, Q5 and Q6 clocks were decommissioned after only operating for a few years. The following year, the Director mentioned that "the two other clocks in operation, Q7 and Q9, are 5 to 10 times less good, so the Observatory's timekeeping relies on a single clock [Q10]."[49] After several years of searching for a new quartz clock, the 1963 annual report stated that "the remarkable quality of the Q10 clock compared with other clocks available on the market has led us to abandon our plan to complete our battery of quartz clocks by subsequent purchases. We believe that the next step will be the acquisition of a permanent rubidium or caesium standard."[50] The triumph of quartz was short-lived, as it was replaced by atomic clocks within a decade.[51]

---

[44] "Les horloges mécaniques ne sont plus utilisées. [...] La précision des horloges à quartz est de 10 à 20 fois plus grande que celle des horloges mécaniques." See "Séance de la Commission de l'Observatoire, de mercredi 23 février 1955," *Séances de la Commission et Procès-verbaux 1949–1955*, AEN, 1EP-352, fol. 3. Guyot explains the difference between quartz and mechanical clocks: Guyot (1955, pp. 1–9).

[45] RD 1959, p. 14 and RD 1961, p. 12. On cultural heritage: Appelbaum (2007); Muñoz-Viñas (2005); De La Torre (2002).

[46] RD 1954, p. 6.

[47] RD 1955, p. 6.

[48] RD 1960, p. 6.

[49] "Les deux autres horloges en fonctionnement, Q7 et Q9, sont 5 à 10 fois moins bonnes, de sorte que la conservation de l'heure de l'Observatoire repose sur une seule horloge [Q10]." RD 1961, p. 6.

[50] "La qualité remarquable de l'horloge Q10 par rapport à d'autres horloges disponibles sur le marché nous a fait abandonner notre projet de compléter notre batterie d'horloges à quartz par des achats ultérieurs. Nous pensons en effet que la prochaine étape sera l'acquisition d'un étalon à rubidium ou à césium à marche permanente." RD 1963, p. 5.

[51] See Riley (2019); Wallis (2018); Cutler (2005); Henderson (2005); Ramsey (2005), Forman (1985).

Changing Practices: From Observer to Operator

The basic principle of the PZT is to use photography to record the transit of the star near the zenith. The light from the star is captured by a tube aimed at the zenith, then reflected by a mercury bath before reaching a photosensitive plate placed at the nadir. The star is photographed twice before and twice after the zenithal transit, forming the four corners of a rectangle on a photographic plate known as a stellar plate. In 1954 Guyot described the operation of the PZT.

The lens has an aperture of 25 cm and a focal length of 3.43 m. At its upper end, the main tube of the instrument carries a part that can rotate 180° around a vertical axis: the turret. The objective and the [stellar photographic] plate are integral with the rotary and rotate at the same time. The device, which works like a robot, is designed so that a single press of a button triggers the automatic exposure cycle. For each star, 4 exposures of 21 s each are taken, separated by an interval of 11 s during which the rotary is rotated 180° each time. As far as possible, the times of the 4 exposures in pairs should be symmetrical with respect to the time of passage of the meridian.[52]

According to Guyot, the instrument works like a "robot," giving the astronomer a degree of flexibility when it comes to triggering the observation cycle. From then on, it is enough to switch on the machine at the right moment and the act of recording takes place automatically, without any direct human intervention. We are witnessing the transition from an active observer to an operator responsible for preparing the instrument, without any direct role in the observation process.

To ensure sharp images, the stellar photographic plate is driven by a carriage to follow the apparent motion of the star. During the measurement cycle, another photographic plate, known as the chronographic plate, records the position of the carriage every two seconds.[53] The comparison of the two plates by means of a measuring device determined the operating state of a reference quartz clock, since the time of the passage of the stars was known thanks to predictive calculations recorded in star catalogues.

With quartz clocks, it was possible to compare the accuracy of the time continuously using frequency comparators, whereas previously it had to be checked once a day with a chronograph. In addition, the presence of Belin chronographs made it possible to record external signals and thus to check the accuracy of the time between different institutions.

---

[52] Guyot (1954, pp. 779–780, 811–812): "Son objectif a 25cm d'ouverture et 3,43m de distance focale. Le tube principal de l'instrument supporte à son extrémités supérieure une partie qui peut tourner de 180° autour d'un axe vertical: le rotary. L'objectif et la plaque [photographique stellaire] sont solidaires du rotary et se retournent en même temps que lui. L'appareil qui travaille comme un robot est construit de telle manière qu'une seule pression sur un bouton déclenche le cycle automatique des poses. Pour chaque étoile, on fait 4 poses de 21 secondes séparées par un intervalle de 11 secondes au cours duquel le rotary tourne chaque fois de 180°. On s'arrange autant que possible pour que les instants des 4 poses soient symétriques deux à deux par rapport à l'heure du passage par le méridien".

[53] Manuel d'utilisation P.Z.T. AEN, 2IND-79, p. 1.

The new operating chain changed many practices at the Observatory and questioned the role and status of the human factor, from that of an active observer to that of a technician responsible for initiating procedures but without direct influence on the production of data.

Error Reduction and Outsourcing

The Neuchâtel Observatory was a small institution without a large staff. Although the introduction of the PZT seemed to reduce the human workload, the preparation, maintenance and processing of the PZT results generated a considerable workload, at least equal to, if not greater than, that of the previously used meridian circle. The PZT shifted staff activity from data production to instrument preparation and measurement of photographic plates and error reduction calculations. From 1954, comparisons of stellar and chronographic photographic plates were first carried out using a Zeiss stereo comparator, before being replaced in 1957 by a Hauser P-324.[54] These instruments were used to accurately determine the coordinates on the plates using a binocular microscope.[55] The measurements were then corrected for errors (environmental and instrumental) and statistical operations were performed to obtain the time data. These time-consuming computations were outsourced to take account of limited staff resources. For this purpose, the Observatory relied on the IBM centre in Mannheim and then on the computing centre in Lausanne.[56] From February 1968, the error reduction calculations were carried out at the computing centre of the University of Neuchâtel on an IBM 1130 computer.[57] Since the star catalogues used for the meridian circle observations could not be used with the PZT—because the observations were made in different parts of the sky—the Observatory asked other scientific institutions to calculate the positions of the stars that could be used (Greenwich Observatory, then Brussels Observatory, and finally the Astronomisches Rechen-Institut in Heidelberg).

The vertical construction of the PZT also made it possible to simplify the statics of the instrument and thus to eliminate the instrumental errors typical of meridian circles, such as collimation, azimuth, irregularity of the trunnions, errors in the bending of the optical tube, etc. Apart from these adaptations, the principle of time determination with the PZT remained the same. Thanks to the reduction of some errors, the accuracy of the data obtained with the PZT had improved by a factor of ten and was now close to one thousandth of a second. However, new sources of error appeared and were evaluated in a study carried out by Walter Schuler, a collaborator with the Observatory.[58] Exposing light spots on a photosensitive plate, developing

[54] RD 1957, p. 5.

[55] Nasim (2019, pp. 164–166).

[56] RD 1956, p. 6; RD 1964, pp. 4–5.

[57] RD 1968, p. 6.

[58] Schuler (1967).

the plate and measuring their positions introduced new problems that limited the instrument's potential. The expansion of the glass plate between exposure at night and measurement in a temperate room, or the development and drying of the gelatine, caused deformations at the periphery of the plate.[59]

In 1959, in order to reduce manpower requirements, the observing cycle was fully automated. It was no longer necessary to have a person on site all night, only to prepare the instrument.[60]

The development of atomic time subsequently thwarted the fortunes of the PZT in Neuchâtel, and astronomical timekeeping was gradually abandoned. However, the PZT remained in use until 1982, as part of the international research programme on latitude variations that had been centred on the Mizusawa Observatory since 1962.[61] In 1982, the Observatory decided to discontinue the service of the PZT, judging that it was no longer necessary either for the institution or for the study of latitudinal variations. It was therefore dismantled and transferred to the Musée International d'horlogerie de La Chaux-de-Fonds (MIH), where it became part of the museum's heritage.

Conclusion

The Neuchâtel Observatory in 1954 represents an intermediate stage between astronomical time and atomic time. It can be seen as an attempt by astronomers to retain control of time measurement in the context of the rapid development of quartz and atomic technologies. The persistence of partly competing and partly complementary technologies enables the institution to remain at the forefront in the field of time measurement. However, the operating chain also enabled the preservation of timekeeping expertise, and in many ways the PZT and quartz clocks represented a step towards an institution dedicated to atomic time. When atomic clocks were being developed in Switzerland, it was natural that the various actors should turn to the Observatory as an institution specialising in time measurement. From this point of view, the operating chain in 1954 can indeed be seen as a period of transition between astronomical and atomic time, at a time of rapid technological development in the post-Second World War context.

In this new configuration, the practices of time measurement evolved and the role of personnel was questioned. From being active observers in the process of producing data, they became more and more like preparers of scientific instruments, which themselves produced accurate time. In this sense, this transformation marked the transition from observer to operator, but also, gradually, from an astronomical observatory to a laboratory dedicated to quantum technologies, as shown by the

[59] Nasim (2019, pp. 167–169), Schuler (1967, pp. 60–64).

[60] Wyser et al. (1959).

[61] Boucher (2015, p. 53); McCarthy and Luzum (1996, p. 623).

resignation of the astronomers at the beginning of the 1970s, as they felt they no longer had a real mission at the Observatory.

References

Appelbaum, B. (2007). *Conservation Treatment Methodology*. Butterworth-Heinemann.

Barrelet, J.-M. (2008). Sydney de Coulon. Industriel, homme politique (1889–1976). In M. Schlup (Ed.), *Biographies neuchâteloises. De 1950 au seuil du XXIe siècle*, vol. 5 (pp. 81–85). Gilles Attinger.

Belteki, D., & Gressot, J. (2023). Introduction. Re-Assembling the history of meridian circle. In D. Belteki, J. Gressot, L. Jeanson, J. Davoigneau (Eds.), *Cahiers François-Viète* 3(14) (special issue *Circles of Precision: Meridian Circles during the Nineteenth and Twentieth Centuries* (pp. 5–20).

Beutler, G. (2005). Methods of Celestial Mechanics, vol. 1: Physical, Mathematical, and Numerical Principles. Springer.

Blaser, J.-P. (1956). Rapport annuel du Directeur sur l'exercice 1955 (RD1955), Archives de l'État de Neuchâtel, 2IND-98.

Bonanomi, J. (1961). Rapport annuel du Directeur sur l'exercice 1960 (RD1960), Archives de l'État de Neuchâtel, 2IND-98.

Bonneuil, C., & Pestre, D. (Eds.). (2015). *Histoire des sciences et des savoirs*, tome 3: *Le siècle des technosciences*. Le Seuil.

Boucher, C. (2015). La genèse du Système International de Référence Terrestre (ITRS). *Revue XYZ, 145*(4), 53–56.

Brooks, R. (1991). Development of micrometers in the seventeenth, eighteenth and nineteenth centuries. *Journal of History of Astronomy, 22*(2), 127–173.

Brooks, R. (2008). Development of scale-reading micrometers in the seventeenth to the nineteenth centuries. *Bulletin of the Scientific Instrument Society, 98*, 8–23.

Canales, J. (2001). Exit the frog, enter the human: physiology and experimental psychology in nineteenth-century astronomy. *The British Journal for the History of Science, 34*, 173–197.

Carter, M., & Carter, W. (2000). Seth Carlo Chandler Jr.: The discovery of variation of latitude. *Polar Motion: Historical and Scientific Problems. ASP Conference Series* 208:109–119

Cutler, L. (2005). Fifty years of commercial caesium clocks. *Metrologia, 42*, 90–99.

Decaux, B. (1948). Les horloges à quartz et leurs applications. Conférence du 29 mai 1948. *Société suisse de chronométrie*: (pp. 266–275).

De La Torre, M. (Eds.). (2002). *Assessing the Values of Cultural Heritage: Research Report*. Getty Conservation Institute.

Forman, P. (1985). The first atomic clock program: NBS, 1947–1954. In *Proceedings of the 17th Annual Precise Time and Time Interval Systems and Applications Meeting, Washington DC*, (pp. 1–18). PTTI.

Gressot, J. (2023). Automating time determination the photographic zenith tube (PZT) of the Neuchâtel Observatory. *Bulletin of the Scientific Instrument Society, 156*, 16–23.

Gressot, J. (2024a). *L'Observatoire cantonal de Neuchâtel et la mesure du temps (1858–1960): Instruments scientifiques, acteurs et espaces savants*. PhD Dissertation, University of Neuchâtel.

Gressot, J. (2024b). Mesurer le temps, construire la précision: Adolphe Hirsch et Mattäus Hipp. In A. Turner & F. Soulu (Eds.), *Hommage à Paolo Brenni et Denis Beaudouin*.

Gressot, J., & Jeanneret, R. (2022). Determining the right time or the establishment of an astronomical culture of precision at the Neuchâtel Observatory. *Journal for the History of Astronomy, 53*(1), 27–48.

Gressot, J., & Jeanneret, R. (2023a). Dialogue interdisciplinaire entre l'objet et l'archive pour l'étude historique et la conservation du patrimoine scientifique et technique: Le cas du tube zénithal photographique (1954–1982) de l'Observatoire de Neuchâtel. *Coré, 5*, 1–18.

Gressot, J., & Jeanneret, R. (2023b). Élaborer le cercle méridien *Ertel & Sohn* de l'Observatoire de Neuchâtel (1858–1861): Besoins scientifiques, possibilités techniques et contraintes financières. *Cahiers François-Viète, 3*(14), 255–287.

Gurfil, P., & Seidelmann, K. (2016). *Celestial Mechanics and Astrodynamics: Theory and Practice.* Springer.

Guyot, E. (1946). Une nouvelle méthode de détermination de l'heure avec la lunette zénithale. *Journal Suisse D'horlogerie, 7–8*, 333–335.

Guyot, E. (1954). À l'Observatoire cantonal de Neuchâtel. De la lunette méridienne à la lunette zénithale photographique. *La Suisse horlogère* 32–33:779–780, 811–812.

Guyot, E. (1955). Sur la précision relative des horloges à quartz et des horloges mécaniques, Annales Françaises de Chronométrie, 4-5 juin 1955, 1-9.

Guyot, E. (1955). Rapport du Directeur de l'Observatoire cantonal de Neuchâtel à la Commission de l'Observatoire sur l'exercice 1954 (RD1954), Archives de l'État de Neuchâtel, 2IND-98.

Henderson, D. (2005). Essen and the national physical laboratory's atomic clock. *Metrologia, 42*, 4–9.

Hirsch, A. (1861–1862). Expériences chronoscopiques sur la vitesse des différentes sensations et de la transmission nerveuse. *Bulletin de la Société des Sciences Naturelles de Neuchâtel* 6:100–114.

Hirsch, A. (1862–1863). Sur les corrections et équations personnelles dans les observations chronographiques de passage. *Bulletin de la Société des Sciences Naturelles de Neuchâtel* 6: 365–372.

Katzir, S. (2016). Variations and combinations: invention and development of quartz clock technologies at AT&T. *Journal of the International Committee for the History of Technology, 22*, 78–114.

Lombardi, M. (2011). The Evolution of Time Measurement, Part 2: Quartz Clocks. *IEEE Instrumentation and Measurement Magazine* (pp. 41–48).

McCarthy, D., & Luzum, B. (1996). Path of the mean rotational pole from 1899 to 1994. *Geophysical Journal International, 125*(2), 623–629.

Muñoz-Viñas, S. (2005). *Contemporary Theory of Conservation.* Elsevier Butterworth-Heinemann.

Nasim, O. (2019). Handling the heavens: Things and the photo-objects of astronomy. In C. Caraffa (Eds.), *Photo-Objects: On the Materiality of Photographs and Photo-Archives in the Humanities and Sciences* (pp. 161–175). EOA.

Oort J. H. (Ed.). (1950). *Transactions of the International Astronomical Union 7* (p. 64). Cambridge: Cambridge University Press.

Ramsey, N. (2005). History of early atomic clocks. *Metrologia, 42*, 1–3.

Richter, Y. et al. (2000). Une institution au service d'une industrie: le Laboratoire Suisse de Recherches Horlogères (LSRH). In T. Perret, A. Beyner, P. Debély, L. Tissot and F. Jeanneret (Eds.), *Microtechniques et mutations horlogères. Clairvoyance et ténacité dans l'Arc jurassien* (pp. 91–134). Gilles Attinger.

Riley, W. (2019). A History of the Rubidium Frequency Standard. *IEEE UFFC-S History* (pp. 1–35).

Ross, F. (1915). *Geodesy.* Government Printing Office.

Schaffer, S. (1988). Astronomers mark time: discipline and the personal equation. *Science in Context, 2*(1), 115–145.

Schmidgen, H. (2003). Time and noise: the stable surroundings of reaction experiments, 1860–1890. *Studies in History and Philosophy of Biological and Biomedical Sciences, 34*, 237–275.

Schmidgen, H. (2009). *Hirn und Zeit. Die Geschichte eines Experiments, 1800–1950.* Matthes & Seitz.

Schuler, W. (1967). Étude théorique et expérimentale de la lunette zénithale photographique (PZT) de Neuchâtel. *Édition Médecine & hygiène.*

Slotten, H. R. (1993). The Dilemmas of Science in the United States: Alexander Dallas Bache and the U.S. Coast Survey. *Isis* 84(1): 26–49

Stephens, C. (2008). Reinventing accuracy: The first quartz clock of 1927. In J. Graf (Ed.), *Die Quarzrevolution: 75 Jahre Quarzuhr in Deutschland 1932–2007* (pp. 12–23). Deutschen Uhrenmuseum.

Sutton, E. (2012). *Observational Astronomy: Techniques and Instrumentation.* Cambridge University Press.

Wallis, E. (2018). Suivre son propre rythme. Alfred Kastler zwischen Physik und Politik 1950–1960. In C. Forster & G. Neuneck (Eds.), *Physik, Militär und Frieden* (pp. 213–228). Springer.

Wyser, C., et al. (1959). L'automatisation de la lunette zénithale photographique. *Publications De L'observatoire De Neuchâtel (Suisse), 7*, 1–10.

Julien Gressot is a historian of science, technology and innovation, specialising in the history of time measurement, scientific instruments and scientific spaces. He is currently a postdoctoral researcher at the University of Neuchâtel, where he is working on the SNSF project *Atomic Clocks at the Neuchâtel Observatory: Time, Quantum Technologies and Innovation (1948–2001).* He is also project manager of an exhibition on the Neuchâtel Observatory involving three museums and the historical site, managed by the *Automates and Merveilles* Association, where he is responsible for the overall coordination, scientific content and scientific curatorship of the exhibitions.

Romain Jeanneret is a conservator specialising in historical metals and technical objects, with expertise in interdisciplinary research applied to complex metallic cultural heritage. He teaches visual documentation as a lecturer at the Haute Ecole Arc Conservation-Restoration (HE-Arc CR) in Neuchâtel. He also runs the restoration workshop for the medieval treasure of the Abbey of Saint-Maurice in Valais, Switzerland, managing restoration and preventive conservation. Since 2019, he has collaborated with the University of Neuchâtel on projects involving the study of materials and the documentation of scientific collections, as well as the analysis of the "*chaîne opératoire*" of the Neuchâtel Observatory.

Part III
Testing Chronometers at the Observatory in the Long Nineteenth Century: Inspection, Performance, and Certification

Chapter 8
The Nineteenth-Century Marine Chronometers of the Royal Astronomical Observatory of the Spanish Navy in San Fernando

Yolanda Muñoz Rey

Abstract The introduction of scientific navigation methods on the ships of the Spanish Navy in the second half of the eighteenth century led to a need among its officers for good instruments, especially marine chronometers. These instruments had to be checked at the Observatory before reaching their destination, where they were entered in the inventory. Despite the economic crisis that the Observatory suffered throughout the nineteenth century, its use as a depository for naval instruments and chronometers was consolidated. In the middle of the nineteenth century, the stock of chronometers amounted to more than 200. At present, there are about 20 marine chronometers on display in the Observatory's collection, which are being restored.

This chapter presents the collection of Marine Chronometers held at the Royal Observatory of the Spanish Navy (*Real Observatorio de la Armada*, hereafter ROA) located in San Fernando (Cádiz), in southern Spain. Although it is of the first magnitude and undoubtedly rich and important, this collection is still largely unknown. Before presenting the collection, however, I will first carry out a historical contextualization of the area of Cádiz where the ROA is located. Secondly, I will provide a brief history of the ROA, the institution which is responsible for the collection. Next, I will explain the acquisition of these chronometers by the ROA, considering the circumstances and reasons that surrounded it, as well as the subsequent evolution of the collection. Finally, I will delve into the pieces by analysing inventories, current stocks, exhibitions in the museum and records in the current catalogue. The objective is to study this group of instruments from both a technical and a historical point of view. Regarding the methodology adopted in this chapter, it relies on a bibliographic and documentary study at the ROA headquarters and the corresponding fieldwork of observation and analysis of the preserved chronometers, as well as interviews with the personnel in charge of them.

Y. Muñoz Rey (✉)
Universidad de Cádiz, Cadiz, Spain
e-mail: yolanda.munoz@uca.es

G. Bernasconi et al. (eds.), *Astronomical Observatories and Chronometry, 18th-20th Century*, History of Physics, https://doi.org/10.1007/978-3-031-95874-8_8

Introduction

The historical context of the ROA collection has been extensively studied by Francisco Javier González González in several books and articles.[1] González shows that the introduction, during the late eighteenth century, of scientific methods of navigation by the Spanish *Armada* gave rise to the need for high quality marine instruments especially chronometers, to apply these navigation techniques with the required precision. The acquisition and use of these instruments was left to the discretion of the ships' commanders who, until well into the nineteenth century, were responsible for acquiring their own instruments, including the chronometers. These were purchased from the best manufacturers in England, France and Germany. In the beginning, the Navy only acquired the instruments necessary for the execution of hydrographic works that were entrusted to commissions designed for that purpose. These instruments, before being used onboard, had to be inspected at the Observatory, where they were registered in the inventory. Despite the economic crisis that the Observatory suffered throughout the nineteenth century, its use as a warehouse for Naval instruments and chronometers was consolidated. This practice was recorded in the Observatory's regulations, which quite regularly included the presence of a watchmaker in charge of the chronometers' maintenance. The Observatory was in correspondence with makers of marine chronometers, signing contracts and testing the chronometers under examination for a time. The Observatory was also obliged to store and preserve in perfect conditions a collection of instruments for each ship.[2] By the mid-nineteenth century, the chronometers purchased could have been around 200. Over time, marine chronometers were replaced by new chronometers or other time control instruments. Damaged chronometers were registered in the inventory as casualties and had been disappearing. Some 17 marine chronometers are currently preserved and on display in the ROA museum collections, and are being restored.

Regarding the geographical context of the ROA's location, one should note that Cádiz is in Andalusia, in the south of Spain. Cádiz and its area (close to Gibraltar), given its position as a meeting point between the Atlantic Ocean and the Mediterranean Sea, as well as Europe and Africa, is a strategic crossroads in commercial and military terms. For this reason, in the eighteenth and nineteenth centuries, Spanish governments established in Cádiz and San Fernando (a city very near Cádiz) the most important military institutions in Spain.

The city of Cádiz, with a long and dense history since its Phoenician origins, has its entire municipal area on a small island and communicates with the nearby city of San Fernando by an arm of land and with the city of Puerto Real by two bridges that cross the Bay. In the eighteenth century, the city experienced an important economic and social Golden Age when the Trade House, which monopolized trading with the Indies (America), moved there from Seville, and the settlement of foreign merchants, who controlled the European and Atlantic wholesale trades from the city, became general. The Enlightenment and the beneficial measures of Carlos III towards the

[1] González González (1992, 1995, 2003, 2004, 2005, 2009, 2011 and 2020).

[2] González González (2005).

Fig. 8.1 The royal observatory of the Spanish Navy in San Fernando. *Source* The author

Bay of Cádiz, as a place of essential military and commercial strategy, culminated the process. The social conformation was what we could call *proto*-bourgeoisie, namely intellectual and cosmopolitan people, with high purchasing power. Urban development was rapid, intense, and very homogeneous, creating an enlightened city of the first order in the European scene (Fig. 8.1).

José Patiño (1666–1736), Secretary of the Navy and the Indies and Prime Minister of King Felipe V (1683–1746), created an Academy of Marine Guards in the Castle of Cádiz in 1717, with a mixed theoretical-practical programme (land-sea), on the model of the English and French systems. The scientist and soldier Jorge Juan y Santacilia (1713–1773), a student of this Academy, with the approval of King Carlos III (1716–1788), later created an astronomical Observatory housed in the tower of this castle. The castle, however, was too small and the instruments suffered from the humidity and salty conditions of the island city. For this reason, King Carlos III moved the Maritime Department, the Astronomical Observatory, the Academy of Marine Guards, the great Arsenal of *La Carraca* and the military urbanization of *San Carlos*[3] to *La Real Isla de León* (the present San Fernando). This meant a huge and rapid development for this city, whose configuration is still determined by it. The population grew very fast in a short time and was absorbed by the cultural and social profile of the city of Cádiz.

[3] Fresnadillo (1989).

The Observatory, based on the project of Gaspar de Molina y Zaldivar (1741–1806) was completed in 1797 on a hill in Torre Alta, the highest area of the city of San Fernando. In its 250 years of history, highly reputed scientists, and military men[4] have worked in the ROA, such as Vicente Tofiño (1732–1795), José de Mazarredo (1745–1812), Jorge Juan y Santacilia (1713–1773), Alejandro Malaspina (1754–1810), Cecilio Pujazón (1833–1891), etc.

Despite the successive moments of crisis in Spain, especially in the nineteenth and first half of the twentieth centuries, the ROA always maintained its scientific activity, acquired the best instruments manufactured in England, France, Switzerland and Germany[5] and participated in important international projects, such as the Malaspina expedition, the Transit of Venus observation and the *Carte du Ciel*, working in collaboration with the main observatories in the world.[6]

Today, the Observatory continues being a centre of scientific activity with the following sections: Astronomy, Ephemeris, Geophysics, Satellite Monitoring, Time Section (in which the official time in Spain is determined), Time and Frequency Metrology, Astrometry and Meteorology. It is also a site to be visited with its historical building, the collection of instruments and research centre, with a library and a very important documentary archive.

The ROA Collection of Historical Instruments is essential to understand an important period in the history of San Fernando, and in the history of Spain and the Spanish Navy, but also more generally in the history of European science from the eighteenth to the twentieth century.[7] These pieces represent an excellent journey through the advances and development of astronomy and navigation, and their social, political, cultural, economic, and scientific interests are undeniable and obvious. Currently around 1,500 instruments are inventoried in the ROA, the only ones still extant.

The Royal Spanish Navy promoted and encouraged scientific and technical studies, was highly engaged in officer training, and was at the forefront of the latest technical advances, hiring eminent scientists and collaborating with non-military Spanish and foreign institutions and scientists.[8] In the case of the ROA, we are faced with an unusual case in Spain of a Scientific Heritage Collection that is preserved almost complete and in its original state and place and contextualized.[9]

In the analysis of the current state of the instruments, we find an acceptable physical state of preservation, with almost all the conservation and museological work still to be done, but also with hardly any economic or personal endowment. Although the efforts of the librarian and the instrument teacher in the last years of the twentieth century and the beginning of the twenty-first century have allowed the

[4] For a list of the ROA Directors, see: https://armada.defensa.gob.es/ArmadaPortal/page/Portal/ArmadaEspannola/cienciaobservatorio/prefLang-es/02InfoGeneral.

[5] González González (1992).

[6] González González and Gálvez (2004).

[7] González González (1992).

[8] González González (1995).

[9] Munoz (2001).

physical recovery of most instruments, this is not enough. Much work is still to do on the historical archive, and the collection of instruments.

About the instrument we are studying, that is, the marine chronometer, we must remember that the eighteenth century marked a complete transformation in the methods of construction of scientific instruments in all areas. Assembling, polishing of lenses and stability of devices were improved. Important instrument makers arose, mainly in England, then in France, Germany, and Switzerland, laying the foundations and creating an industry that became more and more consolidated. Although they inherited a tradition, it was in the nineteenth century that they achieved great prestige and sales volume (González 2003). Clocks were essential auxiliary instruments in the practice of many sciences, to measure as accurately as possible the time of observation and the duration of natural phenomena. The diversity of clock types responds to their specific function and use.

The calculation of longitude and latitude at sea is essential to know the exact location of the ship and to carry out ocean navigation safely. This became a political issue in the struggle for military and commercial hegemony in the Atlantic between the main European powers (Spain, Portugal, England, and France).[10] It would not be until the nineteenth century that the marine chronometer was standardized, England being the origin of its production and its dissemination to other countries. As Moreno reminds us (1995), in the eighteenth century the first prototypes of these instruments were built; however, they corresponded to a rather artisanal production, in which each chronometer was different from the previous one due to the need to constantly introduce technical improvements in their mechanisms. France and England were at the forefront of these prototypes with such outstanding makers as Julien Le Roy, John Harrison, Thomas Mudge and Ferdinand Berthoud.

[10] The calculation of the length was simple and easy for centuries. But the calculation of latitude was more complicated. In the eighteenth century, various techniques were tested by measuring and comparing the position of several stars at different times, and that had to be exact, but this required precision watches and that is why marine chronometers appeared. The use of chronometers to calculate latitude was based on the following principle: the ship had to leave the port with a clock that marked the time of the starting point (the port). At sea, through the position of the stars, the local time of this point where the ship was located had to be deduced. In this way, the difference between this time and the one marked on the clock before (of the port) was the difference in longitude between the ship's position and the starting point. It is considered that the first to propose this solution was Regnier Gemma Frisius in his work *De principiis astronomiae et cosmografiae*, Louvain, 1530. Other authors have proposed that Hernando Colón already discovered it in 1524 (González González 2009). Despite the simplicity of the method, it would take many years for watch technology to deliver watches capable of being used at sea without loss of precision. The work of Christiaan Huygens and Robert Hooke in the seventeenth century laid the foundation for the development of precision watchmaking in the eighteenth century. The definitive application of this method was made possible by the English watchmaker John Harrison, who in 1736 was authorized by the Board of Longitude to test his marine chronometer on a round trip between London and Lisbon. The good results obtained ensured that Harrison received official help to build a new and more sophisticated watch. A few years later, he managed to make a stopwatch that, after 81 days of testing, only accumulated an error of 5 s. In France, meanwhile, attempts were also being made to obtain reliable marine chronometers. In the last quarter century, both Pierre Le Roy and Ferdinand Berthoud tested watches with precision like Harrison's. It was a battle against time to be the first to obtain the best technology to dominate the Atlantic (González González 2011).

Other scholars have insisted on the early involvement of the ROA in the construction of the first chronometers.

Spain, as a maritime and colonial power, could not remain oblivious to the progress that was being carried out in the field of chronometry at sea. That this progress was followed closely at a relatively early stage, is demonstrated by a report, dated 1750, by Jorge Juan to the Marqués de la Ensenada, where he claims to have seen the Harrison clock,[11] something not very easy to get at that time, given the secrecy that surrounded its construction.[12]

In the nineteenth century, also due to imperialist needs, chronometers were considerably improved for navigation by other makers. From 1820 we thus see a second generation of chronometers appear, paving the way for a new typology of chronometers that incorporated relevant technological advances. By the mid-nineteenth century the new typology was standardized, and it would remain stable until after World War II. At that time, electronic chronometers began to appear, giving rise to a third generation of modern chronometers.

The Nineteenth Century Marine Chronometers of the ROA

There are no historic inventories of the scientific instruments in the ROA. A few "Charge Lists" were made in a non-routine and incomplete way, where all types of objects and elements were listed, with scientific instruments scattered throughout (among them chronometers) in very small numbers. Furthermore, their appearance on a list did not correspond to an acquisition or purchase and, therefore, registration, but to the existence at that time of that instrument in that office. That is, some instruments do not appear, others appear in different places and times, etc. and data on the date or place of acquisition or manufacturer are not always mentioned.

Some 300 letters of the internal regime are preserved in the ROA Archive, in boxes 116 and 117, in which all documentation relating to the chronometers is collected. In these letters the director or another ROA official requests of the higher authorities that some chronometer be acquired, without there being documentary evidence that they were actually acquired and became part of the collection. What we can deduce from this extensive correspondence is that there was such a movement of request for and possibly acquisition of these chronometers. These letters also address requests for a review or adjustment of said chronometers to other offices, or their transfer, as well as their assignment to different ships or expeditions. Many of them relate how the instruments arrived damaged by the conditions of the journey and must be returned to the manufacturers for repair. This situation is constantly repeated.

[11] "I have seen the Machine and I would have liked it to be possible to buy it, so that we would not lack it from the beginning." Jorge Juan to the Marquis of La Ensenada; London, February 23, 1750. MN, ms. 2413, Dect. 10, cited by Lafuente and Sellés, (1988, p. 64).

[12] Lafuente and Sellés (1988, p. 63). These authors describe in detail how a great effort was made to select suitable students and send them to train as watchmakers in Berthoud's workshop.

Therefore, currently we can only reliably trust the current inventory of the instruments that have survived to this day and are safely guarded. We can also trust González's bibliographical references that he has been verifying over the years and extracting data about the chronometers from these numerous documentations.

The detailed history of the ROA chronometers from the end of the eighteenth century to the present is undoubtedly a vast study. An indication of its magnitude is the fact that Pablo Bernal Sánchez published a study in 2021 concerning only the watches of the firm of Losada that were acquired in the nineteenth century by the ROA and are currently in the Naval Museum of Madrid.

The objective of this chapter is not to make an extensive and detailed study of all the chronometers that have been acquired and passed through the ROA because it would far exceed the available space. However, we do find it interesting and necessary to make this collection visible and contextualize it in space and time, in the institution and in its historical role, with the challenge always in sight to continue advancing in the future and for years in detailed and more extensive studies that cover the analysis of all the extractable data, at least of the very extensive epistolary documentation and the very numerous Lists of Charges mentioned.

The most interesting revelation of all the preserved communications is not simply that these instruments were acquired, delivered, and used, but that malfunctions, disorders, damage and imperfections were very numerous. Thus, returns to the manufacturers were constant, which meant an evolving cooperation between the latter and institutions such as the ROA in the technical improvement of their construction.

To contextualize this process of acquiring chronometers and their use at the ROA, we can recall that in the last years of the eighteenth century, in the Astronomical Observatory of Cádiz and later in San Fernando, the Marine Guards were already learning the system for calculating moon distance and the use of marine chronometers. The Spanish political, military, and economic hegemony in the Atlantic also allowed the organization of transatlantic scientific expeditions, typical of the enlightened reign of Carlos III, carried out by the ROA and led by Tofiño, Malaspina, Vernacci, Alcalá-Galiano, Jorge Juan and Churruca.[13]

For example, between 1783 and 1789, Tofiño left the Observatory to carry out a hydrographic expedition to map the Spanish Mediterranean and Atlantic coasts. The expedition included a frigate and a brig carrying, among other instruments, eight Berthoud chronometers. The ROA served as the warehouse for instruments for such expeditions and the maintenance workshop for the marine chronometers. It was Jorge Juan who promoted the purchase of Harrison and Berthoud watches in Spain.

Other expeditions follow one another in which we again see Berthoud clocks on board, as well as others, such as the Arnold chronometers (1789) and other types of clocks (longitude clocks, seconds clocks, pendulum clocks, seconds for the carriage): the Malaspina expedition, the expedition to the northwest coast of America, (San Blas station), the expedition to the Strait of Magallanes, the expedition to the coasts of Cuba, the expedition to Yucatan, the expedition to the Antilles. For the huge fleet of the Malaspina Expedition, in addition to instruments from the ROA, 62 new

[13] González González (1995).

instruments were purchased, always by directors of the ROA themselves in London which, before being boarded on the ships, passed through the ROA to be checked and inventoried. Between them, 12 watches were purchased, including four Arnold chronometers. In the 1827 List, only an Arnold and a Pennington chronometer from that series remained in use.[14] On board, the marine chronometers were kept in the chronometer closet, located in the officers' cabin; their number varied according to the importance of the ship and the nature of the mission.[15]

Lafuente and Sellés[16] explain that these chronometers, once at the ROA, underwent several tests to detect errors, defects and problems and were often returned to the manufacturer, in Paris or London, for their repair.[17] What was not done at the ROA was stabilizing the necessary infrastructure for subsequent maintenance. The chronometers needed to be cleaned every two years, and always before a campaign. The efforts made by the directors of the ROA to retain an established and well-trained watchmaker in the Observatory to deal with all the chronometers and watches were constant yet not always successful.

The acquisition and use of these instruments was at the discretion of the ship captains who, until the nineteenth century, were in charge of acquiring their own instruments and chronometers. The Navy only acquired the necessary instruments for the execution of the hydrographic works entrusted to commissions for that purpose. These instruments, before going to their destination, had to be inspected at the ROA.

Eventually, the 1831 regulation established that the ROA must have a mechanician and a watchmaker. As of this date, it was also compulsory for every commander of a warship with more than 12 guns to be in charge of, in addition to a collection of maps, a collection of instruments, among which there must be a chronometer, the acquisition and distribution of which would done be by the ROA. In 1857 a report was sent to the departments of Cádiz, Ferrol, and Cartagena, and to the posts of Havana and the Philippines, requesting a list of the chronometers assigned to the ships so that the observatory would know their actual number. The observatory was still in charge of establishing relations with the chronometer builders, signing the orders, and testing them when they arrived. The 1859 regulation explicitly includes the mission of the ROA as the depository for Navy instruments, and that the Observatory's watchmaker be appointed by Her Majesty at the proposal of the director. In 1864, according to the director of the ROA, the watchmaker was in charge of more than 200 chronometers. In the new Regulation of Cecilio Pujazón of 1873, the depositing and testing of the chronometers is in charge of an officer of the first Section of Instruments and Chronometers of the Navy. In the end, the mechanician and the watchmaker positions became intermittent, definitively disappearing in the 1990s. Currently, the remaining

[14] González González (1995).

[15] Bernal (2020).

[16] Lafuente and Sellés (1988).

[17] This occurred much more often with the first chronometers of the eighteenth century that were prototypes in need of continuous improvements than in nineteenth-century ones, which were by then standardised.

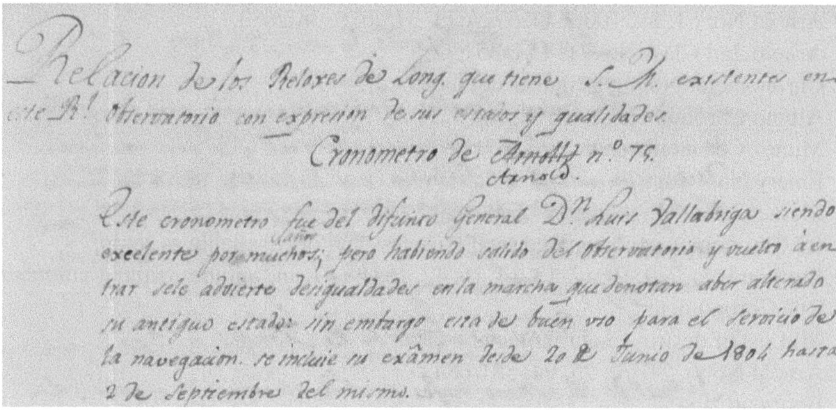

Relacion de los Relojes de Long. que tiene S. M. existentes en este Rl. Observatorio con expresion de sus estados y qualidades

Cronometro de Arnold n.º 75.

Este cronometro fue del difunto General Dn. Luis Vallabriga siendo excelente por muchos años; pero habiendo salido del Observatorio y vuelto á entrar sele advierte desigualdades en la marcha que denotan aber alterado su antiguo estado: sin embargo esta de buen uso para el servicio de la navegacion. se incluie su examen desde 20 de Junio de 1804 hasta 2 de Septiembre del mismo.

Fig. 8.2 Document "Relation of the Long…" mentioned next. *Source* The author

chronometers are inventoried, catalogued, and exhibited in the main building of the ROA, together with the rest of the instrument collection.[18]

About the specific appearance of the chronometers in the Lists of nineteenth century we can read data, as shown in Fig. 8.2, for example.

This chronometer belonged to the late General Don Luis Vallabriga and worked well for many years, but having left the observatory and returned to it, it had some unevenness in its movement that indicated alterations of its former condition: however, it was still good enough for navigation. An examination is included from June 20 September 2, 1804.[19]

This document lists 32 chronometers: 16 from Arnold, 12 from Berthoud, and 4 of unknown make. 10 are in the Commission, 1 is lost and 21 in the Observatory. No acquisition dates appear in this list and the only information that appears is their state of conservation: 80% are deteriorated, unusable or under repair. This is not an official and complete inventory but rather a list that at any given time is requested by higher authorities regarding the existing chronometers in the ROA.

On April 21, 1829, a List of all the instruments kept in the Observatory was sent to the ministry. From it we extract the 44 chronometers that were recorded in his books, as well as their condition (ACM, *Observatorio. Generalidad*. 1831, box 117):

(a) Chronometers that are or may be in use:

Arnold No. 5

Pennington No. 108/499.

(b) More or less damaged chronometers, which are considered sellable, although for much less than their new value:

---

[18] Munoz (2001).

[19] *Relation of the Long. Clocks that His Majesty has existing in this Rl. Observatory with expression of their states and qualities*, Isla de León, September 4 of 1804. Prepared by Julián Canelas, the document is a copy sent by Bermúdez to the ministry. AGM, Observatory. Private affairs. 1797–1804. Box 116.

Arnold No. 61, 89, 100, 344, 372/673, 375/676, 387/683.
Arnold 2nd Class No. 71, 154, 351.
Chronometer No. 8/98 (maker not included).
Albino Chronometer No. 1.
Muñoz Chronometer.
Emery No. 978, 979.
Margetts No. 84.
(c) Chronometers considered totally useless and only sellable as scrap:
Berthoud No. 7, 9, 10, 12, 13, 14, 16, 42, 46, 49, 57, and another without a number.
Rue Chronometer.
(d) Lost or destroyed chronometers:
Arnold No. 6, 71, 72, 75, 94, 381/684.
Berthoud No. 57.
1 Chronometer.
Bonfante No. 1.
Pennington No. 167.
(e) Chronometers delivered:
Pennington No. 180/499.
Arnold No. 56.
Brockbanks.

Up to 1850, we know of 48 marine chronometers owned by the Navy. If we except those manufactured by Berthoud and Arnold in the eighteenth century, which amount to a total of 31 devices, we can see how much less newly acquired material was during the entire first half of the nineteenth century. However, following the reactivation of the Navy in the mid-nineteenth century, this situation changed substantially.

Therefore, the number of chronometers acquired during the second half of the nineteenth century counted by González[20] were: 1850–1858: 74; 1859–1863: 107; 1864–1873: 61; 1874–1904: 141. All of the lists are preserved in the ROA Archive (AROA-097 to AROA-109).

According to González,[21] we can say that in some Lists of the late nineteenth century (1875–1880) we see how the number of chronometers varies year after year. In the case of those assigned to the charge of ships and dependencies, it drops progressively from 168 to 110, assuming wear, deterioration, and loss. In the case of the chronometers deposited in the ROA itself, the number varies between 15 and 30 depending on the year. Starting in 1877, acquisition of chronometers stopped, although eight chronometers were shipped on the Transit of Venus expedition in 1882.

Summarizing, we have seen that, in the middle of the nineteenth century, more than 200 chronometers were in the ROA at some time. Over time, marine chronometers were replaced by new chronometers or other timekeeping instruments. Deteriorated chronometers were registered in the Lists as casualties and gradually disappeared. Currently, there are 17 marine chronometers from the nineteenth century that are

[20] González González (1998).

[21] González González (2005).

included in the new inventory, restored, and exhibited in the collection of the ROA museum.

In the current inventory, registered in the MILES database (intranet of the Ministry of Defence), there are 129 entries for Marine Chronometers, of which 17 are from the nineteenth century and 112 from the twentieth century. None remain from the eighteenth century.

In the following table we can see an extract with the 17 chronometers of the nineteenth century in the current ROA inventory (MILES database, Ministry of Defence intranet accessed on September 10, 2021).[22] We can see the makers and the years of acquisition:

In Fig. 8.3, we can see that all those that are dated belong to the second half of the nineteenth century, especially from the last decade. The manufacturers are 90% English and 10% German, although the firms are very varied, with only 5 Dent chronometers standing out in quantity. We cannot consider this list and its characteristics representative of the collection of chronometers that the ROA has had since they are very low in number in relation to all those that have been registered from the eighteenth to the twentieth century. Nevertheless, they are the ones that have survived until now.

In the current catalogue of the ROA museum, there is a section exclusively dedicated to clocks, with a subsection dedicated to marine chronometers. This catalogue has been recently published (2020) in digital format by Francisco José González González, Francisco López Vico and Mercedes Espada Bordes and can be consulted with free access on the website of the Ministry of Defence.[23] The general order of the catalogue can be consulted in Appendix I. And an example of an inventory entry relating to one of the chronometers can be consulted in Appendix II.

Timepieces at the ROA Today: Heritage and Conclusions

Currently, the measurement of time continues to be an important task at the ROA, mainly because it is the institution that establishes the official time in Spain. The ROA Time Section maintains the pattern of the basic unit of time (the second) and the UTC Coordinated Universal Time (ROA) scale. Royal Decree 1308/1992, of October 23, declares the ROA Laboratory to be a laboratory associated with the Spanish Metrology Centre in the metrological field of time and frequency, and the pattern that is maintained there as the "national standard of the basic unit of time". To carry out this mission, the ROA Time Section is made up of three Services: Time, Electronics and Naval Calibration and Chronometry. Likewise, the Time section

---

[22] The MILES system is only accessible from the Ministry of Defence intranet, that is, from the computers located inside military buildings and with security codes. You can see information about this at: https://ejercito.defensa.gob.es/Galerias/Descarga_pdf/Unidades/Madrid/Ihcm/2020/20200525-guia-archivos-enlaces-web-v1.pdf

[23] General Catalogue of Official Publications: https://publicaciones.defensa.gob.es/media/downloadable /files/links/c/a/catalogo_coleccion_museografica_del_roa_030321.pdf

N° inventory	N° catalogue	Tipe of instrument	Instrument	Maker	City of manufacture	Year	Location in ROA
0036	H1-004	RM	Marine chronometer	French	London	1860	VS-M
0037	H1-017	RM	Marine chronometer	Parkinson & Frodsham	London	S. XIX	SM-A
0042	H1-016	RM	Marine chronometer	Frodsham & Baker	London	S. XIX	VS-M
0067	H1-010	RM	Marine chronometer	Dent	London	1898	SM-A
0070	H1-011	RM	Marine chronometer	Dent	London	1898	SM-A
0074	H1-015	RM	Marine chronometer	Hutton	London	1898	VS-M
0076	H1-002	RM	Marine chronometer	Ehrlich	Bremerhaven	S. XIX	SM-A
0077	H1-005	RM	Marine chronometer	Johannsen	London	1863	VS-M
0079	H1-007	RM	Marine chronometer	Dent	London	1891	SM-A
0082	H1-008	RM	Marine chronometer	Dent	London	1892	SM-A
0213	H1-003	RM	Marine chronometer	French	London	1850-1858	SM-A
0397	H1-001	RM	Marine chronometer	Ehrlich	Bremerhaven	1874	SM-A
0398	H1-019	RM	Marine chronometer	Bruce	Liverpool	S. XIX	SM-A
0399	H1-018	RM	Marine chronometer	Ganter	London	S. XIX	SM-A
0402	H1-020	RM	Marine chronometer	Benson	London	1874-1904	SM-A
0403	H1-006	RM	Marine chronometer	Fletcher	London	1860-1880	SM-A
0581	H1-009	RM	Marine chronometer	Dent	London	1895	SH-R

Fig. 8.3 XIX century chronometers in the ROA inventory (MILES database). *Source* The author

has other assigned tasks that are: advice on watchmaking and chronometry material for use in the Navy; homologation of chronometers and certification of usefulness of the same for the purposes of navigation; acquisition, conservation, recognition, classification, repair and study of the chronometric equipment destined to the service of the Navy and repair and study of the chronometric material of the merchant ships that for this purpose is entrusted to it; design and development of primary time and frequency systems for use on Navy ships.

About the management of the historical heritage, we can say that the Collection of Historical Instruments of the ROA, recognized as a museum collection for

Fig. 8.4 Current showcase with some ROA marine chronometers and other instruments. *Source* The author

the official provision organized by the Defence Museums Network in Spain, has 1245 pieces (inventory of 2020). After a long process of collection, cataloguing and restoration, initiated in the last decades of the twentieth century, this collection has been reorganized.[24]

In this collection of 1245 instruments, the 112 chronometers (17 from the nineteenth century, 95 from the twentieth century) are a numerous subset in comparison to others. After decades of thoughtlessness and dispersion, in 2001, thanks to the first inventories made by Francisco José González González, the nineteenth century chronometers were collected and stored in the Meridian Hall. Later, they were restored and exhibited in a new showcase in the Entrance Hall of the ROA in 2012, where they are on display today. They are in an excellent state of conservation and they are catalogued, and the catalogue includes photographs of excellent quality (Fig. 8.4).

The collection of historical chronometers of the ROA, even though they are a small surviving representation of all those that the institution housed in the past, undoubtedly confirms the crucial role that the measurement of time plays in maritime navigation in the Modern Age for the control of commercial routes and geopolitical and military strategy for the main European powers in competition: France, England, Portugal, Spain, etc. Being in possession of the best time-measurement technology was essential. The ROA was the Spanish institution in charge of this

[24] González González, López and Espada (2020).

service and achieved it by acquiring the best chronometers manufactured in Europe and using them successfully in military actions as well as in the most important overseas scientific expeditions of that time.

Nowadays, at the ROA, and next to the new building of the Time Section and its facilities, the old neoclassical building is still in use and, in addition to housing the headquarters of other current scientific research units, it houses the historical archive, the library and the collection of scientific instruments that, distributed in cabinets present in almost all the rooms and spaces of the building, is coextensive with the library itself. The chronometers that have survived over the years, now safeguarded, monitored, inventoried, and catalogued, are currently in a very good state of conservation, and exhibited in the main entrance hall. Since the eighteenth century, when the first clocks arrived at the ROA, the institution has been measuring time continuously for more than 250 years and its presence, active or heritage-exhibitive, continues impressing visitors and positions itself at the forefront of scientific work.

Acknowledgements I want to thank the University of Neuchâtel, particularly the organizers of the international workshop *Astronomical Observatories and Chronometry: Time, Science and Instruments (18th–20th Century)* held from December 1 to 3, 2021, at the Musée international d'horlogerie in La Chaux-de-Fonds, Switzerland, for the financing of my attendance.

Appendix I

General order of the catalogue of museum collection of ROA (Gonzalez et al., 2020):

1. Astronomical observation instruments.

Glasses and telescopes.
Accessories and other astronomical instruments.
Instruments and materials for astrophotography.

2. Clocks for navigation and science.

Marine chronometers and record clocks.
Astronomical pendulums, quartz oscillators and atomic patterns.
Chronographs and other devices for measuring time.

3. Geophysical observation instruments.

Instruments of geodesy and hydrography.
Meteorological instruments.
Instruments of magnetism and seismology.

4. Devices for technical, scientific and educational use.

Nautical navigation and astronomy instruments.
Slide rules, calculators and computers.
Physical and electrical apparatus, workshop machines and other instruments.

5. Other ROA equity funds.

Paintings, engravings and other funds considered historical-artistic heritage.
Engraving of matrices with astronomical images.
Funds deposited in other Navy units.

Appendix II

Example of an inventory entry relating to one of the chronometers (Gonzalez et al., 2020) (Fig. 8.5):

Marine Chronometer
Johannsen, A., No. 428.
London, 1863.
Sphere (diameter): 11.5 cm. Box: 19 × 18 × 18 cm.
Brass, steel, glass, wood.

Fig. 8.5 Marine chronometer Johannsen in ROA catalogue. *Source* MILES database from ROA Library on September 15, 2021 by author

Registry: A. Johannsen, 3 Lower Ashby St. London No. 428.
Location: Library, Room 8, Showcase.
Inv: 0077/B8.

Marine chronometer in a mahogany case with a double lid, hinges, corner pieces and lacquered brass handles. Bronze mortar mounted on cardan suspension. The engraved silver-plated brass dial indicates the hours, minutes and seconds. It also has a graduated sector to indicate the state of the winding (from 0 to 56 h). It is a traditional type chronometer (circular brass plates, with snail and chain power transmission; flywheel mounted on the plate and chronometer escapement). It has a normal open bimetallic flywheel, cylindrical spiral and nuts for thermal and gear adjustment (oscillation).

This chronometer was purchased under an R.O. July 29, 1862. It arrived in San Fernando in November 1863. It was used on the frigates Gerona (1866–1879 and 1881–1898) and Numancia (1898–1902). It is currently on display in the showcase of Room VIII of the Library. It is in good condition and works.

References

Bernal Sánchez, P. (2020). *José Rodríguez Losada y la armada española a través del Real Instituto y Observatorio de la Armada en San Fernando.* Ministerio de Defensa.

Fresnadillo, R. (1989). *El Castillo de la Villa de Cádiz (1467–1947).* Excelentísimo Ayuntamiento de Cádiz.

González González, F.J. (1992). *El Observatorio de San Fernando (1831–1924).* Ministerio de Defensa.

González González, F. J. (1995). *Instrumentos Científicos del Observatorio de San Fernando (Siglos XVIII, XIX y XX).* Ministerio de Defensa.

González González, F.J. (2003). La colección de instrumentos antiguos del Real Instituto y Observatorio de la Armada: Piezas incorporadas entre 1995 y 2002. In *Actes de la VII Trobada de d'Historia de la Ciencia i de la Técnica*, pp. 273–284. Societat Catalana d'Historia de la Ciencia i de la Técnica.

González González, F.J., & Gálvez Fernández, F.J. (2004). Expediciones científicas españolas para la observación de los Tránsitos de Venus. In *Tribuna de Astronomía y Universo*: 36–43.

González González, F.J. (2005). *El Real Observatorio de la Armada.* Ministerio de Defensa.

González González, F.J. (2009). *Instrumentación:Cronómetros y Sextantes. Conference presented at Seminario Navegación Astronómica y GPS*. Universidad Internacional Menéndez Pelayo.

González González, F.J. (2011). *Instrumentos Náuticos:Evolución histórica. Conference presented at Seminario Navegación Astronómica.* Semana Naval de la Armada en la Escuela de Guerra Naval.

González González, F.J., López Vico, F.R. & Espada Bordes, M. (2020). *La Colección Museográfica del Real Instituto y Observatorio de la Armada. Catálogo de Instrumentos Científicos y Patrimonio Cultural.* Ministerio de Defensa.

Lafuente, A., and Sellés, M. (1988). *El Observatorio de Cádiz (1753–1831).* Ministerio de Defensa.

Moreno, R. (1995). *José Rodríguez de Losada: Vida y obra.* Fundación Juanelo Turriano.

Muñoz Rey, Y. (2001). *La Colección de Instrumentos Antiguos del Real Instituto y Observatorio de la Armada. Análisis del Estado Actual y Programa de Actuación Conservadora,Restauración y Museología.* Ministerio de Defensa and ROA.

Yolanda Muñoz Rey (Madrid, 1973), Doctor of Arts and Humanities, Master in Historical-Archaeological Heritage, Master in Innovation and Investigation, Degree in Fine Arts and Degree on Art History. She is professor at University of Cádiz (Spain), researcher at the University Research Group HUM726, academician at the Royal Academy of San Romualdo of Sciences, Letters and Arts. She has research awards, distinctions and fellowships, done research stays (Sant Petersburg, Faro, TetuánLevanger and Cambridge), and published extensively (books, articles, papers) on history, museology and conservation of scientific-historical instruments.

Chapter 9
The Certification of Chronometers at the Neuchâtel Cantonal Observatory. Setting up and Renewing Equipment and the Legal Framework for Certification (1858–1912)

Amandine Cabrio

Abstract The Cantonal Observatory in Neuchâtel was founded in 1858 to monitor chronometers manufactured in the region. To ensure this delicate mission, suitable apparatus and instruments were used, improved, changed. The certification was also based on a legal framework. Both have evolved over time to meet scientific and economic needs. This article addresses this evolution between 1855 and 1912.

On 18 May 1858, the decree was issued for the creation of a Cantonal Observatory in Neuchâtel, "suitable for the scientific determination of time and for the publication of tables for the adjustment of precision clocks and marine chronometers presented by clockmakers or manufacturers resident in the Canton of Neuchâtel."[1] This marks the official birth of a scientific institution dedicated to the development and quality of precision watchmaking, responding to a need expressed by political and industrial circles since the early decades of the nineteenth century.[2] The Neuchâtel Cantonal Observatory remained in operation until 2007, when it was closed by the canton. Some of its activities were then taken over by the University of Neuchâtel and the Swiss Centre for Electronics and Microtechnology (CSEM).

The certification of chronometers and the determination and distribution of time were entrusted directly to the Institute by the cantonal authorities when it was set up. It was an instrument of intervention by the scientific establishment, and therefore

[1] Official Gazette of the Republic and Canton of Neuchâtel (1858).

[2] In particular in the report written by the Neuchâtel experts on their return from the Paris Universal Exhibition in 1855: Richard et al., (1856); also Petitpierre (1871, p. 252); Tissot-Daguette (undated [1827–1829]).

A. Cabrio (✉)
Manuscript and Private Archives Service, Bibliothèque publique de Neuchâtel, Neuchâtel, Switzerland
e-mail: amandine.cabrio@unine.ch

G. Bernasconi et al. (eds.), *Astronomical Observatories and Chronometry, 18th-20th Century*, History of Physics, https://doi.org/10.1007/978-3-031-95874-8_9

by the State, on behalf of the watchmaking industry. The certification issued by the Neuchâtel Observatory is therefore a certificate of compliance based on specific rules and technical equipment.[3]

This service included an observation period for the timekeepers. This consisted of a series of tests that varied according to the type of chronometer. In general, they include changes in position, temperature, atmospheric pressure, etc. The aim was to record the rate readings so that the accuracy of the timepiece could be calculated. This first part of the certification process resulted in a certificate in the form of a running report signed by the director of the scientific establishment. From 1865, it also included the classification of timepieces by means of annual competitions. They intended to reward the best pieces and the prizes were based on the results of the first phase.

Initially reserved for Neuchâtel watchmakers, the certification of chronometers provided them with a significant commercial advantage, as it enabled them to obtain a certificate issued by a scientific institution. These attestations of compliance increased the confidence of buyers and distinguished them from their competitors. Certification was subsequently extended to timepieces made by watchmakers in the cantons of Berne in 1905 and Vaud in 1928. This development brought watchmaking production in the "industrial district"[4] of the Jura Arc within the scope of the Observatory's activities and signalled the introduction of a regional policy in favour of the watchmaking industry. The scientific establishment, in particular through its certification mission, and the watchmaking industry of the Jura Arc influenced each other in a constant interplay. The proximity of the scientific and watchmaking players largely explains this permeability, which led to the development of "a whole series of institutional measures to promote the dissemination of knowledge."[5]

This network of institutions, in turn, influenced the development and success of the watchmaking industry. Béatrice Veyrassat has shown the emergence of a "pattern of diffuse authority" in the canton of Neuchâtel, thanks in particular to "the birth of a series of institutions for industrial regulation."[6] Indeed, in the second half of the nineteenth century, the creation of the Cantonal Observatory, followed by the watchmaking schools and the creation of umbrella organisations such as the Swiss Watchmaking Chamber, enabled the watchmaking industry to rely on bodies for the dissemination of knowledge. At various levels, these organisations helped to standardise the production and quality of watchmaking products.

The certification mission undertaken by the Cantonal Observatory of Neuchâtel took place against the background of a growing number of observatories. Throughout Europe in the nineteenth century, these scientific institutions were created and financed by governments, in the context of a strong development in positional astronomy, but also in practical astronomy, including geodesy, metrology and chronometry, which brought together disciplines central to the establishment of

[3] Couret et al. (1995).

[4] Bettini (1979); Pasquier (2008).

[5] Pasquier (2008, p. 27).

[6] Veyrassat (1997, pp. 216–217).

state governance and economic policies based on scientific expertise. In this context, several observatories were responsible for the certification of chronometers, including Greenwich, Kew,[7] Besançon,[8] as well as Geneva[9] and Washington.

The context in which the observatories were created, and the history of these scientific spaces have been the subject of numerous studies over the last twenty years, dealing in particular with the emergence of a new relationship between science and the state.[10] However, the importance of the instrumental heritage and its spatial layout, environment and location,[11] and particularly the study of the certification of chronometers and the technical and regulatory apparatus on which it was based remains marginal. Nevertheless, it is worth highlighting specific studies of certain devices and instruments, including the analysis of certain parts dedicated to the certification of chronometers,[12] as well as the study of the role of observatories in the development of marine chronometry[13] and that on the economic history of chronometry.[14]

The aim of this paper is to analyse the establishment and subsequent development of the technical apparatus and the regulations governing the observation of chronometers, between the first official request for an observatory in Neuchâtel and the completion of a new building for scientific observations in 1912, through three key periods. The first period concerns the design and implementation of the technical and regulatory framework between 1855 and 1865. The second will focus on the major revision of the regulations between 1889 and 1901, which brought the Neuchâtel regulations into line with those of other chronometric observatories. The third will focus on the renewal of the instruments and the reorganisation of the premises between 1910 and 1912, which made it possible to reorganise the services of the observatory.

This chronology makes it possible to study the introduction and evolution of the technical and regulatory framework on which certification was based, and to observe the tensions between the various players. The changes made are often the result of compromises between scientific requirements and the needs of the watchmaking industry.

---

[7] MacDonald (2015).

[8] Saint-Martin (2008).

[9] Fallet (1995); Giovannini (2007).

[10] Lamy (2007); Aubin et al. (2010); De la Noë (2011).

[11] Schmidgen (2003); Boistel (2005); Aubin (2015).

[12] Babey (2007); Baldi (2013); Derigny (2016); Gressot and Jeanneret (2020).

[13] Fallet (1995).

[14] Donzé (2023).

Designing and Implementing Chronometer Certification (1855–1865)

On 10 September 1855, the Neuchâtel State Council appointed a cantonal commission to visit the second Universal Exhibition in Paris and draw up a report on the industries present. This type of event was to become unmissable for all export industries, as they were "instruments of economic and industrial development, [and] places where the state and its institutions are represented."[15] On their return, these experts sent a report to the committee that had commissioned them, the result of an analysis of the watchmaking, machine tool and railway industries.[16] In this report, the Neuchâtel experts concluded that, among a series of measures to support Neuchâtel industry, it was necessary to set up an observatory to issue setting tables for chronometers manufactured in the region in order to give them a commercial advantage. The scope of the scientific establishment was defined and included basic instruments for determining time, but not specific equipment for certifying chronometers.[17]

The conclusions of the experts' report were echoed in political circles, and in January 1856 Henri Ladame (1807–1870), who taught chemistry, physics and meteorology at the Gymnase and then at the Neuchâtel Academy, and who founded the Neuchâtel Society of Natural Sciences with Louis Agassiz (1807–1873), was commissioned to carry out a study of the project. Although Ladame's conclusions do not seem to have been preserved, the presentation of a full history of the project by Aimé Humbert (1819–1900), State Councillor for Public Education, to the Grand Council at its meeting on 17 May 1858 provides a record of its content.[18] In his history, Humbert notes that Ladame, after visiting and studying the instruments and operation of the observatories in Geneva, Zurich and Bern, proposed the creation of a technical apparatus for Neuchâtel. This apparatus consisted of a meridian telescope, a parallax telescope, a barometer, a clock, several thermometers and an apparatus for chronometric observations at different temperatures. In contrast to the technical apparatus proposed by the experts in their 1856 report, which was solely focused on determining time, Ladame included instruments for observing the influence of external factors on the chronometers' variation. Ladame's draft was submitted to the Department of Education in May 1856. The discussions were suspended between 1856 and 1857, following the attempted overthrow of the Neuchâtel Republic by a royalist coup.[19] The observatory project was then taken up again by the State Commission for Industrial Schools, which was asked to provide an opinion on Ladame's project in August 1857 (AEN 1IP-606). On 20 February 1858, the commission presented a report to the Council of State based on an analysis of the instruments, architecture and operation of the Geneva observatory (AEN 1TP-792).

---

[15] Debluë (2015, p. 31).

[16] Richard et al. (1856).

[17] Richard et al. (1856, p. 11).

[18] Bulletin officiel des délibérations du Grand-Conseil de la République et Canton de Neuchâtel (1857, p. 605–626.

[19] Barrelet (1993).

However, the Director of Public Education did not consider the commission's project to be sufficiently developed to be submitted to his department's consultative committee. The objective was to present a comprehensive draft at the Grand Council session in May 1858. The Director of Public Education therefore requested the assistance of Professor Rudolf Wolf (1816–1893), Director of the Bern Observatory and Professor of Astronomy at the Swiss Federal Institute of Technology in Zurich (AEN 1TP-792). The current state of research makes it impossible to ascertain whether Wolf was in fact commissioned to conduct a study. The pivotal role of the report submitted in March 1858 by Adolphe Hirsch (1830–1901), a German astronomer, in the establishment of an observatory in Neuchâtel is evident.

Adolphe Hirsch was born on 21 May 1830 in Halberstadt. He studied astronomy at the Universities of Berlin and Vienna.[20] Hired by the astronomer Urbain Le Verrier (1811–1877), director of the Paris observatory from 1854 to 1877, the circumstances surrounding Adolphe Hirsch's arrival in Neuchâtel and his mandate to draw up a project for the Neuchâtel observatory in 1858 are the subject of various accounts.[21]

On 31 March 1858, Adolphe Hirsch presented a new project to the authorities stating:

> It is first necessary to determine the scale on which an observatory should be built in order to ensure that it fulfils the purpose for which it is intended. Once this has been done, the next step is to identify the most suitable site, the construction plan and finally the number and nature of the scientific instruments with which it should be equipped.[22]

In his project, the astronomer proposed a technical apparatus and a layout of spaces inspired by the Greenwich Observatory[23] and Pulkova.[24] This therefore necessitated the allocation of a dedicated space to each scientific practice, with the instruments determining the precise arrangements and constituting the elements around which the architectural structure was to be designed.[25] Following his appointment as director of the cantonal observatory, Hirsch was responsible for overseeing the construction of the building and initiating the procurement of the requisite instruments for astronomical research and the certification of chronometers. For instruments dedicated to positional astronomy, the primary method of acquisition was to purchase them from specialist craftsmen and manufacturers, with the majority based in Germany. A substantial corpus of correspondence between Hirsch, Aimé Humbert and Georges Guillaume (d. 1896), who succeeded Humbert as head of the Department of Public Instruction, offers invaluable insights into this period (AEN 2IND-83).[26]

Conversely, the acquisition of instruments and equipment for the certification of chronometers appears to have been relatively limited. The ice-house, which was

---

[20] Schaer (2001).

[21] Legrandroy (1900/1, pp. 3–35); Fischer (2001, pp. 41–47); Schaer (2001, pp. 181–184); Burgat-Greller and Schaer (2001).

[22] Hirsch (1858, p. 3).

[23] Smith (1991); Rooney and Nye (2009).

[24] Aubin et al. (2010, pp. 33–57).

[25] Aubin (2015, p. 62).

[26] Gressot and Jeanneret (2020).

necessary for the observation of chronometers in cold temperatures, was constructed directly in the cellar of the building (AEN 2IND-83). The structure was relatively simple in design, comprising a converted space that was closed off by a door and insulated by layers of straw, coal and straw matting.[27] The ice for this cold room was delivered to the cantonal observatory on a regular basis, possibly by the town. In June 1858, with the objective of procuring two precision astronomical clocks, one of which would serve as a regulator for the comparison of chronometers while the other would be utilised for astronomical observations (AEN 2IND-83), Adolphe Hirsch, in collaboration with the political authorities, initiated a competition, which was disseminated to the region's clockmakers. In the spring of 1860, five clocks were submitted to the cantonal observatory by various Neuchâtel watchmakers.[28] Two clocks were selected as the most exemplary: the first was that of the Association ouvrière du Locle, and the second was that of the late Mr Alexandre Houriet.[29] The Association ouvrière clock was installed in the chronometer observation room and remained in service there until 1912.[30]

The observatory building was inaugurated in April 1860. The report on the visit by members of the Neuchâtel Society of Natural Sciences on 25 May 1860 provides a detailed account of the areas and instruments that had already been installed at the time.

The members of the Society were greeted by the director and subsequently taken to the meridian room. [...] The Society was then led into the adjoining room, which contained the chronometers under observation and the electrical apparatus. In the chronometer room, a Fortin barometer is located. In terms of meteorological instruments, the observatory still has a normal thermometer and several thermometrographs.[31]

The technical inventory required to certify chronometers was incomplete, lacking an oven for observing timepieces at elevated temperatures. This oven was not constructed until 1862 by Matthias Hipp.[32] The cooler and oven enabled the Cantonal Observatory to certify chronometers, taking into account temperature changes during the running of the chronometers, specifically the expansion of the materials. Despite the necessity for a number of adjustments and revisions, notably to the ice box,[33] this technical apparatus remained stable until 1878, when the cold room was replaced by

---

[27] Hirsch (1860–1861, p. 6).

[28] Hirsch (1860–1861, p. 4, 9–10).

[29] Hirsch (1860–1861, p. 9).

[30] Département de l'Instruction publique (1912, p. 130).

[31] Bulletin de la Société des Sciences Naturelles de Neuchâtel (1858–1861, pp. 229–238).

[32] Hipp was a German-Swiss watchmaker and electrician who served as the manufacturing director of the telegraph network in Bern between 1852 and 1860. He then went on to found and direct the telegraph and electrical equipment factory in Neuchâtel. He was a pioneer of electric clock-making and collaborated closely with Hirsch to establish a system for distributing time by telegraph from the cantonal observatory. Additionally, he developed a number of devices for the scientific community, including an electric clock and chronographs.

[33] Hirsch (1860–1861); Hirsch (1861–1862).

an ice cabinet manufactured by the Société des instruments physiques de Genève.[34] The design and installation of the technical apparatus were conducted in accordance with scientific requirements, while taking into account the constraints of Neuchâtel. At the outset, the Neuchâtel observatory was constrained by limited resources, including space, budget, and staff, in comparison to larger observatories such as Paris and Greenwich.

Simultaneously with the establishment of the observatory and its services, Adolphe Hirsch collaborated with Georges Guillaume to draft two sets of regulations pertaining to the observation of chronometers and the internal operations of the institution. In 1858, he undertook a visit to Greenwich with the intention of drawing inspiration from the organisation of the chronometer service. The establishment of this legal framework represents a pivotal moment in the process of establishing and operationalising the observatory, as it situates its activities at the core of the state apparatus. The regulations are incorporated into the Canton's compendium of laws and decrees, thereby conferring legal authority upon them. Moreover, they establish compliance criteria that will contribute to standardise watchmaking production over the long term.

The initial regulations for the certification of chronometers, published in August 1860,[35] delineated the observation periods and deposit fees, yet did not specify the criteria for obtaining a rate bulletin.[36] The director of the observatory was thus afforded complete freedom regarding the criteria for the certification of compliance. The observatory's service regulations were formally adopted on 22 January 1861, as set forth in the Règlement pour le service de l'Observatoire cantonal 1861. These regulations delineated the duties of the observatory personnel and the procedure for certifying the chronometers, which were required to be compared on a daily basis with the reference time. In particular, the Director was responsible for scientific observations, determining and maintaining time, supervising the electrical equipment used to transmit time, and publishing the results of the chronometer certification process. Additionally, he oversaw astronomical and meteorological observations. The role of the assistant, which was not filled until 1864, was to provide assistance to the director. He was responsible for the maintenance of the building and the instruments, and could be entrusted with observation and calculation tasks.

The first significant alteration to chronometer testing was implemented in 1864.[37] This amendment introduced the possibility of certifying watches fitted with an anchor escapement and reduced the duration of the tests, thereby increasing the number of deposits. As illustrated below, the decree also introduced oven observation periods for pocket chronometers (Fig. 9.1).

[34] Hirsch (1878).

[35] Feuille officielle (1860).

[36] The Cantonal Observatory accepted three categories of timepieces: marine chronometers, regulators and pocket chronometers.

[37] Decree amending the testing service and the tariff for timepieces submitted to the cantonal observatory for examination (1864).

Category	Observation time	Positions	Températures
Marine Chronometer	3 months	flat	room temperature
Pocket Watch	1 month	falt and hanging	room temprature and 32°
Watch with anchor escapement	15 days	flat	room temperature

Fig. 9.1 Regulations for chronometer testing in 1864

This modification was a response to the expressed need by several watch manufacturers to be able to certify lever escapement watches and to reduce the duration of certain tests (BPUN OBSE102-4).[38] The effects of utilising the lever escapement to set chronometers had not yet been empirically investigated.[39] The potential for manufacturers to have this type of timepiece certified motivated them to refine their settings.[40] The observatory was to play an important role in validating not only innovations but also the use of certain components ~~employed in the adjustment~~ of chronometers (e.g. escapement, balance wheel or balance spring[41]) through certification.

In its early years, the observatory operated a certification system that permitted the observation/expertise of chronometers. The certification consisted in the issuance of a certificate of compliance in the form of a running report. However, from the outset of the observatory's activities, Hirsch aspired to introduce an annual competition designed to reward the most accurate timepieces and to encourage the quality of watchmaking production.[42] He returned to this project on several occasions. In his report for the year 1864–1865,[43] Hirsch proposed an award model with a prize of 200 francs for the best marine chronometer and three prizes for the best pocket

[38] Hirsch (1864).

[39] Houriet (1888).

[40] The distinction between chronometers and escapements was abolished with the inauguration of annual competitions on 20 June 1866. This change was based on the promising results obtained by the use of the lever escapement in chronometer settings, as outlined by Hirsch in 1865.

[41] The escapement is a mechanical device that regulates the beating of the watch's mechanical system. The escapement is situated between the energy source (e.g., spring, weight) and the resonator (e.g., balance wheel, pendulum). The balance wheel is a moving component that regulates the mechanism of a watch. It is regulated and perpetuated by the escapement. The balance spring, or spiral spring, is responsible for determining the frequency of oscillation, which in turn serves as the basis for the precision of the resonator.

[42] Hirsch (1860–61, pp. 17 and 22).

[43] Hirsch (1864–65, p. 9).

chronometers, with values of 150, 100 and 50 francs, respectively.[44] The funding for these prizes was largely derived from the revenue generated by taxes on the observation of chronometers.[45] The expertise provided by the observatory thus became a service that was sold to watch manufacturers, as well as an instrument for classifying watch production and creating a quality scale.

On 5 December 1865, the Conseil d'État allocated the sum of 500 francs to distribute five prizes to the best chronometers observed at the cantonal observatory (AEN 1CE-240, pp. 1026–1027). The inaugural regulations for the annual competitions were thus enacted on 20 June 1866.[46] All chronometers that had obtained a "*bulletin de marche*" were eligible to participate and were rewarded according to their accuracy. The objective of these competitions was to foster emulation, stimulate research, and enhance the quality of Neuchâtel's timepieces. Certification enables standardisation of production, as the results for specific components (e.g., escapement, balance spring, and balance wheel) will either encourage the watch industry to utilise them on a large scale or to eliminate them. Furthermore, certification bestowed a commercial advantage upon watch manufacturers, as it enhanced market confidence in regional production. In 1870, Hirsch observed the increasing value of the bulletins issued by the observatory, which had acquired a degree of "authority."[47]

Adaptation and Standardisation of the Rules Governing the Observation of Chronometers (1889–1901)

From 1878/9 onwards, the watchmaking community called for a reform of the rules governing the observation of chronometers at the cantonal observatory in Neuchâtel. These demands were based on comparisons made with the regulations of other observatories at the Paris Universal Exhibition.[48] In particular, the regulations of the Kew and Geneva observatories took into account a wider range of criteria than those of Neuchâtel when calculating chronometer variations.[49]

[44] In 1864, the tax levied on the certification of a marine chronometer was 20 francs per piece, which represented a percentage of the final sale price ranging from 2.5 to 6%. The fees for pocket chronometers and anchor watches were 10 francs and 5 francs, respectively. A total of 65 chronometers were registered between 1864 and 1865, resulting in an estimated revenue of 650 francs.

[45] Hirsch (1864–65, p. 9).

[46] No. 78 "Règlement pour la distribution des prix alloués aux meilleurs chronomètres et montres présentés à l'observatoire cantonal" (20 June 1866, 1862, pp. 495–498).

[47] Hirsch (1870, p. 11).

[48] Clerc et al. (1879).

[49] In the chronometer rate readings, only the diurnal variation is taken into account (Regulations for the observation of chronometers at the Cantonal Observatory and for the distribution of prizes awarded to the best marine and pocket chronometers emerging from the annual competition, 27 December 1872).

From 1874, the rate was calculated in Geneva according to a formula that includes the diurnal variation, the compensation error and the mean rate.[50] For watch manufacturers, these criteria, which were also adopted by the Kew, Yale and Besançon observatories in 1875, enabled the results of observations to be compared internationally and were more representative of the rate of a chronometer. In response to the requests made in the report on the 1878 Universal Exhibition,[51] Hirsch initiated a process of consultation with manufacturers with the objective of implementing significant changes to the existing regulations. This process was completed 23 years later, after the death of Adolphe Hirsch in 1901.

The reform was not subjected to any form of study until 1887, when a subcommittee was established. Due to Hirsch's health problems between 1889 and 1890, the subsequent stage of the process was uncertain and even appeared to have been abandoned until 1897, which provoked the discontent of several watch manufacturers who had deposited their chronometers (BPUN OBSE-102-4.10).[52] In 1897, the committee responsible for revising the regulations commenced its work anew, as evidenced by several letters attesting to meetings between Hirsch and the manufacturers (AEN 1IND-341). It is likely that the new regulations were finalised in early 1899. However, Hirsch and the manufacturers elected not to implement the revised regulations in the light of the Universal Exhibition scheduled to take place in Paris in 1900 (AEN 1IND-341). This decision was taken in order to avoid the possibility of chronometers displaying rate bulletins based on the previous regulations being used alongside those based on the revised regulations.

The draft revision of the Neuchâtel regulations was presented to the Congress of Chronometry, which coincided with the Universal Exhibition in Paris.[53] In comparison to the regulations in place in Geneva, Kew and Besançon, the prospective Neuchâtel regulations were still the subject of criticism from other chronometric observatories. In particular, there were calls for the establishment of a unified classification system for the annual competitions, as well as for the unification of the tests and criteria for obtaining "*bulletins de marche*."[54] Following the International Congress of Chronometry and the Universal Exhibition, work was resumed on revising the Neuchâtel regulations, taking into account the proposals made by the watchmakers and astronomers present (AEN 1IND-341). The demise of Adolphe Hirsch on 16 April 1901, following a protracted illness, further postponed the implementation of the regulations. Louis Arndt (1861–1940), who had been appointed acting director,

---

[50] Geneva Observatory Regulations (23 October 1874).

[51] Clerc et al. (1879).

[52] A review of the collection dedicated to the observatory, held by the Neuchâtel Public Library (BPUN 102OBSE-4), reveals that certain letters indicate a reluctance on the part of the director of the scientific establishment to change the regulations. It is not possible to prove this or to ascertain Adolphe Hirsch's precise motivations, but it is nevertheless important to emphasise the slowness of the process.

[53] Gautier (1902).

[54] Gautier (1902, pp. 154–159).

Category	Class	Observation time	Positions	Températures
Marine Chronometer	-	63 days	flat	18° - 32° - 4° - 11° - 25°
Deck Watch	-	65 days	5 positions	18° - 32° - 4° - 11° - 25°
Pocket Watch	1st	46 days	5 positions	18° - 32° - 4°
Pocket Watch	2nd	29 days	2 positions	18° - 32° - 4°
Pocket Watch	3rd	15 days	2 positions	18° - 32°

Fig. 9.2 Regulations for chronometer testing in 1902

assumed the role. The regulations, which came into force on 1 January 1902,[55] introduced a number of important changes, as shown (Fig. 9.2).[56]

The observatory thus expanded its certification scope to include deck chronometers[57] and precision clocks,[58] in addition to the existing categories. Each category is distinguished by the duration of the tests and the conditions for obtaining a running report, as outlined in the Regulations for the Observatory of Chronometers at the Cantonal Astronomical Observatory and for the distribution of prizes awarded to the best chronometers emerging from the 1901 competition. The most significant alteration/novelty pertains to the methodology employed to ascertain the rate of a chronometer.

> Art.13—The 4 main criteria for the operation of a stopwatch are: 1. the average deviation from daytime operation [...]. 2. The average deviation corresponding to a change of position [...]. 3. Compensation. This is judged on the basis of the thermal coefficient and the average error in compensation. [...] 4. The restart [...].[59]

These significant alterations standardised the Neuchâtel chronometer observation criteria, aligning them with those employed in other chronometer observatories.[60] This enabled a more comprehensive consideration of the diverse parameters influencing chronometer performance and facilitated transnational comparisons of watch

[55] Regulations for the observatory of chronometers at the Cantonal Astronomical Observatory and for the distribution of prizes awarded to the best chronometers emerging from the 1901 competition.

[56] The five positions are: vertical, hanging from the top; vertical, hanging to the left; vertical, hanging to the right; horizontal, dial down; horizontal, dial up. The two positions are: horizontal, dial up and vertical, hanging from the top.

[57] Estelle Fallet elucidates that the on-board chronometer is a large watch (45 to 65 mm in diameter) that is typically placed in a case without suspension, which facilitates its transportation. The chronometer is employed to ascertain the status of the ship following a comparison with the primary marine chronometer. See Fallet (1995, p. 21).

[58] It should be noted that the clocks are not included in the table, as there is no information available regarding the length of time they were observed and the tests they underwent. This information will only be available with the 1905 regulations.

[59] Regulations for the chronometer observatory at the Cantonal Astronomical Observatory and for the distribution of prizes awarded to the best chronometers in the 1901 competition.

[60] De Vanssey et al. (1902).

manufacturers' results. This led to an adaptation of the technical apparatus, which was completed in 1912. In addition to the criteria for assessing the rate at which chronometers ran, the categories of chronometers accepted, the duration of tests and the changes of position were standardised to correspond with the practices of the chronometer observatories in Geneva and Besançon.[61] This approach can be explained by the fear of the State Council of Neuchâtel and the scientific establishment that watchmakers would reduce their deposits at the cantonal observatory. However, it is also necessary to consider the broader context of globalisation and standardisation that affected astronomy at the time,[62] as well as the watchmaking industry. On the one hand, the large-scale circulation of data between observatories and the need to be able to compare and reuse them accelerated this standardisation movement.[63] Conversely, the Swiss watch industry was witnessing a phenomenon of mechanisation, which led to the introduction of new manufacturing standards that enabled the production of greater quantities of higher-quality products.[64]

The Renewal of Spaces and Equipment for the Observation of Chronometers (1910–1912)

Following the death of Adolphe Hirsch in 1901, the arrival of Louis Arndt at the head of the cantonal observatory initiated a series of changes. Upon assuming his role, Louis Arndt observed a number of structural deficiencies in the edifice and the necessity to renew certain apparatus utilized for certification purposes, primarily to align with the observation regulations currently under consideration.[65] Consequently, several projects were devised with the objective of transforming the building and purchasing new equipment.[66] On 22 March 1905, the observatory commission approved a budget of 7,500 swiss francs for the renovation of the building and the purchase of new equipment (AEN 1EP-352). The construction of the residential building on the observatory site also permitted the chronometry service to be relocated to the main building (AEN 1TP-794). The refurbishment of the premises permitted the installation of a new refrigeration unit in the basement, ordered from Ernst Schneider & Cie, which was intended to enhance the safety of the chronometers under observation. The refrigeration appliance and the spaces in which the chronometers were located had been designed in such a way that they were able to protect the

61 De Vanssey et al. (1902).

62 Pestre (2015).

63 Porter (1995); Pestre (2015).

64 Donzé (2023).

65 Arndt (1902).

66 This renewal also concerned the time service. In point of fact, Adolphe Hirsch bequeathed a considerable fortune to the State, with the intention of providing financial support for the advancement of astronomical research at the observatory.

chronometers from being damaged, while also reducing the risk of accidents caused by moving parts.

In order to avoid the great inconvenience of having to transport the chronometers for comparison to another room and thus avoid the condensation of humidity on the parts, the coolers were built large enough to allow the observer to enter.[67]

From 1909 onwards, the project to construct a new building, the future Hirsch pavilion, and the refurbishment of the old building enabled the chronometry service to be redeployed (AEN 1TP-794). The chronometry service's activities, which included determining the time and certifying chronometers, would remain in the original building, while scientific observations were brought together in the new building. The work conducted in the original building permitted the space to be reorganised in a more rational manner.[68] The chronometry service and the scientific service of the cantonal observatory were now settled into separate buildings, which had been adapted to the equipment, processes and practices they housed.

The publication in 1909 by the assistant astronomer Hermann Stroele (1880–1921), who was in charge of chronometer certification, of an article detailing the process of observing chronometers in the Bulletin de la Société des sciences naturelles de Neuchâtel[69] marked a pivotal moment in the transformation of the process of observing chronometers. This publication, which was made without the authorisation of the observatory's director, criticised the continued use of the eye and ear method for observing pocket chronometers.[70] As a result, it caused a stir within the cantonal administration as well as criticism from the watchmaking world, both parties fearing that the cantonal observatory would lose all credibility as a result.[71] Furthermore, Stroele failed to mention the trials conducted by Louis Arndt since 1905 on the use of a counter to automate the observation of pocket chronometers.[72]

In the early months of 1910, the observatory commission resolved to procure a chronograph. This type of device, which was widely used in observatories,[73] enabled the simultaneous recording of data from the seconds hand of the astronomical clock and the electrical impulse triggered by an observer or a micrometer recorder when the seconds hand of a chronometer passed by at certain times, on a paper ribbon using two ink pens activated by an electrical impulse. This method has the significant advantage of reducing the workload of the observer, minimising errors and providing a documented record of the observation, thereby facilitating verification.

---

[67] Département de l'Instruction publique (1912).

[68] Département de l'Instruction publique (1912).

[69] Stroele (1908–1909).

[70] The cantonal observatory had been utilising a recording chronograph to compare marine chronometers since 1882 (BPUN 102OBSE-4.5). The observatory had possessed a recording chronograph for timekeeping since 1864. The various references to a chronograph in Hirsch's annual reports do not permit the determination of the precise chronology of its use for the certification of marine chronometers.

[71] Letter from Georges Favre-Jacot to the Department of Commerce and Industry (1st June 1909, AEN 1IND-343).

[72] Arndt (1906, pp. 13–14).

[73] Lamy and Soulu (2015).

Its intended purpose was solely for the observation of chronometers, yet the price and fragility of the item ultimately prevented its purchase (AEN 1EP-352). The spike chronograph, which was employed for astronomical observations, was then favoured (AEN 1EP-352) and its use became widespread for all chronometers in stock during 1910.[74] The utilisation of this apparatus by the observatory personnel responsible for the comparison of chronometers serves to illustrate the significance of the production of certification that was designed to minimise the influence of human error. Furthermore, it furnishes proof of the observations made, thereby facilitating enhanced monitoring of the work carried out by the observers.

The standardisation of data, the definition of quality criteria and the growing volume of information to be produced by observatories were leading to a transformation in the field of certification. This standardisation was characterised by the automation of observation, and thus influenced by the ideology and techniques of work rationalisation that emerged in the late nineteenth and early twentieth centuries.[75]

The relocation of the chronometry service to new premises resulted in the acquisition of a seismograph manufactured by Bosch in Strasbourg.[76] Installed in July 1911 in the basement of the chronometry building, its use enabled the astronomers to better identify the causes of the variations in the rate of astronomical clocks owned by the observatory and thus eliminate external factors in the calculation of the rates of clocks submitted by the watchmakers. In this way, this approach enabled the exclusion of poor-quality clocks from observations and the denial of a running report that they would have otherwise received. This was in accordance with the recommendations set out in the report by the Director of the Cantonal Observatory to the Inspection Commission, concerning the new installations for observing chronometers and clocks, dated 1 November 1904 (AEN 1IND-345).

The modifications to the technical apparatus employed for the certification of chronometers between 1905 and 1912 were a consequence of the emergence of new scientific requirements. The segregation of specific procedures necessary for certification and the incorporation of automated processes demonstrate a commitment to maintaining the dependability of the expertise.

Conclusion

A critical examination of the design, implementation and subsequent alterations to the technical and regulatory framework for the certification of chronometers reveals that the proposed changes were formulated at the nexus of scientific, political and industrial considerations.

The period between 1855 and 1865 saw the development of the technical apparatus, during which the necessity for the watch industry to obtain certification that

[74] Arndt (1911).

[75] Gardey (2008).

[76] Arndt (1913).

would validate the technical quality of chronometers became apparent. In order to achieve this, Adolphe Hirsch devised a technical and regulatory apparatus that guaranteed the reliability of his expertise and helped to build confidence in certified watchmaking products. Hirsch's vision was based on the model of the astronomical and chronometric observatories of the time and proposed a "scientific space"[77] adapted to various practices. Furthermore, Hirsch advocated for the necessity of establishing a connection between positional astronomy and practical astronomy, with the objective of aligning the Neuchâtel observatory within a European network.

The second period under analysis, spanning from 1889 to 1901, illustrates the delay of the cantonal observatory in the regulatory framework of certification. In contrast to several European chronometer observatories, the Neuchâtel observatory considered only one criterion in its assessment. However, this stance was untenable in the face of mounting pressure from the watch industry and other observatories. In fact, the increased exchange of data and the possibility of comparing them were leading to standardisation at various levels. Consequently, the Neuchâtel observatory began standardising its regulations approximately three decades after the other chronometric observatories. This significant transformation also necessitated the replacement of the technical equipment, which was required to comply with the established regulations.

The final period under consideration in this paper, spanning from 1910 to 1912, demonstrates that the automation of the data collection process did not merely impact positional astronomy, but also extended to the various branches of practical astronomy. The observatory sought to reduce the impact of personal bias and external influences in order to provide high-quality expertise. These endeavours demonstrated the pivotal role of confidence in the certifying institution, which in turn facilitated the development of confidence in the certified products.

In considering the specific characteristics of the watchmaking industry in the Jura Arc, it is important to acknowledge the broader global movements that were shaped by the Neuchâtel observatory. The objective is to demonstrate that smaller observatories can offer equally compelling case studies, notwithstanding their size. Conversely, the limitations of a constrained workforce, limited space and budgetary constraints serve to illustrate the adaptability of the 'observatory' model.

The specific case of the certification of chronometers also illustrates the pivotal role played by these scientific institutions in the economic and social advancement of a region. The objective of this preliminary investigation is to stimulate further inquiry into the Neuchâtel cantonal observatory and analogous observatories.

[77] Lamy (2007).

References

Manuscripts

Archives de l'État de Neuchâtel (AEN):

Fonds du Département de l'Économie publique (EP), Série "Économie publique," cotes AEN 1EP-352 à AEN 1EP-368; Série "Économie publique," AEN 2EP-576 à AEN 2EP-1151.

Fonds du Département de l'Industrie (IND), Série "Industrie I," cotes AEN 1IND-340 à 1IND-352; Série "Industrie II: Observatoire cantonal," cotes AEN 2IND-1 à 2IND-99.

Fonds du Département de l'Instruction publique (IP), cotes AEN 1IP-624, AEN 1IP-640A et AEN 1IP-641A.

Fonds du Département des Travaux publics (TP), cotes AEN 1TP-792 à 1TP-794, puis AEN 1TP-995 à 1TP-1003.

Printed texts

Arndt, L. (1902). *Rapport du directeur de l'Observatoire cantonal à la commission d'inspection sur l'exercice 1901*. E. Sauser.

Arndt, L. (1906). *Rapport du directeur de l'Observatoire cantonal de Neuchâtel à la commission d'inspection sur l'année 1905 suivi du Rapport spécial sur le Concours des chronomètres observés en 1905*. E. Sauser.

Arndt, L. (1911). *Rapport du directeur de l'Observatoire cantonal de Neuchâtel à la commission d'inspection sur l'année 1910 suivi du Rapport spécial sur le Concours des chronomètres observés en 1910*. E. Sauser.

Arndt, L. (1913). *Rapport du directeur de l'Observatoire cantonal de Neuchâtel à la commission d'inspection sur l'année 1912 suivi du Rapport spécial sur le Concours des chronomètres observés en 1912*. E. Sauser.

Aubin, D. (2015). L'observatoire. Régime de spatialité et délocalisation du savoir. In D. Pestre (Eds.), *Histoire des sciences et des savoirs* (vol. 2, pp. 55–72). Seuil.

Aubin, D., Bigg, C., & Sibum, H. O. (2010). *The heavens on earth: Observatories and astronomy in nineteenth century science and culture*. Duke University Press.

Babey, V. (2007). L'observatoire chronométrique de Neuchâtel. Évaluation et évolution d'une société de services à travers ses instruments scientifiques de la deuxième moitié du 19e à la première moitié du 20e siècle. In H.-J. Gilomen, M. Müller and L. Tissot (Eds.), *Dienstleistungen: Expansion und Transformation des dritten Sektors (XV.-XX. Jahrhunderts)/Les services: essor et transformation du secteur tertiaire (XVe-XXe siècle)* (pp. 329–344). Chronos.

Baldi, R. (2013). *Une approche pluridisciplinaire intégrée pour l'étude et la conservation de la collection d'objets de l'Observatoire de Neuchâtel. Rapport sur la recherche historique*. Université de Neuchâtel, Institut d'histoire de l'art et de muséologie/Musée international d'horlogerie de La Chaux-de-Fonds (unpublished manuscript).

Barrelet, J. M. (1993). *Histoire du Pays de Neuchâtel*, tome 3: *De 1815 à nos jours*. G. Attinger.

Bettini, G. (1979). Dal settore industriale al distretto industriale: Alcune considerazioni sull'unita' di indagine dell'economica industriale. *Rivista di Economia e Politica Industriale, 5*(1), 7–21.

Boistel, G. (ed.). (2005). *Observatoires et patrimoine astronomique français*. ENS Éditions.

Bulletin de la Société des sciences naturelles de Neuchâtel. (1843–1897).

Bulletin de la société neuchâteloise des sciences naturelles. (1897).

Bulletin officiel des délibérations du Grand-Conseil de la République et Canton de Neuchâtel. (1857).

Burgat-Greller, M., & Schaer, J.-P. (2001). Adolphe Hirsch (1830–1901), directeur de l'Observatoire de Neuchâtel de 1858 à 1901. *Bulletin De La Société Neuchâteloise des Sciences Naturelles, 124*, 23–39.

Cabrio, A. (2020). Observer et certifier les chronographes et les compteurs de sport à l'observatoire cantonal de Neuchâtel (1924–1932). In L. Tissot and T. Busset (Eds.), *Sports et entreprises: La performance en jeu* (Vol. 11, pp. 97–111). Éditions CIES.

Clerc, C., Grossmann, J., & Ducommun, P. et al. (1879). *Rapport présenté au Conseil d'État par les experts horlogers délégués à l'exposition universelle de 1878*. Imprimerie L.-A. Borel.

Couret, A., Igalens, J., & Penan, H. (1995). *La certification*. Presses Universitaires de France.

Daston, L. (2008). On scientific observation. *Isis, 99*, 97–110.

Daston, L., & Galison, P. (2007). *Objectivity*. Zone Books.

De La Noë, J., & Soubiran, C. (eds.). (2011). *La (re)formation des observatoires astronomiques sous la IIIe République: Histoire contextuelle et perspectives actuelles*. Presses universitaires de Bordeaux.

Debluë, C.-L. (2015). *Exposer pour exporter: Culture visuelle et expansion commerciale en Suisse (1908–1939)*. Alphil.

Derigny, C., & Jeanneret, R. (2016). *Projet OBS: Approche pluridisciplinaire intégrée pour l'étude et la conservation de la collection d'objets de l'Observatoire chronométrique de Neuchâtel*. Haute École ARC Neuchâtel (unpublished manuscript).

Donzé, P.-Y. (2023). *Histoire sociale et économique de la chronométrie*. Alphil.

Fallet, E. (1995). *La mesure du temps en mer et les horlogers suisses*. Institut L'Homme et le Temps.

Fischer, G. (2001). Adolphe Hirsch (1830–1901): L'astronomie et les sciences de la Terre. *Bulletin De La Société Neuchâteloise des Sciences Naturelles, 124*, 40–47.

Garde, D. (2008). *Écrire, calculer, classer: comment une révolution de papier a transformé les sociétés contemporaines (1800–1940)*. Éditions la Découverte.

Gautier, R. (1902). Rapport de la commission chargée de l'étude des épreuves et concours pour les chronomètres dans le but d'obtenir une unification des épreuves dans les observatoires. In *Congrès international de chronométrie: Compte-rendu des travaux, procès-verbaux, rapports et mémoires publiés sous les auspices du bureau du congrès* (pp. 153–156). Gauthier-Villars.

Geneva Observatory Regulations. (1874).

Giovannini, M. (2007). *L'Observatoire de Genève de 1928 à 1975: Un établissement scientifique face aux enjeux de son temps*. Mémoire de licence présenté à la Faculté des Lettres de l'Université de Fribourg (unpublished manuscript).

Gressot, J. (2023). Automating time determination: The Photographic Zenith Tube (PZT) of the Neuchatel observatory. *Bulletin Scientific Instrument Society*, 16–23.

Gressot, J., & Jeanneret, R. (2020). The Ertel Meridian Circle (1858–1912) of the Neuchâtel observatory: Material analysis and culture of precision. In *XXXIX Scientific instrument symposium, London (conference paper)*.

Hirsch, A. (1858). *Rapport de M. le Dr Hirsch sur le projet de fonder un Observatoire cantonal à Neuchâtel*. [Neuchâtel].

Houriét, C. (1880). *Résumé d'un cours public d'horlogerie donné à la Chaux-de-Fonds*. École d'horlogerie de La Chaux-de-Fonds.

Journal suisse d'horlogerie. (1876–1918). Journal suisse d'horlogerie et inventions-revue pour l'horloger. (1918–1921). Journal suisse d'horlogerie et de bijouterie. (1921–1935). Journal suisse d'horlogerie et de bijouterie et Bulletin de la Foire suisse de l'horlogerie à Bâle. (1936–1947). Musée international d'horlogerie de La Chaux-de-Fonds.

Lamy, J. (2007). *L'observatoire de Toulouse aux XVIII et XIXesiècles: Archéologie d'un espace savant*. Presses universitaires de Rennes.

Lamy, J., & Soulou, F. (2015). L'émergence contrariée du chronographe imprimant dans les observatoires français (fin 19e-début 20e). *Annals of Science, 72*(1), 75–98.

Legrandroy, E. (1900–1901). Adolphe Hirsch. *Bulletin de la Société Neuchâteloise Des Sciences Naturelles, 29*, 2–35.

MacDonald, L. T. (2015). *Making kew observatory: The royal society, the British association and the politics of early Victorian science.* Cambridge University Press.

Pasquier, H. (2008). *La "recherche et développement" en horlogerie: Acteurs, stratégies et choix technologiques dans l'arc jurassien suisse (1900–1970).* Alphil.

Pestre, D. (ed.). (2015). *Histoire des sciences et des savoirs,* vol. 2: *Modernité et globalisation.* Seuil.

Petitpierre, A. (1871). *Un demi-siècle de l'histoire économique de Neuchâtel: 1791–1848.* J. Sandoz.

Porter, T. (1995). *Trust in numbers: The pursuit of objectivity in science and public life.* Princeton University Press.

Seiler, A. (1848–1973). *Recueil des lois, décrets et autres actes du gouvernement de la République et Canton de Neuchâtel.*

Regulations for the Observatory of Chronometers at the Cantonal Astronomical Observatory and for the distribution of prizes awarded to the best chronometers emerging from the 1901 competition.

Richard, L., Favre, D., Tixier, P., Courvoisier, A., & Combe, É. (1856). *Rapport présenté au comité du canton de Neuchâtel pour l'Exposition universelle de 1855, à Paris.* Imprimerie Leidecker & Combe.

Rooney, D., & Nye, J. (2009). "Greenwich observatory time for the public benefit": Standard time and Victorian networks of regulation. *British Journal for the History of Science, 42*(1), 5–30.

Saint-Martin, A. (2008). *L'office et le télescope: Une sociologie historique de l'astronomie française, 1900–1940.* Thèse Université Paris IV Sorbonne (unpublished manuscript).

Schaer, J.-P. (2001). Adolphe Hirsch, astronome, fondateur et directeur de l'Observatoire de Neuchâtel (1830–1901). In Schlup (Eds.), M*Biographies neuchâteloises: Tome 3: De la Révolution au cap du XXe siècle* (pp. 181–184). Attinger.

Schaffer, S. (1988). Astronomer mark time: Discipline and the personal equation. *Science in Context, 2*(1), 115–145.

Schmidgen, H. (2003). Time and noise: The stable surroundings of reaction experiments, 1860–1890. *Studies in History and Philosophy of Biological and Biomedical Sciences, 34,* 237–275.

Smith, R. (1991). A national observatory transformed: Greenwich in the nineteenth century. *Journal for the History of Astronomy, 22,* 5–20.

Stroele, L. H. (1908–1909). L'observation des chronomètres et le calcul des résultats à l'Observatoire de Neuchâtel. *Bulletin de la Société Neuchâteloise des Sciences Naturelles, 36,* 3–29.

Tissot-Daguette, D. (undated [1827–1829]). *Coup d'œil sur l'état actuel de la manufacture d'horlogerie en Suisse.* Fonds de la Société d'émulation patriotique, Bibliothèque publique de Neuchâtel.

Trueb, L. (2012). *L'observatoire de Neuchâtel: son histoire de 1858 à 2007.* Éditions Institut l'Homme et le Temps.

Veyrassat, B. (1997). Manufacturing flexibility in nineteenth-century Switzerland: Social and institutional foundations of decline and revival in calico-printing and watchmaking. In C. F. Sabel (Eds.), *World of possibilities: flexibility and mass production in western industrialization* (pp. 216–217). Cambridge University Press.

Whitley, R. (2006). *The intellectual and social organization of the sciences.* Oxford University Press.

Wise, M. N. (1995). *The values of precision.* Princeton University Press.

Amandine Cabrio is a doctoral candidate in the history of science and technology at the University of Neuchâtel, where she is preparing a thesis on the certification of chronometers at the Neuchâtel Cantonal Observatory. She also works as an assistant in the Manuscripts and Private Archives Department of the Neuchâtel Public Library. She holds a master's degree in museum studies and a postgraduate diploma in archival studies.

Chapter 10
Astronomy and Chronometry: Institutions for Testing Chronometers After the Foundation of the German Empire in 1871

Günther Oestmann

Abstract After the foundation of the German Empire in 1871, the infantry general Albrecht V. Stosch called for the establishment of a self-sustaining production of nautical instruments and chronometers and introduced national testing standards. In conjunction with these ambitious efforts and growing demand, a new industry of precision mechanics gradually developed in Germany. In this context, the Imperial Admiralty planned an observatory specifically for the purpose of examining and testing chronometers, which was completed in 1883. The *Deutsche Seewarte* in Hamburg set up a chronometer department, and in 1877 the first annual competition for the testing of chronometers was organised. A third important institution for testing chronometers was the Marine Observatory in Wilhelmshaven, built between 1874 and 1878, which also had other functions (geomagnetic and meteorological observations, tide prediction).

With the creation of the North German Confederation, the constitution of which came into force on 16 April 1867, a new era began for the shipping of the German coastal territories. Article 54 stipulated that the merchant ships of all the federal states were to form a federal navy. The times of particular regulations for the German territories came to an end, and after the foundation of the German Empire in 1871 the institutional framework as well as the development perspectives of the Prussian navy changed fundamentally.

Already on 25 September 1861, the "Hydrographic Bureau" (since 1879 "Hydrographic Agency"), subordinate to the Department for Equipment in the Prussian Navy Ministry, had been founded by a Supreme Cabinet Order. The main task of this office was initially the correction of nautical charts according to newspaper reports and communications from the Foreign Office. Summaries appeared in the *Nachrichten für Seefahrer*, which were primarily addressed to the navy's warships, but excerpts

G. Oestmann (✉)
Technischen Universität, Berlin, Germany
e-mail: oestmann@nord-com.net

G. Bernasconi et al. (eds.), *Astronomical Observatories and Chronometry, 18th-20th Century*, History of Physics, https://doi.org/10.1007/978-3-031-95874-8_10

were also published in newspapers via the Ministry of Trade and Commerce. The Hydrographic office was also responsible for processing the results of overseas expeditions, procuring navigational instruments and making the first cartographic surveys of domestic waters.[1] Another important topic was the unsolved problem of compass compensation, which had received little attention until then. It required an immediate solution in the transition from wooden to iron shipbuilding. Moreover, the instrumentation of the navy was not up to date. Most of the instruments were procured by the shipyards, and there were hardly any checks and examinations according to scientific principles. In 1876, it was decreed that in future the procurement, examination and acceptance of nautical instruments had to be carried out by the Hydrographic office.[2] Instrument depots were set up at the imperial shipyards in Wilhelmshaven, Kiel and Danzig.

The first head of the newly founded Imperial Admiralty in 1872 was Albrecht von Stosch (1818–1895), General of the Infantry.[3] It was his ambition to make the German navy independent of foreign (especially English) precision instruments by promoting domestic industry. In connection with the introduction of a state examination system and the growing demand for nautical, precision mechanical and optical instruments, a new branch of industry developed in the German Empire.

In this endeavour, Stosch was motivated by personal scientific interests, which he also wanted to promote within the navy. On the other hand, he saw his efforts as an opportunity to influence the merchant navy: shipowners and merchants could be offered the opportunity to profit from the knowledge of the Imperial Navy in order to minimise the loss of ships, crews and cargo as well as to determine economic trade routes. Stosch was an opponent of any particularistic efforts and hoped that this would link the Hanseatic cities more closely to the German Empire and win them over to the Imperial Navy. The Hydrographic office seemed to him to be the first step towards a naval ministry encompassing both the navy and the merchant navy, but it was never established.[4] On 28/30 March 1889, the *Reichsmarineamt* replaced the *Kaiserliche Admiralität* as supreme administrative authority, and in 1893, it took over the supervision of navigation marks from the Ministry of the Interior. The Hydrographic office was renamed *Nautische Abteilung des Reichsmarineamts*, which went along with an expansion of its areas of responsibility: the Nautical Department now cared for nautical cartography and navigation marks, the processing of sailing instructions and beacon directories, as well as nautical instruments and instrument testing.[5]

---

[1] Freiburg im Breisgau, Bundesarchiv/Militärarchiv: RM 1/1800, fol. 147r-164v, 168r-179v; RM 1/658, fol. 4r-6v; Anon. (1875–2); Festschrift Hydrographisches Bureau (1911, p. 6).

[2] Freiburg im Breisgau, Bundesarchiv/Militärarchiv: RM 1/1800, fol. 174r.

[3] Sieg (2005).

[4] Sieg (2005, pp. 391–396).

[5] Festschrift Hydrographisches Bureau (1911, pp. 14–15).

The *Deutsche Seewarte*

The *Deutsche Seewarte* was responsible for nautical-scientific matters and instrumentation for the merchant navy. This institution emerged from the *Norddeutsche Seewarte*, founded in 1867 by Wilhelm v. Freeden (1822–1894), director of the navigation school Elsfleth, and was subordinated to the Imperial Admiralty in 1874 (Fig. 10.1). The *Seewarte* began its work as an imperial institute on 1 February 1875,[6] and the geophysicist Georg Balthasar Neumayer (1826–1909) was appointed as director.[7] Neumayer's relations with the Admiralty dated back to 1872; in that year he had come to the attention of Stosch on the occasion of a lecture on magnetism on iron ships held at the Geographical Society in Berlin and had been strongly recommended by the astronomer Wilhelm Foerster (1832–1921) for the post of "Hydrographer to the Admiralty".[8] In this position, Neumayer had already developed extensive activities in the fields of nautical instrumentation, surveying and cartography in a short time.[9] In 1873 he founded the *Hydrographische Mittheilungen* (since 1874 *Annalen der Hydrographie und Maritimen Meteorologie*). Due to his geophysical research and services to polar research, Neumayer achieved international fame. He finally retired in 1903 at the age of 77. However, the Imperial Navy did not want to see a scholar but a naval officer as head of the *Seewarte* and appointed Rear Admiral Alfred Herz (1850–1936) as his successor. After his promotion to staff officer, Herz had started off as first officer of the cruiser *Marie*, then commander of the gunboat *Mücke* and the ironclads *Oldenburg*, *Württemberg* and *Deutschland*. Afterwards he held various shore positions, retired from active service in 1894 and was appointed coastal inspector for the Elbe and Weser area.[10] Herz remained director of the *Seewarte* until 1911. During his term of office, the regulations for the creation of the "German chronometer" were enforced, which had been passed under Neumayer's aegis in 1898.[11]

Initially, the *Seewarte* had three departments (I: Maritime Meteorology and Oceanography, II: Procurement and testing of nautical instruments, scientific publications, III: Weather telegraphy and storm warnings). After the Imperial Navy had set up chronometer observatories in Wilhelmshaven and Kiel, the director of the Hamburg Observatory, George Rümker, successfully lobbied Mayor Gustav Heinrich Kirchenpauer (1808–1887) for the creation of a chronometer testing facility for the merchant navy, which was then attached to the *Seewarte* as its fourth department.

Rümker (1832–1900) was the son of the Hamburg observatory director Charles Rümker and had already been introduced to the activities at the observatory by him

[6] Berlin, GStA PK: I. HA Rep. 89; Winterhalter (1889, pp. 234–241).

[7] On the machinations with which Neumayer "disembarked" his competitor. See Freeden see Horn (1972, pp. 64–69).

[8] Foerster (1911, pp. 131–132).

[9] Wislicenus (1909); Heidke (1926, p. 10).

[10] Anonymous (1930, p. 161).

[11] Oestmann (2012 and 2014).

Fig. 10.1 Building of the *Deutsche Seewarte* (inaugurated on 14 September 1881) in neo-Renaissance style, according to a design by Gustav Jacob Kirchenpauer (1847–1914) and Ludwig Hermann Philippi (1848–1908). Photograph of the nineteenth century, in possession of the author

as a youth.[12] In 1851 he went to Berlin to study and two years later was appointed as an observator at Durham Observatory. To support his sick father, George Rümker returned to Hamburg in 1855 as an adjunct at the observatory. After his father's retirement to Lisbon, he was entrusted with the management of the Hamburg observatory, first on an interim basis in 1857 and then permanently from 1867. In addition to nautical science (in 1871 Rümker was appointed chairman of the examination commissions for the mariners' and helmsmen's examinations and *Reichsprüfungs-Inspektor* for the nautical examinations of the merchant navy for the eastern part of the German Empire) his special attention was devoted to precision timekeeping. The great importance of the chronometer in an era of shortening shipping routes and travel times through scientific route guidance was clearly recognised by Rümker:

> The tendency of modern times is to shorten sea voyages as much as possible, and apart from the changes in the construction of ships, the sailing instructions first established by Maury on the most scientific basis possible have contributed to this in the most essential way. In order to be able to use them practically, however, it is absolutely necessary that the seaman is given the means to determine the ship's location at sea with the greatest possible accuracy, as often as he has the opportunity, since only when he is reassured about the latter and certain that he is not too close to dangerous places marked on the nautical charts, will he be able to

[12] Schorr (1900).

> actually follow the regulations and advice given to him in the instructions and to continue his voyage according to them without delay.[13]

Already in 1875, the *Annals of Hydrography* had published the announcement of the establishment of a fourth area of activity of the *Seewarte*:

> In addition to the various branches of activity of the new institute, which have been developed in more detail here, another will in all probability be added in the near future, the cultivation of which is of outstanding importance for navigation. This is the supervision of the examination of chronometers, preferably before their purchase, on the part of the state or the private sector. This examination will not be carried out within the Seewarte, but probably at the observatory in Hamburg, which will be equipped with some special facilities for this purpose. Until recently, there was no chronometer testing institute in Germany, as such institutes have been active in other countries, e.g. in England, for many years for the benefit of the nautical public and for the improvement of chronometer production in particular. No matter how many times this serious deficiency has been pointed out, the remedy has encountered considerable difficulties; these difficulties can be easily overcome by a suitable combination of the activities of the two institutes in Hamburg, the Seewarte and the observatory. The Royal Observatory in Kiel, which has already been carrying out chronometer examinations for the navy according to the latest principles for almost two years, and had also taken over the function of a chronometer testing institute in the aforementioned sense on an interim basis, will be considerably relieved by the intended institution as soon as it comes into being, so that it will be able to dedicate itself to the interests of the navy in conjunction with the observatory in Wilhelmshaven with regard to the chronometer question.[14]

In the summer of 1876, Rümker visited the Special Loan Collection of Scientific Instruments in the South Kensington Museum in London and was commissioned to take a look at the observatories in Greenwich and Liverpool (Bidston near Birkenhead) that were entrusted with chronometer testing. The experience gained there was to be used in the planning of the Hamburg facility.[15]

Not far from the Hamburg observatory at *Millerntor* a small house of about 3.5 × 5.5 m floor space was located, in which Adolf Repsold (1806–1871) had once set up his dividing engine, which had been acquired by the Hamburg Admiralty. This building was extended in 1876. The ice cellar, 2 m long and 1.86 m wide, had double walls for insulation and a ceiling made of zinc sheeting. Above the ceiling was a space that was filled with ice. This caused the air under the ceiling to cool and sink downwards. A constant temperature of + 5° C could be maintained in summer.

---

[13] "Die Tendenz der Neuzeit ist darauf gerichtet, die Seereisen so viel wie möglich abzukürzen, und dazu haben, abgesehen von den Veränderungen in der Konstruction der Schiffe, auf das Wesentlichste die zuerst von Maury auf möglichst wissenschaftliche Basis begründeten Segelanweisungen beigetragen. Um aber diese praktisch verwerthen zu können, ist durchaus erforderlich, daß dem Seemanne die Mittel gegeben werden den Schiffsort auf See mit möglichster Genauigkeit, so oft sich ihm die Gelegenheit darbietet zu bestimmen, da er erst dann wenn er über letztere beruhigt ist und sicher ist, sich in nicht zu großer Nähe gefährlicher auf den Seekarten verzeichneter Stellen zu befinden, im Stande sein wird die ihm in den Anweisungen erteilten Vorschriften und Rathschläge auch wirklich zu befolgen und seine Reise nach denselben ohne Verzug weiter fortzusetze." Letter to the Mayor of Hamburg, Gustav Heinrich Kirchenpauer, (June 1874); Hamburg, State Archive: Best. 111–1, 12,538, nr. 1.

[14] Anonymous (1875a, pp. 101–105).

[15] Neumayer (1878, pp. 8–9).

The chronometers were placed in lockable zinc boxes on shelves. At the bottom, the boxes had a drawer into which chlorine calcium could be poured to absorb moisture.

From the heating chamber of the building, which was equipped with air heating, a duct ran under the floor into the room for observation at high temperatures. Here there was a chronometer heating apparatus consisting of a box with double sheet metal cladding, the spaces between which were filled with slag wool for insulation. At the top, the box had a lid with double glazing to allow easy access to the chronometers placed on slatted shelves for winding. Under the roof-shaped sheet metal floor was the heating duct, but it proved more practical to place four gas burners here on each side, above which an intermediate plate was mounted. This prevented the parts of the cover plate directly above the flame from overheating. The exhaust air did not reach the upper room with the chronometers, but was discharged through openings on the sides. During the test, the chronometers were compared with uncompensated reference timekeepers, so-called “thermochronometers”. These were chronometer nr. 20 by Gebrüder Eppner and the nr. 108 by Christian Friedrich Tiede (1794–1877),[16] whose original bimetallic compensation balance had been replaced by a simple brass balance in 1878 by Wilhelm Bröcking (1834–1897).

The first competitive trial in accordance with the common procedures of the Royal Observatory at Greenwich and Liverpool Observatory was held in 1877. For the observation of the chronometers in medium temperatures of 15–20 °C, the large room of the testing institute was provided, which also served as a working room for the observatory director and the department assistant. Both here and in the cold room there was a telegraphic connection to the Hamburg observatory, through which time transmission was ensured by means of a chronograph and an electric slave clock. The room also contained escapement models, an air pump to study the behaviour of chronometers at different air pressures, a device to study the influence of magnetism and a work table to enable watchmakers to make any necessary adjustments to the chronometers they had sent in.[17] Furthermore, a specialist library was available and the Institute held the most important watchmaking journals of the time.[18] A packing room located in the basement, a messenger room and a room for the departmental assistant completed the facilities.

Towards the end of the 1880s, however, the institute’s premises proved to be increasingly inadequate, as it was neither possible to examine chronometers at a constant temperature nor to carry out commissioned tests and the annual competitive tests in parallel, as only one heating apparatus was available. A reconstruction of the existing building or a new building seemed urgently needed to adequately accommodate the growing demands on the facility.[19] Moreover, extensive road construction work was planned nearby: a ring road was to be built on the *Stadtwall* from the *Holstentor* to the *Millerntor*, as well as a further road connection to St. Pauli in the immediate vicinity of the testing Institute. Due to these disturbances and expected

[16] Munich, Deutsches Museum: 67,127; 1935 donated by the Seewarte.

[17] Neumayer (1878, pp. 29–31); Gelcich (1892, pp. 177–178).

[18] Neumayer (1878, p. 48).

[19] Jahresbericht über die Tätigkeit der Deutschen Seewarte (hereafter, JBDS), nr 12 (1889).

continual vibrations its relocation had to be initiated. At the same time, the necessary expansion was considered.[20]

Nearby the building of the *Seewarte* at the *Stintfang* (completed in 1881) a new facility with a floor space of 173 square metres was erected in 1892.[21] Here, too, was a cold room with an ice chamber in the cellar as well as a gas-fired heating apparatus for ordinary examinations of chronometers and one for the competitive trials. The latter had a rectangular floor space of 3.5 × 1.3 m and could accommodate up to 60 chronometers. A special room was reserved for the testing of precision pocket watches. Here, another small heating apparatus was set up, and furthermore, in addition to journals on astronomy, hydrography and chronometry, there was a collection of instruments and models,[22] drawings of chronometer compensation devices, a watchmaker's workbench and a chronograph with accessories. Inside the institute building, there was a 5 × 4 m room with double walls for constant temperature and three brickwork pillars, which were isolated from the foundations of the building. Two of these pillars were used to suspend precision pendulum clocks: a pendulum clock with electric winding by the Hamburg clockmaker Carolus Arnold, the other clock (nr. 55) made by Adolph August Kittel (1846–1921) had been purchased in 1889.[23] The third pillar carried a special chamber that was used to examine chronometers both at different atmospheric pressure and altered humidity.[24] The time determinations required for the Institute were still carried out on the meridian instruments of the Hamburg observatory, and there was a wire connection for the transmission of time signals. There was also a connection with the *Seewarte*, so that a pendulum clock by Theodor Knoblich (1827–1892) located there in the basement could be connected to the chronograph. A total of five precision pendulum clocks were used: Tiede nr. 375 (in the basement of the observatory), Kittel nr. 25 (in the telegraph room of the observatory), Kessels nr. 1284 (in the meridian room of the observatory),[25] Knoblich nr. 2090 (in the basement) and Kittel nr. 55 (in the middle room of the chronometer testing institute).

In 1883, a specially constructed machine ("Combe's Apparatus") driven by a gas motor was set up in the atrium of the *Seewarte*. This was used to examine anemometers, but could also be equipped with a rocking device for testing chronometers to study the effect of ship movements (rolling and pitching) on the rate (the device was internally called a "merry-go-round" by the staff).[26] A series of tests conducted in 1883, 1886 and 1903 showed only a relatively slight acceleration of the rate

---

[20] JBDS (1890, pp. 41–42).

[21] Hamburg, State Archive (111–1, 12,539). See Kirchenpauer (1882).

[22] JBDS (1899, pp. 26–32).

[23] Oestmann (2016).

[24] JBDS (1893, pp. 24–26).

[25] Of this clock only the compensating pendulum is extant Kassel, Mathematical-Physical Cabinet; Oestmann (2011, pp. 78–9).

[26] Neumayer (1885, pp. 13–15); Anonymous (1905). It had been built "according to instructions given by the director" by the mechanics W. Ritter in Altona and Frank von Liechtenstein in Hamburg.

during rolling and pitching, which, however, occurred particularly strongly with a combination of both movements.[27]

Since 16 September 1876, a time ball had been in operation on the tower of a harbour warehouse on the *Kaiserkai*, which was operated initially by a push-button and from 1899 by the contact signal of a clock.[28]

The observatory was becoming increasingly busy due to the annual chronometer tests, and Rümker therefore suggested that the testing institute should be assigned to the *Seewarte* in future. This happened after Rümker had to resign from all offices due to illness on April 1, 1899,[29] and Carl Stechert (1856–1925) took over its management. Stechert had studied astronomy, mathematics, physics and philosophy in Berlin since 1878, received his doctorate in Kiel in 1884 and had been an assistant at the *Seewarte* since 1891.[30]

In 1900, an octagonal, double-walled wooden hut with a meridian gap and an inner diameter of 2.41 m was built next to the institute building. This served to house a passage instrument made by Carl Bamberg (nr. 1263). An electrical connection for time transmission was established between the institute building and Arnold's clock, regulated to sidereal time, was hung in the hut. For this purpose, a Strasser & Rohde type A pendulum clock (nr. 219), also regulated for sidereal time, was purchased, which was equipped with a nickel steel pendulum according to Ludwig Strasser's design and an electrical contact device.[31]

From 1909 onwards, a pendulum clock made by Johann Josef Ludwig Nieberg (1804–1884), which had been transferred from the *Seewarte* to the Chronometer Department, served as a registration clock in the daily comparisons.[32] The following year, a Riefler pendulum clock (nr. 223; with airtight tank and pendulum type J1, nr. 731) was added to the inventory, but it was not set up and put into operation until 1911.[33] Improvements or extensions to the heating and cooling equipment for chronometer testing were also made in those years. In 1908, a cooling system and freezer for testing chronometers at low temperatures were installed in a cellar room of the Institute facing north-east.[34] In 1910, three chronometers for the German South Polar Expedition could be tested at temperatures as low as -10° C in this room.[35] At the beginning of 1912, a new six-part chronometer test box for higher temperatures was built and an old, damaged test box was replaced. The test boxes were equipped with an automatic regulating device for gas heating.[36]

---

27 Anon. 1889–1; Rottok (1905).

28 Mahnkopf and Schorr (1924, pp. 476–477); Machoczek (2011).

29 Hamburg, Staatsarchiv 111–1, 12,540.

30 Mahnkopf (1925).

31 JBDS, 23 (1900, pp. 58–60).

32 JBDS, 32 (1909, p. 42).

33 JBDS, 33 (1910, p. 45) and 34 (1911, p. 27).

34 JBDS, 31 (1908, p. 41).

35 JBDS, 33 (1910, p. 45).

36 JBDS, 35 (1912, p. 25–26).

A recurring theme in the annual reports of the *Seewarte* was the low participation of the merchant navy in the chronometer examinations.[37] The situation of the shipping companies was very difficult in the wake of the long period of economic depression from 1873 to 1896 and new expenditure was limited to the bare essentials, which also had a detrimental effect on the manufacture and sale of chronometers. The second cause for complaint was the inadequate keeping of chronometer journals. As early as 1881, the "*unsystematic compilation of observations on the behaviour of chronometers at sea, which had never been tested or had been tested too seldom*" was criticised and a "*strictly scientific journal keeping*" was demanded, as this was the only way to eliminate deficiencies in the design and to give the manufacturers advice for improvements.[38] However, this demand proved to be a "pious hope" two years later.[39] As late as 1903, the compulsory introduction of chronometer journals proposed by the Maritime Professional Association was rejected on the grounds that the "requirements for careful keeping of journals are too high in view of the varied activities and demands made on officers."[40] In 1906, it was finally possible to report that the keeping of chronometer journals on the ships of the merchant navy had "made gratifying progress both with regard to their distribution and their careful keeping."[41]

In the early years, the Chronometer Department was heavily used for examinations of precision timepieces for scientific purposes. In 1882, 41 chronometers were handed over to the Institute by shipmasters of the merchant navy and 60 were handed in by watchmakers and scientific institutes. 20 chronometers would be sold by the manufacturers to shipping companies and scientific institutions mainly through the intermediation of the department, the remaining 40 were for the most part sent in for detailed examination for use on German expeditions in the context of international polar research and for the observation of the transit of Venus in 1882. Furthermore, 5 pendulum clocks were tested on behalf of the Commission for Polar Expeditions and the Brunswick Polytechnic Institute,[42] and in the following year 33 chronometers, 10 precision pocket watches and 2 pendulum clocks were tested on behalf of the German Polar Commission and the Commission for the Observation of the Transit of Venus.[43]

Due to the simultaneous competitive trials in Kiel and Wilhelmshaven, the number of timekeepers delivered to Hamburg in 1885 decreased considerably, and the purchase of chronometers by the merchant navy still left much to be desired. In contrast, the number of precision pocket watches examined had increased considerably since the first trial in 1882, and a larger-scale trial was ordered from the end of 1886.[44]

---

37 JBDS, 1 (1878, p. 134); 2 (1879, p. 76–77); 3 (1880, p. 93).

38 JBDS, 4, (1881, p. 47).

39 JBDS, 6 (1883, p. 43).

40 JBDS, 26 (1903, p. 42).

41 JBDS, 29 (1906, pp. 36–37).

42 JBDS, 5 (1882, pp. 36–37).

43 JBDS, 6 (1883, p. 41).

44 JBDS, 8 (1885, pp. 50–51).

In 1887, a modification of the test specification was introduced. This had been proposed six years earlier by Carl Börgen, according to which temperature compensation, rate jumps and rate acceleration should be represented separately and used for quality assessment. Börgen's procedure was preferred to the representation of the rate by a derived formula, as favoured by C. F. W. Peters. From the thirteenth competitive trial (1889/90) onwards, the chronometers submitted were divided into four quality classes.

The Chronometer Department continued to be used only to a limited extent by shipmasters and shipowners, however.[45] Apparently, the construction of the new free port after the accession of Hamburg to the customs union (*Zollanschluß*) also had a negative effect: the port facilities were now located at a greater distance, and the transport of chronometers became more cumbersome.[46] In 1896, a number of larger shipping companies (Hamburg-Amerika-Linie, Laeisz, and Jebsen) handed over the chronometers of their ships that had been in port for a longer period of time for examination,[47] but the number decreased again the following year. The "water clerks" employed by chronometer manufacturers went on board immediately after the arrival of a ship to take over the chronometers, thus avoiding transport to the *Seewarte*. This practice could only be countered by providing a motor launch to facilitate transportation and to convince shipowners of the useful service provided by the *Seewarte*.[48]

In the period from 1876 to 1891, 489 chronometers from captains of the merchant navy, 391 from watchmakers, scientific institutes and explorers, 162 pocket watches (outside the regular examinations) and 26 pendulum clocks were tested in the Chronometer Department of the *Seewarte*, and 18 chronometers were observed on behalf of the Imperial Navy.[49]

Nothing remains of the *Seewarte*; the building was destroyed in the 206th air raid on Hamburg on 8 April 1945 between 22:23 and 22:44 p.m. and listed as "totally burnt out."[50]

Kiel Chronometer Observatory

In 1873, the Imperial Admiralty entered into negotiations with the Prussian Ministry of Spiritual, Educational and Medical Affairs on the suggestion by Georg von Neumayer to take the chronometers that had been stored at the Imperial Dockyard into the Kiel Observatory, which was still under construction. The astronomer Carl Friedrich Wilhelm Peters (1844–1894) was to check the clocks and watches and

---

[45] JBDS, 12 (1889, p. 39, 18; 1895, p. 25).

[46] JBDS, 15 (1892, p. 36, p.18; 1895, p. 25).

[47] JBDS, 19 (1896, pp. 26–27).

[48] JBDS, 10 (1897, pp. 36–37).

[49] JBDS (1891, p. 31).

[50] Hamburg, Staatsarchiv: 135–2 II Statistisches Landesamt II, Nr. 2354, 5.

ensure that they were stored appropriately. On 3 October 1873, 28 chronometers were transferred from the Imperial Dockyard to the observatory and checked. It turned out that only very few chronometers were faultless and temperature compensation of almost all was insufficient. In order to obtain replacements for bad chronometers and to be able to expand the stock if necessary, chronometer makers were invited to submit timekeepers. In 1874, 67 chronometers were submitted for examination, of which 42 belonged to the Imperial Navy. 25 chronometers had been submitted by Nieberg, Knoblich, Tiede and Eppner.[51] In the following year, there were already 50 chronometers belonging to the navy at the observatory, and 41 watches submitted for examination were added. In 1878, the Imperial Dockyard also handed over the 41 existing pocket chronometers to the Kiel Observatory for testing and storage.[52] However, their facilities for chronometer testing soon proved to be inadequate. There was an ice house the temperature of which could be kept at 5 °C, but in the warming room there was only a simple oven for heating, with which, of course, no defined temperature conditions could be established. The examination room for medium temperatures was exposed to considerable daily heat fluctuations due to solar radiation.[53] After the observatory had applied for the construction of a building with an improved heating system as early as 1874 and this application was repeated five years later, the Imperial Admiralty approved the construction of a new chronometer observatory in 1880, in which up to 120 chronometers could be tested.

A site was chosen on the grounds of the former Düsternbrook shipyard, southeast of the former Naval Academy (today the Reventlou pier is located there). This site was favourable because there was no interference from vibrations or magnetic influences to be feared and the building could face north while enjoying good light conditions to facilitate the temporary generation of low temperatures. The three-storey, flat-roofed brick building, completed at the end of 1883, had a square ground plan with a side length of 12.24 m and was executed in neo-Renaissance style. A round tower with a wooden superstructure was attached to the north-eastern corner.

In the basement there was an air heating system for the building, a central hot-water heating system for the chronometers' heat boxes and a workshop room for setting up scientific experiments and for minor repairs to the chronometers. The ground floor consisted of four rooms: a chronometer observation room, the director's study, a keeper's room and a room for receiving or delivering timekeepers.

In the chronometer observation room there were two heat boxes connected to the hot water heating system, each of which could hold 48 chronometers. The temperature could be regulated as desired by means of air shafts and flaps leading to the outside, and the chronometers could be easily read through the with glass windows in the lids. Shelves attached to the walls held the chronometers for testing at normal temperature. A maximum of 129 chronometers could be examined at the observatory. The "Great Test" took place between the end of October and the beginning of March and lasted 120 days. The temperatures in the chronometer boxes were lowered from 30° C to 5°

[51] Rottok (1900, p. 790).

[52] Peters (1875, pp. 343–344).

[53] Rottok (1900, p. 791).

C in ten-day intervals. This lowest value was then held for 20 days, increased to 30° in decades and finally lowered again to room temperature. There was no ice cellar, so the temperature reduction was dependent on the weather—after all, the minimum of the test cycle fell at the end of December. Occasionally, however, the required lowest temperature of 5° could not be reached because of mild weather.[54]

A pier insulated from the masonry of the building ran through all the floors in the tower and ended in two separate piers at the top of the observation room. On one of these pillars a universal instrument could be set up, the other served as a foundation for a passage instrument by Carl Bamberg in Berlin (1847–1892) with which time determinations were made to check the observatory's two precision pendulum clocks. The master clock had been made by Ferdinand Dencker (1837–1917), and the comparison clock by the Kiel chronometer maker Johann Christoph Schnoor (1837–1902).[55] The latter had a mercury contact for transmitting the second beats to a chronograph. A second stylus of the chronograph was actuated by a Morse key, which the observer pressed synchronously with the chronometer to be tested. In this way, the second signals of the comparison clock and chronometer could be conveniently compared on the paper strip. This "chronographic comparison mode" was supposed to be accurate to 1/100 s with some practice. Also, the passage instrument was electrically connected to the chronograph.

In the director's room was another isolated masonry pillar for examining sextants and other angle-measuring instruments.

To transmit the midday time signal to the ships moored in the harbour, a small, manually operated time ball had initially been mounted on the observatory. However, this device had the disadvantage of poor visibility, so that in 1883 it was decided to erect a time ball at the Imperial Dockyard, which was to be released electrically from the observatory. As a transitional measure, a 13 m high flagpole was erected at the chronometer observatory, to which a time ball with a diameter of 1 m and a drop height of 3.5 m could be attached and released manually by means of a wire hoist. On the roof of the artillery magazine of the Imperial Dockyard on the eastern shore of the *Kieler Förde*, the new system was then put into operation on 24 January 1886. At a height of 30.7 m above the ground, the time ball was now clearly visible. The electrical supply line was 6.2 km long and was laid over land, as an underwater cable would have been too expensive. The time ball was probably executed by Carl Bamberg, who had already built similar installations in Neufahrwasser, Hamburg, Cuxhaven and Bremerhaven.[56]The daily manual release process is described in detail in a contemporary article in the *Illustrirte Zeitung* from 1890.[57] With the fall of the time ball at 12:00 mean Kiel local time, a gun was fired on a guard ship in the harbour, a process popularly known as the "soup shot" (Fig. 10.2).

C. F. W. Peters was in charge of the *Seewarte* until 11 April 1888. In this year he was appointed as director of the observatory in Königsberg and his successor became

[54] Freiburg im Breisgau, Bundesarchiv/Militärarchiv: RM 3/10774, fol. 119v.

[55] Freiburg im Breisgau, Bundesarchiv/Militärarchiv: RM 3/10772, fol. 220r.

[56] Loewenherz (1880, pp. 40–50).

[57] Lühning (2007, pp. 428–430).

Fig. 10.2 The firing of the noon cannon in the harbour of Kiel. Top, left to right, the chronometer observatory outside and inside, and the electrically released time ball on the Imperial Shipyard. Xylograph in *Leipziger Illustrirte Zeitung*, 1890, after a drawing by Ferdinand Lindner (1847–1906). Original in possession of the author

Corvette Captain Hans Nees v. Esenbeck.[58] In 1899 Corvette Captain Carl Rottok (1854–1931), who joined the Imperial Navy in 1871 and had attended the Kiel Naval Academy in 1887/88, took over. He had been a teacher at the School for Midshipmen (*Deckoffiziersschule*).[59]

In addition to chronometer examinations, the activities of the Kiel Chronometer Observatory extended to the storage and administration of all the chronometers of the Baltic Naval Station (including the school and observation chronometers), time determinations and time ball service, the revision of the chronometer journals delivered by the ships of the Baltic Naval Station and the solution of scientific questions related to clocks and chronometers and aimed in particular at their improvement.

Due to the extensive destruction of Kiel during the Second World War, no remains of either the chronometer observatory or the time ball installation have survived. It is also not known whether or how long the observatory was in operation after the First World War.

---

[58] Biographical data and information on his military career could not be ascertained.

[59] Data according to information from the *Deutsche Dienststelle (WASt) für die Benachrichtigung der nächsten Angehörigen von Gefallenen der deutschen Wehrmacht*, Berlin, 12 June 2012.

Wilhelmshaven Naval Observatory

The Wilhelmshaven Naval Observatory was established in the urban district of Alt-Heppens in 1874–78. In contrast to the institution in Kiel, which only served to test chronometers, its tasks were laid out more broadly.

Neumayer played a major role in the founding and establishment; together with his friend Carl Nicolay Jensen Börgen (1843–1909), he made careful magnetic measurements at several locations around the town and selected the most suitable site for the building. The review of the construction plans for the new institute as well as the selection and testing of the instruments was almost entirely in Neumayer's hands, as Börgen was then frequently occupied by his participation in the Venus expedition.[60]

Although chronometer testing and timekeeping were part of its duties, geomagnetic and meteorological observations as well as tidal research and forecasting were added.[61] In 1879, a tide table was compiled for the North Sea, and the following year for all European coasts.

Börgen was appointed director of the observatory. After studying physics and astronomy in Copenhagen, Kiel and Göttingen he had been an assistant at the observatories in Göttingen and Leipzig and taken part in the Second German Arctic Expedition under Carl Koldewey (1837–1908) in 1869/70. In 1874–76 he led the German expedition to the Kerguelen Islands on SMS *Gazelle* to observe the passage of Venus. During his long 34 years of service, Börgen built up the geomagnetic observation service, time and tide service. After his death, Corvette Captain Hans Capelle (1864–1948) was appointed as his successor in 1909, who later (from 1919 to 26) headed the *Deutsche Seewarte.*

The two-winged building of the observatory in Wilhelmshaven had two observation towers, and on one of those towers the time ball installation was mounted. This was put into operation on 21 July 1878. The time ball fell at 12:00 mean local time and a second time at 12:32 35 s to indicate London local time. After the legal abolition of local times and the general introduction of mean solar time of the 15th meridian east of Greenwich in 1893, the time ball was released at 12:00 and at noon mean solar time of the Greenwich meridian at 13:00. On 21 March 1910, after a three-year trial period, the Norddeich coastal radio station began regular transmission of radiotelegraphic time signals controlled by the Naval Observatory. In the park grounds of the institute was an iron-free building for geomagnetic investigations.

The first inspection report published by Börgen shows that there was neither a heating apparatus nor an ice cellar. The chronometer room was heated to 30° by means of the existing hot-water heating system, and at the turn of the year one hoped (as in Kiel) for correspondingly low temperatures.[62] Depending on the meteorological conditions, it was sometimes not possible to comply with the conditions required by the examination regulations. For example, Börgen noted in 1888:

---

60 Stechert (1910, p. 30).

61 Werner (1894, pp. 83–84).

62 Börgen (1880, pp. 448–449).

> It was not quite possible to obtain the desired lowest temperature of 5°, because the temperature of the air outside, which was let into the room by opening the windows, did not drop low enough to make it possible on the days when the temperature of 5° was needed.[63]

In Kiel, Wilhelmshaven and Hamburg, a large number of special investigations were undertaken into the behaviour of chronometers under various conditions. Carl Rottok analysed 823 of the chronometer journals kept on ships of the Imperial Navy since 1874 and investigated whether shocks, vibrations or the firing of guns could cause chronometer malfunctions. It turned out that in most cases no rate changes had occurred, and if they coincided with shocks or vibrations, these could not be proven with certainty to be the cause.[64] Rottok also investigated the influence of atmospheric pressure,[65] the behaviour of chronometers during transport over land and in different positions, and the influence of the type of compensation on the quality class of the chronometers.[66]

C. F. W. Peters studied in detail the sudden changes in the rate of chronometers. After ruling out the possibility of rough transport and magnetism, Peters believed that the change in humidity was the cause: if a chronometer was brought from the dry testing room onto the ship, water vapour could condense on the surface of the balance spring. The resulting slight mass increment increased the moment of inertia and thus slowed down the rate.[67] Peters thought he could quantify the phenomenon by means of corresponding humidity coefficients and advocated the construction of airtight chronometer cases. At the Hamburg Trade and Industry Exhibition and the Berlin Trade Exhibition in 1889 and 1896, Ferdinand Dencker exhibited such cases constructed in collaboration with Peters.[68] However, tests carried out at the *Seewarte* in 1888 showed that constant behaviour of chronometers in relation to the degree of humidity of the surrounding air could not be proved.[69]

Until 1919, the Naval Observatory was an independent authority directly subordinate to the Reichsmarineamt and was then affiliated with the Wilhelmshaven Naval Shipyard.[70] In spring 1941, the observatory was moved to Greifswald, and the files stored at the naval shipyard of Wilhelmshaven fell victim to an air raid. The observatory building was largely destroyed and the ruins demolished on 1 February 1955 (Figs. 10.3, 10.4 and 10.5).

---

63 "Es gelang nicht ganz die gewünschte niedrigste Temperatur von 5° zu erhalten, weil die Temperatur der Luft im Freien, welche durch Oeffnen der Fenster in das Zimmer hinein gelassen wurde gerade in den Tagen, wo die Temperatur 5° gebraucht wurde, nicht so tief sank, um diese zu ermöglichen." Freiburg im Breisgau, Bundesarchiv/Militärarchiv: RM 3/10774, fol. 142r.

64 Rottok (1905).

65 Rottok (1905).

66 Rottok (1910).

67 Peters (1886, 1887).

68 Voller (1889, p. 370); Gohlke (1896, p. 347).

69 Anonymous (1889b).

70 Festschrift Marinewerft Wilhelmshaven (1931, pp. 96–98).

Fig. 10.3 Kiel Chronometer Observatory, not far from the Naval Academy (foreground right). Contemporary postcard; photograph by Arthur Renard (?) Kiel, Schleswig-Holsteinische Landesbibliothek

Fig. 10.4 Wilhelmshaven Naval Observatory (Postcard, c. 1900; Wilhelmshaven, Stadtarchiv, Best. 5882)

Fig. 10.5 Time ball on the *Kaiserkai* in Hamburg (postcard, c. 1900; in possession of the author)

References

Anonymous. (1875a). Die deutsche Seewarte in Hamburg. *Annalen der Hydrographie und Maritimen Meteorologie, 3*, 101–105.

Anonymous. (1875b). Das Hydrographische Bureau der Kaiserlichen Admiralität. *Annalen der Hydrographie und Maritimen Meteorologie, 3*, 145–150.

Anonymous. (1889a). Bericht über die im Lichthofe der Deutschen Seewarte in der Zeit vom 16. Mai bis 24. Juli 1886 mit dem Chronometer-Schaukel-Apparate ausgeführten Untersuchungen zur Ermittelung des Verhaltens von Marine-Chronometern auf bewegter Unterlage. *Annalen der Hydrographie und Maritimen Meteorologie, 17*, 225–235.

Anonymous. (1889b). Bericht über die im Lichthofe der Deutschen Seewarte in der Zeit vom 26. Januar bis 4. Juni 1888 ausgeführten Untersuchungen zur Ermittelung des Verhaltens von Marine-Chronometern bei verschiedenen Feuchtigkeitsgraden der atmosphärischen Luft. *Annalen der Hydrographie und Maritimen Meteorologie, 17*, 107–118.

Anonymous. (1905). Der Combesche Apparat in seiner Anwendung als Chronometer-Schaukelapparat. *Deutsche Uhrmacher-Zeitung, 29*, 120–121.

Anonymous. (1930). Zum 80. Geburtstage des Konteradmirals a. D. und Direktors a. D. der Deutschen Seewarte Alfred Herz. *Annalen der Hydrographie und Maritimen Meteorologie, 58*, 161–162.

Berlin, Geheimes Staatsarchiv Preußischer Kulturbesitz: I. HA Rep. 89.

Börgen, C. (1880). Die Chronometer-Prüfungen auf dem Observatorium zu Wilhelmshaven in den Jahren 1878–79 und 1879–80. *Annalen der Hydrographie und Maritimen Meteorologie, 8*, 448–455.

Festschrift Hydrographisches Bureau. (1911). *50 Jahre vom Hydrographischen Bureau des Königlich Preußischen Marine-Ministeriums zum Nautischen Departement des Reichs-Marine-Amts 1861–1911*. E. S. Mittler & Sohn.

Festschrift Marinewerft Wilhelmshaven. (1931). *75 Jahre Marinewerft Wilhelmshaven 1856—25. Juni 1931*. Stalling.

Foerster, W. J. (1911). *Lebenserinnerungen und Lebenshoffnungen*. Georg Reimer.

Freiburg i. Br., Bundesarchiv/Militärarchiv: RM 1/1800; RM 3/10774.

Gelcich, E. (1892). *Geschichte der Uhrmacherkunst von den ältesten Zeiten bis auf unsere Tage* (5th ed.). Bernhard Friedrich Voigt.

Gohlke, E. (1896). Die Berliner Gewerbe-Ausstellung. *Allgemeines Journal der Uhrmacherkunst, 21*, 211–212, 234–237, 258–259, 280–283, 346–348 (continued in supplement I), 390–392.

Hamburg, Staatsarchiv: 111–1, 12539; 12540; 135–2 II Statistisches Landesamt II, Nr. 2354.

Heidke, P. (1926). Neumayer als Deutscher und Gelehrter. *Zum hundertsten Geburtstage des Gründers der Deutschen Seewarte Georg von Neumayer* (= Beilage zum Juni-Heft der *Annalen der Hydrographie und Maritimen Meteorologie,* 54) (pp. 1–17)

Horn, W. (1972). Die Anfänge der Deutschen Seewarte: Zur 150. Wiederkehr des Geburtstages von Wilhelm Ihno Adolf v. Freeden. *Zeitschrift des Vereins Für Hamburgische Geschichte, 58*, 64–69.

JBDS. (1878–1912). *Jahresbericht über die Tätigkeit der Deutschen Seewarte.*

Kirchenpauer, A. (1882). Das neue Dienstgebäude der deutschen Seewarte in Hamburg. *Centralblatt der Bauverwaltung, 2*(62–63), 70.

Loewenherz, L. (Ed.). (1880). *Bericht über die Wissenschaftlichen Instrumente auf der Berliner Gewerbeausstellung im Jahre 1879*. Springer.

Lühning, F. (2007). *"...Eine ausnehmende Zierde und Vortheil": Geschichte der Kieler Universitätssternwarte und ihrer Vorgängerinnen 1770–1850. Zwei Jahrhunderte Arbeit und Forschung zwischen Grenzen und Möglichkeiten*. Wachholtz.

Machoczek, D. (2011). Der Hamburger Zeitball—ein Zeitsignal für die Schifffahrt. In G. Wolfschmidt (Eds.), *Hamburgs Geschichte einmal anders—Entwicklung der Naturwissenschaften, Medizin und Technik,* tome 3 (pp. 78–89). Tredition.

Mahnkopf, H. (1925). Carl Stechert†. *Astronomische Nachrichten, 225*, 143–144.

Mahnkopf, H., & Schorr, R. (1924). Die deutsche Seewarte zu Hamburg. *Deutsche Uhrmacher-Zeitung, 48*(456–457), 476–480.

Neumayer, G. (1878). Erster Jahres-Bericht über Organisation und Thätigkeit der Deutschen Seewarte umfassend den Zeitraum vom 1. Januar 1875 bis Schluss des Jahres 1878 (= Aus dem Archiv der Deutschen Seewarte, 1, 1878, Nr. 1). Hammerich & Lesser.

Neumayer, G. (1885). Die Deutsche Seewarte: I. Beschreibung der Zentralstelle in Hamburg (= Aus dem Archiv der Deutschen Seewarte, 7, 1884, Nr. 2). Hammerich & Lesser.

Oestmann, G. (2011). Heinrich Johann Kessels (1781–1849): Ein bedeutender Verfertiger von Chronometern und Präzisionspendeluhren. *Biographische Skizze und Werkverzeichnis* (= Acta Historica Astronomiae, 44). Harri Deutsch.

Oestmann, G. (2012). *Auf dem Weg zum "Deutschen Chronometer": Die Einführung von Präzisionszeitmessern bei der deutschen Handels- und Kriegsmarine bis zum Ersten Weltkrieg* (= Deutsche Maritime Studien, 21). Hauschild.

Oestmann, G. (2014). Towards the "German chronometer": The Introduction of Precision Timekeeping in the German Mercantile Marine and Imperial Navy in the Nineteenth Century. *Antiquarian Horology and the Proceedings of the Antiquarian Horological Society, 35*, 949–963.

Oestmann, G. (2016). "Von jeher ein Querulant": Zum Leben und Werk des Altonaer Chronometermachers Adolph August Kittel (1846–1921). *Klassik Uhren, 5*, 10–19.

Peters, C. F. W. (1875). Die Chronometer-Beobachtungen auf der Sternwarte Kiel und die bisher aus ihnen gewonnenen Resultate. *Annalen der Hydrographie und Maritimen Meteorologie, 3*, 343–348.

Peters, C. F. W. (1886). Untersuchungen über den Einfluß der Feuchtigkeit der Luft auf den Gang der Chronometer. *Annalen der Hydrographie und Maritimen Meteorologie, 14*, 505–512.

Peters, C. F. W. (1887). Ueber den Einfluss der Luftfeuchtigkeit auf den Gang der Chronometer. *Vierteljahrsschrift der Astronomischen Gesellschaft, 22*, 284–292.

Rottok, C. (1900). *Das Chronometer-Observatorium in Kiel. Marine-Rundschau, 11*, 789–798.

Rottok, C. (1905). Haben Stöße, Erschütterungen, welche das Schiff erleidet, oder starke Schiffsbewegungen Einfluß auf den Chronometergang? *Annalen der Hydrographie und Maritimen Meteorologie, 33*, 212–216.

Rottok, C. (1910). Verhalten der Chronometer mit der Zeit in Bezug auf Güte; welchen Einfluß hat die Art der Kompensation auf die Güte der Chronometer? *Annalen der Hydrographie und Maritimen Meteorologie, 38*, 297–304

Schorr, R. (1900). Todesanzeige [George Friedrich Wilhelm Rümker]. *Astronomische Nachrichten, 152*, 127–128.

Sieg, D. (2005). *Die Ära Stosch: Die Marine im Spannungsfeld der deutschen Politik 1872 bis 1883.* Dieter Winkler.

Stechert, C. (1910). Georg von Neumayer. *Vierteljahrsschrift der Astronomischen Gesellschaft, 45*, 10–42.

Voller, C. A. (1889). Die Feinmechanik auf der Hamburgischen Gewerbe- und Industrie-Ausstellung. *Zeitschrift Für Instrumentenkunde, 9*, 354–372.

Walle, H. (1979). *Der Einfluß meteorologischer Navigation auf die Entwicklung der deutschen transozeanischen Segelschiffahrt von 1868 bis 1914 (PhD thesis, Rheinische Friedrich-Wilhelms-Universität Bonn).* [Reprinted in 2009 with the title: *Rahsegler in Deutschland: Von der Seewarte zur Gorch Fock.* Dieter Winkler].

Werner, B. V. (1894). *Die Kriegsmarine, ihr Personal und ihre Organisation.* Friedrich.

Winterhalter, A. G. (1889). *The International Astrophotographic Congress and a Visit to Certain European Observatories and Other Institutions (= United States Naval Observatory: Washington Observations for 1885.* Government Printing Office.

Wislicenus, G. (1909). Georg v. Neumayers Wirken für die deutsche Marine. *Marine-Rundschau, 20*(2), 840–844.

Günther Oestmann (b. 1959) was trained as a watchmaker and studied art history and the history of science at Tübingen and Hamburg. In 1992 he received a PhD with a study on the astronomical and astrological significance of the clock in Strasbourg Cathedral (awarded with the Philipp-Matthäus-Hahn-Prize in 1993) and in 2001 he completed a postdoctoral thesis (Habilitation) on Heinrich Rantzau and his attitude towards astrology. In 2002 he received a master craftsman's diploma from the Federal Technical College for Clock- and Watchmakers in Karlstein, Austria. From 2006–08 he worked as a curator of the Maritime Collection at the Deutsches Museum in Munich and since 2010 he is lecturer at Technical University, Berlin. In 2013 the Musée international d'horlogerie (La Chaux-de-Fonds) awarded him the "Prix Gaïa," and in 2014 he was elected as corresponding member of the International Academy of the History of Science (Paris). Three years later he was appointed as extraordinary professor of history of science at Technical University Berlin.

Part IV
Stories of Instruments: Acquisition, Use, and Impact

Chapter 11
On the Accuracy of Celestial and Terrestrial Clocks: The Merit of Felix Schmeidler's Half-Century Monitoring of Two Riefler Clocks at Munich Observatory

Christiaan Sterken

Abstract We present a preliminary analysis of the recorded differences between clock time and radio-signal time, along with the temperature data that were recorded by Felix Schmeidler during almost half a century. We furthermore deal with the impact of astronomical clock readings on the interpretation of the wavering frequencies of specific types of pulsating variable stars. Considering that almost no observatories in the world had access to Riefler time standards covering five decades or more, and that many observatories worked until the early 1980s with time standards that were even orders of magnitude lower than those of state-of-the-art mechanical clocks, we outline the merit of some interpretations put forward to explain cyclic and quasi-cyclic behavior of pulsating stars in terms of the presence of one or even multiple companion stars.

The Sound of Light

Stars are born, live and die: their life cycles are of the order of millions to billions of years. During their lifespan, stars appear to be in quasi-equilibrium, but some events and processes may provoke sudden changes that can even lead to explosive cataclysms. Some properties of gaseous spheres are not unlike those of musical instruments: stars also possess natural frequencies as well as overtones. Physical processes bring forth sound waves in stellar interiors, and such waves make the star resonate as if it were a violin, a kettledrum or an organ pipe. Since the speed of sound depends on the temperature and on the density of the ambient gas, wave fronts become curved, and the propagating sound waves reflect at the outer boundaries. This causes surface gravity waves (like the ripples we can see on the surface of a pond) and

C. Sterken (✉)
Vrije Universiteit Brussel, Pleinlaan 2, 1050 Brussels, Belgium
e-mail: Christiaan.Sterken@vub.be

G. Bernasconi et al. (eds.), *Astronomical Observatories and Chronometry, 18th-20th Century*, History of Physics, https://doi.org/10.1007/978-3-031-95874-8_11

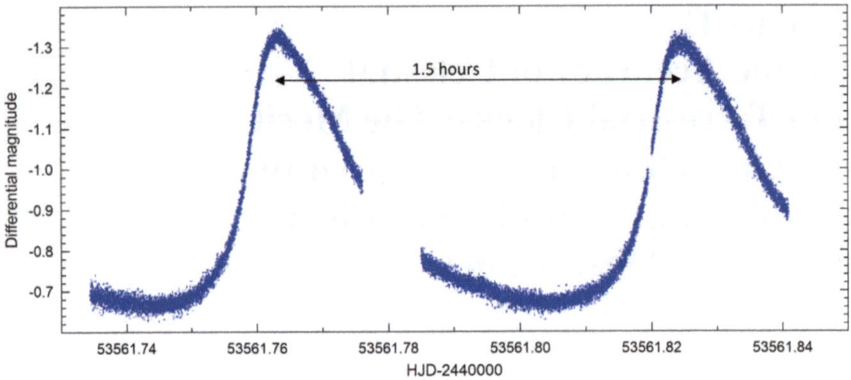

Fig. 11.1 The high time-resolution (dots are 5 s apart) light curve of CY Aquarii. The length of the pulsation period is indicated. Adapted from Tuvikene et al. (2010)

acoustic waves in the stellar interior. Just as in a closed organ pipe—i.e., a column open at one end and closed at the other—waves in a gaseous sphere reverberate and manifest themselves as Chladni-type patterns formed by the node lines along which there are no pressure changes.

Some stars possess a mechanism that generates and maintains steady standing waves inside their interior, and that mechanism makes the star expand and contract, and renders the star cyclically brighter and fainter. The frequency of such stellar pulsations depends on the size and on the density (thus mass) of the star, whereby the brightest stars have the longest vibration periods. The periods of such oscillations range from minutes to years to decades, and the frequencies of these "cosmic sounds" fall entirely outside the audible frequency range of the human ear. Some of these variations have been observed over several centuries as cyclic changes in the light output, like the one illustrated in Fig. 11.1.

Asteroseismologists measure the light (and/or velocity) of pulsational phenomena, and record the time of observation. Both observables, time and the selected aspect of light, underwent substantial improvements in accuracy and precision since the middle of the nineteenth century. However, the importance of accurate timings is often overlooked: whereas errors of even a minute or longer would barely hamper the outcome of an analysis of, for example, χ Cygni (a 450-day pulsator for which data have been available since 1686), timing errors become increasingly troublesome when studying short-period pulsators with cycles of hours or minutes (and even less) over time spans of decades.

From such light curves so-called *timings* can be extracted, i.e., times of maximum light or times of minimum light. To each such timing, a cycle number E is allocated, and assuming that the cycle numbers have been correctly assigned, the corresponding time series leads to the derivation of the period by simple linear regression, as illustrated in Fig. 11.2. That relationship looks very tight, but our eye is misled because for such a short-period star (6000 cycles per year), the E-number range becomes very large. A perfect correlation would result in small deviations from the fit, with a

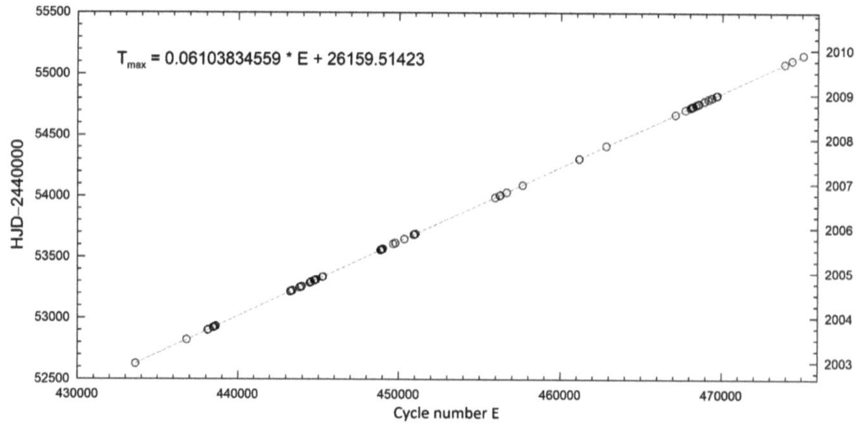

Fig. 11.2 Ephemeris (2003–2010) for CY Aquarii: Heliocentric Julian Date as a function of cycle number *E*

random pattern—a so-called white noise. This can only happen when the pulsating star behaves like a perfect pendulum clock, and when the observational data, *i.e.*, the timings, have modest gaussian errors.

What happens in practice is very different, though. First and foremost, just as a pendulum is subjected to outside forces like draughts from a fan or vibrations by street traffic, not one single pulsating star behaves like a perfect pendulum: there are internal processes as well as external influences (such as binarity) that interfere with the pendulum's steady swing. And, there are the observational uncertainties, which are of various kinds:

1. Photon statistics, visible as scatter in Fig. 11.1, which is entirely caused by the properties of the detector (electronic device, photographic emulsion, or even the human eye). If the object had been, say, ten times brighter, the scatter would become a factor of about three less.
2. The time resolution of the data: one could easily combine the dots in Fig. 11.1 into bins of 30–40 s and obtain a smoother curve, but then the definition, especially of the time of maximum, would become much less reliable.
3. The accuracy of the time standard that has been used to record the data: while this is no longer a problem today since atomic clock times are available over the internet, this problem can be huge when even going a couple of decades back in time.

It is the uncertainties of the third kind that define the realm of this contribution.

Felix Schmeidler (1920–2008) and the Riefler Pendulum Clocks

More than a century ago, Munich Observatory acquired two Riefler pendulum clocks, viz., R23 (1894) and R33 (1910). These clocks were state of the art in accuracy, though they differed from other pendulum clocks in three essential ways: the implementation of a compensated pendulum, the mechanism of the motive apparatus, and the arrangement for constant atmospheric pressure due to the entire clock being mounted inside a sealed cylindrical glass case.

Dr. Felix Schmeidler (1920–2008) came as a student to the Munich Bogenhausen Observatory in 1938. After graduating in 1941, he was hired as an assistant-calculator, and started his lifelong observing project of measuring absolute declinations of fundamental stars at the vertical circle (the *Askania Vertikalkreis* delivered to the Munich Observatory in 1928, see Figs. 11.3 and 11.4. From 1958 till shortly before his death, he carried out meticulous measurements with that instrument while recording,[1] on a weekly basis, the difference between both Riefler-clock times and the time signal of Radio Munich. As one of the most remarkable staff members at the observatory, he tackled the unsolved problem of the systematic differences of 0.2 to 0.4 arcseconds in declination between the fundamental catalogues of the northern and southern hemisphere. His approach was to measure.

1. the same stars,
2. at similar latitudes north and south of the equator,
3. with the same meridian instrument,
4. by the same observer, and
5. using exactly the same data reduction procedures.

To this end, Schmeidler organized in 1952 the relocation of the Bogenhausen *Vertical Circle* to Mt. Stromlo near Canberra in Australia. In the course of more than 15 months, he worked off a list of 1,630 stars in 339 nights.[2] After the instrument was reinstalled at Bogenhausen, Schmeidler measured the same stars over again during an entire year—again the same observer using exactly the same data reduction procedures.

After tedious data reduction it turned out that the systematic differences between the declinations of the northern and southern fundamental stars had vanished. Schmeidler went on observing as long as he could, and as such he spent 66 years of his life determining the exact position of a selection of stars.

The by-product of his labour is a log of about 5,000 entries representing the Riefler clock times versus the time signals broadcast by the *Bayerischer Rundfunk*.[3] Schmeidler's recordings are very valuable for documenting and for understanding the accuracy of recorded observing times at observatories during the last century.

---

[1] http://www.usm.lmu.de/Geschichte/Klassik/06_Schoenberg/01_Astro/index.php (last accessed April 30, 2023).

[2] Wiedemair (2020).

[3] Schneider (1963).

Fig. 11.3 Munich observatory, Bogenhausen. Schmeidler's 7-zoll (18-cm) Vertical Circle (*Askania Vertikalkreis,* 1927). Photo C. Sterken

Figure 11.5 shows the daily clock rate of R23 as a function of time, and the corresponding ambient temperature. The graph shows that the daily clock rate of R23 was quite stable since about 1975, and reveals that the room temperature fluctuations dramatically decreased after 1968, and even became very stable during the last ten

Fig. 11.4 Munich observatory, Bogenhausen. Riefler pendulum clock R23, delivered in 1894 to the Königliche Sternwarte München. Photo C. Sterken

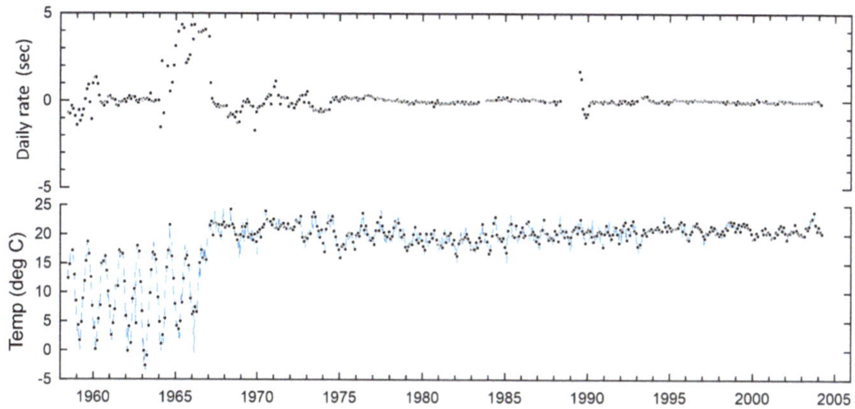

Fig. 11.5 Top: Daily rate of R23 as a function of time. The daily rate data have been truncated at 5 s. Bottom: Atmospheric pressure and room temperature as a function of time in the clock room. Similar data are also available for R33, though less frequent

years of the recording. Atmospheric pressure also improved a lot from 1975 on, but left much to be desired in terms of stability.

The Real Problem of CY Aquarii

Figure 11.6 shows the residuals *O–C* (observed *minus* calculated[4]) of CY Aquarii over about 80 years. The shape has a pattern that suggests cyclicity, as indicated by the model curve (covering about 75 years) for a pulsating star in a binary system, where the observed time of arrival of the light depends on the position of the star in the binary orbit (the so-called *light-time effect*). The graph clearly shows that the scatter around the mean *O–C* curve seems to have decreased from about 1983 on. Part of this decrease is due to improvement in detector performance: older data are visual estimates by eye, or photographic magnitudes, later recordings were made with photoelectric detectors (photomultipliers and charge-coupled devices).

A Systematic Approach

Since 2010 we have attempted to pursue a systematic approach by freezing observing procedures, and by implementing exactly the same data reduction procedures carried out by one and the same person. We used the instrumental setup at the *Cusanusgymnasium* in Bruneck (Italy), where its *Astrogruppe* built a small observatory on a dark

[4] Sterken (2005).

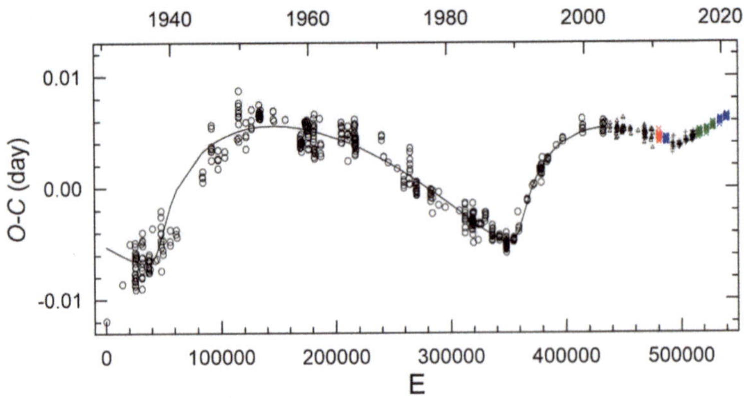

Fig. 11.6 Residuals *O–C* (observed *minus* calculated) for all data collected over about 80 years. The model curve covers about 75 years and stands for a pulsating star in a binary system in which the observed time of arrival of the light depends on the position of the star in the binary orbit[5,6]

site in the mountains near Bruneck (Figs. 11.7 and 11.8).[7] The team is led by Christof Wiedemair, teacher of mathematics and physics at the gymnasium.

The first condition was not met since the optical system was upgraded, the second condition was, unfortunately, also abandoned, though the reduction process followed a very similar workflow. The result can be seen in Fig. 11.6: not only has the scatter during the last decade become absolutely minimal, even the expected large-scale fluctuation in the *O–C* curve has not come true. The latter result suggests that the data that led to the fitted model could as well be represented by a series of linear stretches, representing sudden period jumps (that is, except for the observational scatter). Another decade of monitoring is needed to find out which of the models is right.

Conclusions

Felix Schmeidler's lifelong pursuit of clock monitoring provides us with an unequalled view of attainable clock accuracy in an active professional astronomical observatory setting. Considering that almost no observatories in the world had access to Riefler time standards covering five decades or more, and that many observatories worked until the early 1980s with time standards that were even orders of magnitude lower than those of state-of-the-art mechanical clocks, the merit of some of the interpretations put forward to explain the cyclic and quasi-cyclic behaviour of

[5] Fu and Sterken (2003).

[6] See footnote (1).

[7] See https://www.cusanus-gymnasium.it/bildungsangebot/wahlangebot/astrogruppe/ (accessed on 16 October 2024).

Fig. 11.7 The very first 25-cm reflecting telescope of the Cusanusgymnasium, Bruneck. Photo C. Wiedemair, with permission

pulsating stars in terms of the presence of one or even multiple companion stars is open to question. The full analysis of Schmeidler's recordings will be of great use in evaluating the reliability of published stellar magnitude time series obtained since the onset of precise timekeeping at other observatories.

Our own approach to the almost century-long attempt to understand the enigmatic period behaviour of CY Aquarii clearly illustrates the need for strict observational procedures to be followed, as well as the necessity of data verification by outsiders, in addition to a standing commitment and enthusiasm.

Fig. 11.8 The successor of the 25-cm instrument, a 40-cm reflecting telescope. Photo C. Wiedemair, with permission

At least another decade of conscientious monitoring of CY Aquarii will be needed to find out whether the star's period changes gradually, or abruptly. In other words, whether the phrase *Natura non facit saltus* (Nature does not make jumps), commonly attributed to the Swedish biologist and philosopher Carl Linnaeus (1707–1778), is applicable to this stellar system.

Acknowledgements The author is indebted to Reinhold Häfner for introducing him to Felix Schmeidler in February 2004. Christof Wiedemair is thanked for proofreading the manuscript.

References

Fu, J. N., & Sterken, C. (2003). Long-term variability of the SX phoenicis star CY Aquarii. *Astronomy & Astrophysics, 405*, 685.

Schneider, R. (1963). Zeitzeichen und Normalfrequenz. *Technische Hausberichte.*

Sterken, C. (2005). The O–C diagram: Basic procedures. In C. Sterken (Eds.), *The light-time effect in astrophysics. Proceedings of ASP conference series* (Vol. 335, pp. 3–24).

Tuvikene, T. et al. (2010). New times of maximum of CY Aquarii. *The Journal of Astronomical Data, 16*(2).

Wiedemair, C. et al. (2020). CCD photometry of CY Aquarii VI. The 2019–2020 seasons. *The Journal of Astronomical Data, 26*(1).

Christiaan Sterken obtained an MSc degree in Mathematics at the University of Ghent, his PhD in Astronomy at the University of Brussels, and his Habilitation degree at the University of Liège. He is Emeritus Research Director at the Belgian Fund for Scientific Research, and Emeritus Guest Professor at the University of Brussels. His principal field of research is the photometry of variable stars (luminous blue variables, massive binaries, pulsating main sequence stars, and cataclysmic variables), and comets. He taught courses in observational astronomy, and on the history of natural sciences at the Universities of Brussels and Ghent. Christiaan Sterken was Sarton Chair holder 2006–2007 at the University of Ghent, and was the 2021–2024 President of Commission C3 History of Astronomy of the International Astronomical Union.

Chapter 12
"Who Put that Hole in the Telescope?" The Development of the World's Most Accurate Timepiece

Bart Fried

Abstract While the astrometric and theoretical work at Greenwich Observatory and the United States Naval Observatory was the focus of its early existence, other problems were being studied at the same time. The recent discovery of an extraordinary and completely forgotten Brashear 20.3 cm lens, made with a large centre hole, has shed new light on one of these other problems. That is, the astrometric and geodetic problem of the apparently random variation of latitude – a problem that has plagued all observatories of the past. It would take the better part of a century and the development of highly specialised and exceptionally accurate telescopes for astronomers to complete the puzzle of this short-term variation, but in the end these telescopes, designed for astrometric observations to solve what was essentially a geodetic problem, also proved to be the world's most accurate timepieces.

Introduction

In the history of astronomical observatories, few if any have been more important in the development of precision time keeping than Greenwich Observatory and the United States Naval Observatory. Both are rooted in providing star charts, ephemerides, and improved methods of keeping time to aid maritime navigation. In many ways both observatories were at the forefront of the "culture of precision" and the establishment of professionalism in astronomy. And while the astrometric and theoretical work done at both places had been the focus of their early existence, other problems were being studied concurrently. The recent discovery of an extraordinary and completely forgotten Brashear 20.3 cm lens, produced with a large centre hole in it has cast new light on one of those other problems. That is, the astrometric and geodetic problem of the seemingly random Variation of Latitude—a problem that plagued all observatories of the past. It would take 272 years and the development of

B. Fried (✉)
New York, USA
e-mail: oldscope@gmail.com

G. Bernasconi et al. (eds.), *Astronomical Observatories and Chronometry, 18th-20th Century*, History of Physics, https://doi.org/10.1007/978-3-031-95874-8_12

highly specialized and exceptionally precise telescopes for astronomers and geodesists to complete the puzzle of this short-term variation—this wobble of the Earth's axis of rotation. And in the end, these telescopes, designed for astrometric observations to solve what is essentially a geodetic problem, proved also to be the world's most accurate timepieces. This article discusses the problems and the evolution of these unusual telescopes; the theory that first solved the Variation of Latitude, and the step-by-step developments that led to these telescopes' change of purpose to that of timekeeping.

The Problem, Simply Stated…

With the advent of positional astronomy, called Astrometry, it became possible to increase the precision of both clocks and geodetic instruments by comparison with the motion of the heavens. Prior to the advent of Cesium atomic clocks, it was the essential method for regulating our time keeping and for fixing positions on Earth. Star positions—the frame of reference—are charted by observations from "fixed" locations—astronomical observatories—and the precise locations of all observatories are calculated by using known star positions that are relatively stable, and from regular transit timings, local time can be set. Stability of this system is key to its precision. Proper motion of reference stars; observational effects; instrument idiosyncrasies, environmental factors and fluctuations of the earth's rotation are among the many factors that need to be accounted for, to make accurate time-keeping possible. And therefore, *If the latitude or longitude of the observatories varies in an irregular pattern, the whole system becomes inaccurate.*

Between March and September 1728, the English Astronomer James Bradley (1692–1762) made reasonably accurate latitude measurements for his observatory, and he immediately noticed what appeared to be variations in his measurements. In an analysis of Bradley's methods, Chandler found that Bradley made the following note in his logbook for September 19th:

> I judge this difference cannot arise from the uncertainty of hanging of the wire truly in the same place, but may possibly proceed from some small alteration in the tube itself, etc., if the succeeding observations shall be found to confirm the small disagreement that those already made of *Υ Drac.* seem to show.[1]

Bradley thus attributed the problem of a seemingly varying latitude to errors within his instruments. He was the first of many to doubt their instruments.

The next significant astronomer to take up the problem was the English astronomer Sir George Biddell Airy KCB PRS, (1801–1892). Airy was a brilliant mathematician who graduated from Trinity College, Cambridge in 1823. Not long after, he became Lucasian Professor of Mathematics at Cambridge in 1826 and then Plumian Professor of Astronomy and director of the Cambridge Observatory in 1828. In 1835

---

[1] Chandler (1901a, p. 71).

he was appointed England's Astronomer Royal, a position he held until 1881. By 1843, Airy was prepared to tackle the Variation of Latitude problem head on. The Royal Greenwich Observatory had state-of-the-art instruments for making transit timings, yet the observatory was also plagued with irregular latitude determinations which were, again, attributed to instrumental error. So Airy designed an improved altazimuth transit instrument (Fig. 12.1) that was very rigid, and unlike the large transit instrument at the United States Naval Observatory (USNO), the ability of Airy's new telescope to rotate in azimuth made it more versatile though arguably not as accurate.

Fig. 12.1 Airy's Altazimuth Telescope of 1847. Credits: G. Dolan, "…Where East Meets West." Published online by the Royal Observatory Greenwich: http://www.royalobservatorygreenwich.org/articles.php?article=1091 (accessed on 16 October 2024)

Airy certainly must have felt that it wasn't accurate enough. It wasn't long after the altazimuth telescope was installed that he was planning the next iteration of a vertical, fixed telescope with improvements to ensure the greatest accuracy.

Next, the Reflex Zenith Tube ...

Conceived in 1843 and installed in 1847, the altazimuth transit was soon to be replaced by several instruments including a completely novel telescope invented by Sir George in 1848. Now known as the Airy Reflex Zenith Tube,[2] it had the following improvements:

1. A Permanent, stationary vertical tube assembly
2. A 12.7 cm Achromatic objective
3. Self-levelling pool of Mercury
4. 90-degree diagonal prism.

The highly rigid iron tube, which was permanently fixed and no longer rotatable, eliminated the separate optical tube assembly and its horizontal pivots, along with the circumferential azimuth ring of the earlier altazimuth transit. It was only able to view the zenith and the trade-off for the highly restricted number of stars available to observe was that Airy had eliminated two major sources of error. The term 'reflex' refers to the reflecting pool of liquid mercury placed approximately halfway down the light cone, which reflected the converging cone back up and through the objective lens, where it was picked off by the centred 90-degree prism. The micrometer eyepiece was mounted to the side, out of the light path. The purpose of the mercury was to ensure that the optical path remained vertical even if there were any minor changes to the verticality of the tube itself. The mercury was a self-levelling reflector and remained undisturbed below the observing deck (Fig. 12.2).

The Airy Reflex Zenith Tube (RZT) was fitted with a 12.7 cm lens made by the celebrated lens maker Edward Troughton (1753–1835). Troughton died about sixteen years before the RZT was installed and the lens was already an antique well before his death, having been produced in 1793. Fortunately, glass doesn't change with age and the venerable achromat was used for many decades to come. In order to control operator error in making timings, the upper portion, holding the lens and eyepiece/micrometer section, was able to rotate. An observer working rapidly was able to make a timing, rotate the observation station 180 degrees and make a second timing. In theory this reduced the problem of the Personal Equation, but it was still a source of potential error. The other source of potential error was the fact that the light path made a double pass through the objective lens, so that any possible issues with the glass, such as any inhomogeneity or striae, would be doubled as well. In his pursuit of more accurate measurements for the determination of the observatory's latitude, Airy employed his Reflex Zenith Tube for 31 years, from 1851 to 1882,

[2] Airy (1854, pp. FI–FXV).

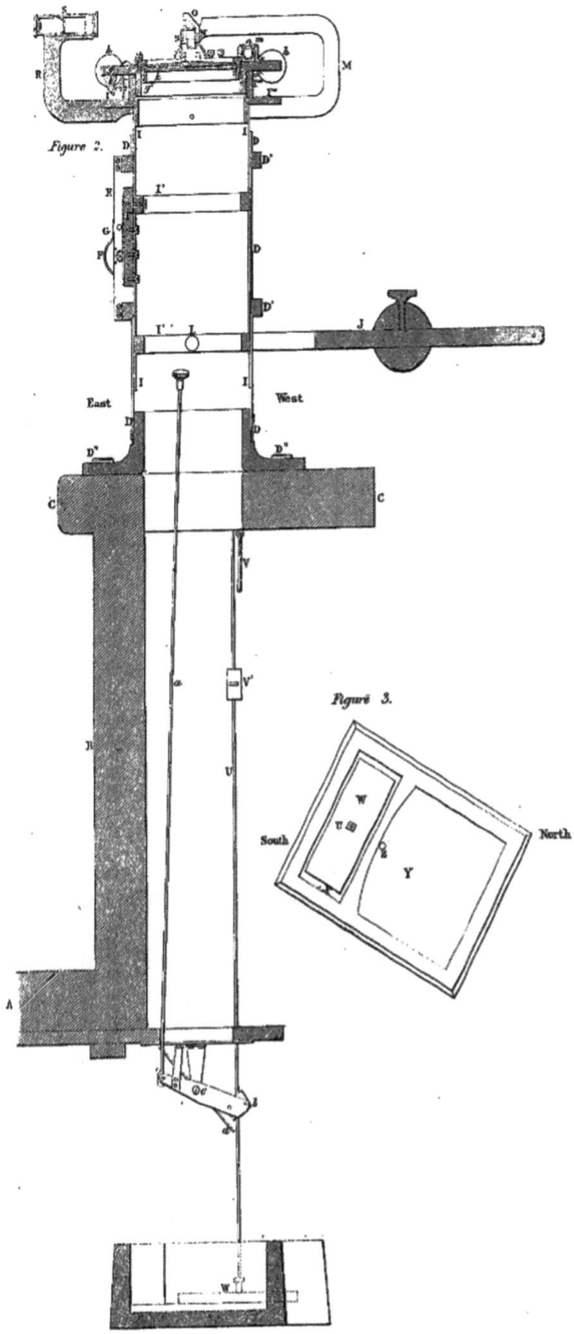

Fig. 12.2 The Airy Reflex Zenith Tube. Credit: Airy GB (1854) Appendix I. Description of the Reflex Zenith Tube of the Royal Observatory, Greenwich. In: Greenwich Observations in Astronomy, Magnetism and Meteorology made at the Royal Observatory, Series 2, vol. 16, pp. FI-FXV http://www.royalobservatorygreenwich.org/articles.php?article=1091

Fig. 12.3 Seth Chandler's original Almucantar design. Credits: *National Academy of Science, Biographical Memoirs*, vol. 66, p. 51

to make measurements of only one star Gamma Draconis. However, "Persistently anomalous results which were presumed to be the result of undiscoverable sources of instrumental error lead to its abandonment. The problem however was not with the telescope, but the interpretation of the results."

Airy, however, was not working in a vacuum. The problem was apparent to every observatory doing quality astrometric work, whether in the pursuit of accurate time-keeping or for producing ephemerides. One astronomer in particular, the American Seth Carlo Chandler, Jr. (1846–1913) became interested in the problem in the late 1870s and he invented two instruments that he hoped would give the necessary accuracy required for useful latitude measurements. One was his patented Altitude-Instrument, also called a Chronodeik.[3] This was a self-levelling instrument with a telescope that operated via pendulum action. There had been numerous earlier pendulum zenith sectors invented, but Chandler's unit was able to provide accurate solar observations giving 1 s accuracy of local time. However, he found that it was not stable or substantial enough for stellar work and it was put aside in favour of another invention of his, the Almucantar (Fig. 12.3). With this instrument, Chandler revived the self-levelling aspect of Airy's mercury pool, but with a twist. Rather than a reflecting pool, the Almucantar employed a self-levelling altitude bearing which floated on mercury. And rather than a fixed meridian transit instrument, the Almucantar rotated in azimuth.

---

[3] Chandler (1881).

Chandler first used his new Almucantar to study the local latitude at the Harvard College Observatory, from May 1884 through June 1885.[4] He detected a pattern of variation in his latitude measurements and could not identify any instrumental or Personal Equation causes. But he did not identify a period, and so he was hesitant to declare a polar motion cause. Then, in the late 1880s, he examined measurements from various other observatories that had published latitude measurements, and after analysing all the results, he arrived at a 427-day period and published his results in 1891. This confirmed George Biddle Airy's much earlier results. Sadly, vindication came too late for Airy, who passed away in 1892 just months after Chandler published a teaser of his discovery.[5] Fittingly, the effect carries the name of its discoverer and is known as the Chandler wobble.

Chandler was successful in characterizing the periodicity of the variation of latitude, yet he was no closer to understanding its cause than any other astronomer or geodesist, with the exception of American astronomer Simon Newcomb (1835–1909), who hinted at the possible cause as "being the consequence of the 'fluidity of the oceans' and the 'elasticity of the Earth'.[6] Yet there wasn't nearly enough high-quality data to explain changes in the periodicity of the Earth's wobble. This need gave rise to the International Latitude Observatories, a set of observing stations on the same parallel around the globe organized by the International Latitude Service, in order to more accurately measure the Earth's wobbly rotational axis. Understanding this phenomenon was deemed critical if there was to be any increase in time-keeping precision that modern societies required. The astronomical community also needed a deeper understanding of the issue, as all fundamental astrometry depended upon accurately knowing how to compensate for the variation. Thus, six initial stations were installed near the parallel of 39° 08' North latitude. These were:

- Gaithersburg, Maryland, United States
- Cincinnati, Ohio, United States
- Ukiah, California, United States
- Mizusawa, Iwate, Japan
- Charjui, Turkestan
- Carloforte, Italy.

The RZT Reborn ...

Chandler's Almucantar inspired other, similar instruments with one in particular utilizing some novel concepts. The Reverend G. A. Fargis, S.J. (1854–1916) at Georgetown University in Washington, D.C. developed a 'Photo-Chronograph' in 1892.[7] It used a mercury bearing system to self-level the optical path, similar in

[4] Carter and Carter (1995).

[5] Chandler (1891, pp. 59–61).

[6] Carter and Carter (1995).

[7] Fargis (1892, p. 89).

concept to Chandler's Almucantar. But Fargis added a photographic system with a time mark inserter, which eliminated the issue of the Personal Equation and measuring the plates was easy, since it could be done at leisure in the comfort of an office during the day. It's unclear why Fargis's photographic improvements didn't immediately catch on in the greater astronomical community, but they were a step in the right direction.

During this same period, starting in 1888, American astronomer Charles Leander Doolittle (1843–1919) was conducting Variation of Latitude studies at Lehigh University, continuing there until 1895. Then he took up the directorship of the new Flower Observatory of the University of Pennsylvania. He brought along his highly capable son, Eric Doolittle (1870–1920), who shared his father's interest in the problem. The new observatory was fully equipped with top quality Warner & Swasey (W&S) instruments including a fine 102 mm Zenith telescope that allowed the Doolittles to immediately jump back into their latitude variation studies. Unfortunately, while it was not too far from the 39° 08' parallel, the Flower Observatory was not close enough to participate directly in the ILS programme. However, Charles Doolittle had in his mind to improve upon the original Airy Reflex Zenith Tube which was, at that time, still the only one ever built. Following a paper by Chandler in 1901 in which he sang the praises of the Airy Reflex Zenith Tube,[8] it was brought back into use in 1902, again for variation of latitude observations until 1910, after which no further observations were made with it at Greenwich. Later, it was used off and on in other locations until November 1929 and then it was finally donated to the Science Museum, London in 1937. It resides in storage there today. However, Chandler's recommendation for its continued use must have been well understood by Charles Doolittle, who set about raising the funds to build his improved version. While the original RZT had been state-of-the-art in its day, there were still deficiencies that were well known and that could be improved. USNO astronomer William Markowitz commented:

> Only bright stars could be observed with Airy's Reflex Zenith Tube, which was used chiefly to observe 7 Draconis. This star transits near the zenith at Greenwich. [...] Eddington[9] analysed the results and concluded that the deficiencies of the reflex zenith tube were its small aperture, the difficulty in determining the focal length, and the second passage of light through the objective. The reflex zenith tube was not used (at Greenwich) after 1910.[10]

The elder Doolittle set about to cure these deficiencies. His design first addressed the issue of determining the focal length, i.e., maintaining the proper focus with an instrument that has no focusing mechanism at the eyepiece. This was accomplished by a clever solution, simply adding or subtracting small amounts from the mercury pool changes the focal plane position accordingly, and only a minute amount needs to be adjusted as any change in the height of the pool is doubled due to reflection. This was apparently easier than introducing an elevating system for the mercury pool.

---

[8] Chandler (1901b, p. 122); see Markowitz (1961).

[9] Eddington (1911, p. 541).

[10] Markowitz (1961, pp. 88–90).

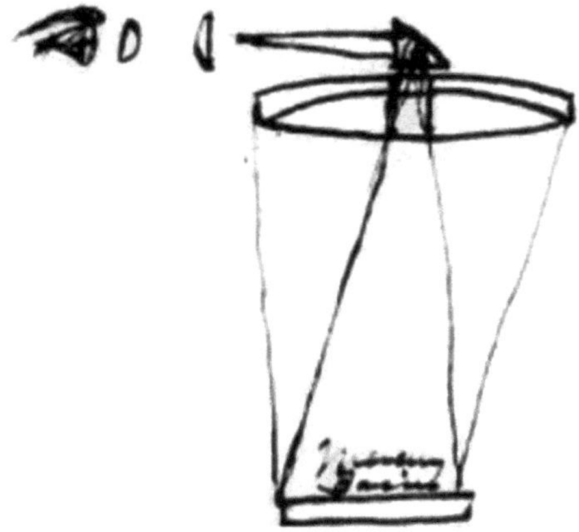

Fig. 12.4 Brashear's sketch of the Wharton RZT objective, included in his letter to Hale. Credit: California Institute of Technology Archives and Special Collections, George Ellery Hale Papers

In order to increase the usefulness of the telescope, a much larger objective lens of 20.3 cm was ordered from the John A. Brashear Co. of Allegheny, PA. This is the unusual lens revealed in the introduction. It has a focal length of 381 cm, yielding a focal ratio of f/18.8, and a 3.8 cm centre hole. The larger lens allowed the Doolittles to measure a greater number of stars which passed over their zenith, increasing the efficiency of the programme. The centre hole was the elder Doolittle's solution to the issue of a double pass through the objective. The cone of light from the f/18.8 system is just narrow enough to pass cleanly through the hole, up to the diagonal system above it. The rotating eyepieces and diagonal system were also improvements as they incorporated a dual observing station to increase the speed of the measurements and increase the overall accuracy.

In 1903, the University received a monetary gift from the founder of the Bethlehem Iron Company, philanthropist Joseph Wharton (1826–1909), for the construction of their RZT. The lens was ordered and completed the same year by John A. Brashear (1840–1920) and the instrument was installed in 1904, 12 feet due East of the 102 mm W & S zenith telescope. Brashear, writing to his friend George Ellery Hale (1868–1938), described it to him, "We have just finished an 8-inch objective for Professor Doolittle for a vertical telescope, for variation of latitude observations. It was a difficult task as we had to figure it with an inch and a half hole through the center. The instrument is to be used as per diagram"[11] (Fig. 12.4). Eric Doolittle described Brashear's successful effort for an article for *Popular Astronomy*. "Through the exact center of the objective a hole 1.5 inches in diameter is bored; the difficult and dangerous work of drilling this hole in a figured lens was performed by Mr. Brashear who made all of the optical parts of the instrument, the mounting being constructed by Warner and Swasey" (see Figs. 12.5, 12.6, 12.7, 12.8).[12]

[11] Brashear to Hale, 2 January 1904; California Institute of Technology Archives and Special Collections: George Ellery Hale Papers.

[12] Doolittle (1906, p. 530).

Fig. 12.5 The Wharton lens, after its recent discovery at the home of a University of Pennsylvania Professor. The cast iron cell was provided with three sets of leaf springs along the inner wall surface to maintain the centricity of the two elements with respect to each other and the cell. The two elements, Crown glass and Flint glass are separated by a metal spacer ring. Brashear's craftsmanship producing this lens was superb. It was tested and determined to have ~15 nm RMS; which translates to a 97% Strehl ratio (100% = optical perfection in terms of theoretical performance). Credits: Photo courtesy of Dr. Roger Ceragioli

The Wharton reflex zenith tube was used until July of 1911. In 1912, C. L. Doolittle published his results, comparing observations made with both the Wharton tube and the W. & S. Zenith telescope. He wrote.

> The Wharton Instrument has proved in some respects less satisfactory than was hoped. Very appreciable systematic differences are sometimes present depending on the order of bisection E-W or W-E. This might not be a serious matter if the differences were constant but such is not the case. After continuing with apparent constancy, perhaps for months, they sometimes disappear entirely, or even change signs. Much care has been given to this matter and it is hoped that the effect has been eliminated from the final result. During the last year a reversing prism was employed, making the direction of apparent motion of the star the same in both bisections. We can hardly suppose that this was a disadvantage, but the probable error was quite appreciably greater after its introduction than before.[13]

Although Doolittle was less than happy with his results, in reality the Wharton RZT greatly improved the accuracy of its predecessor, the Airy RZT. It was glowingly characterised by Variation of Latitude investigator Frank E. Ross. "The Wharton instrument proved to be a vast improvement over the Greenwich instrument," wrote

[13] Doolittle (1912, p. 133).

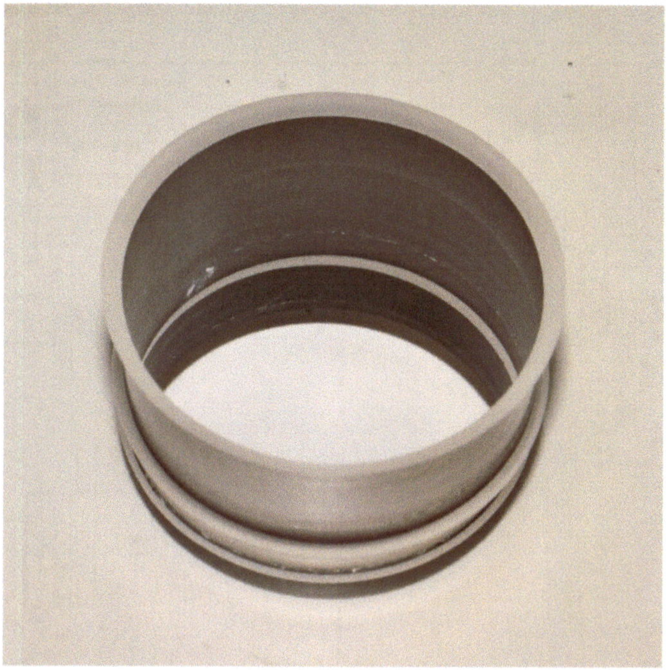

Fig. 12.6 Closeup of the cored and chamfered hole drilled out of the centre of the doublet objective lens. Credits: Photo courtesy of Dr. Roger Ceragioli

Ross.[14] However, the Doolittles moved on from their Variation of Latitude research and the Wharton instrument was abandoned. The Brashear lens of the Wharton RZT was saved by Penn's Prof. Robert Koch and it now resides in Arizona, at The John E. Greivenkamp Museum of Optics; the rest of the telescope was unceremoniously left to rust outside on the lawn near the observatory, and eventually it was trashed. Thus only the lens remains from what, at that time, was almost certainly the most precise visual telescope in the world.

Frank Elmore Ross (1874–1960) and the Photographic Zenith Tube

American mathematician, astronomer and optical designer Frank Ross began working for the Nautical Almanac Office in 1902 and was introduced to astronomy by Simon Newcomb. Then from 1903 to 1905, he worked as a research assistant at the Carnegie Institution. Next, from 1905 through 1915, Ross was the director of the ILO Gaithersburg, MD station and it was here that his interest in developing an improved

[14] Ross (1915, p. 11).

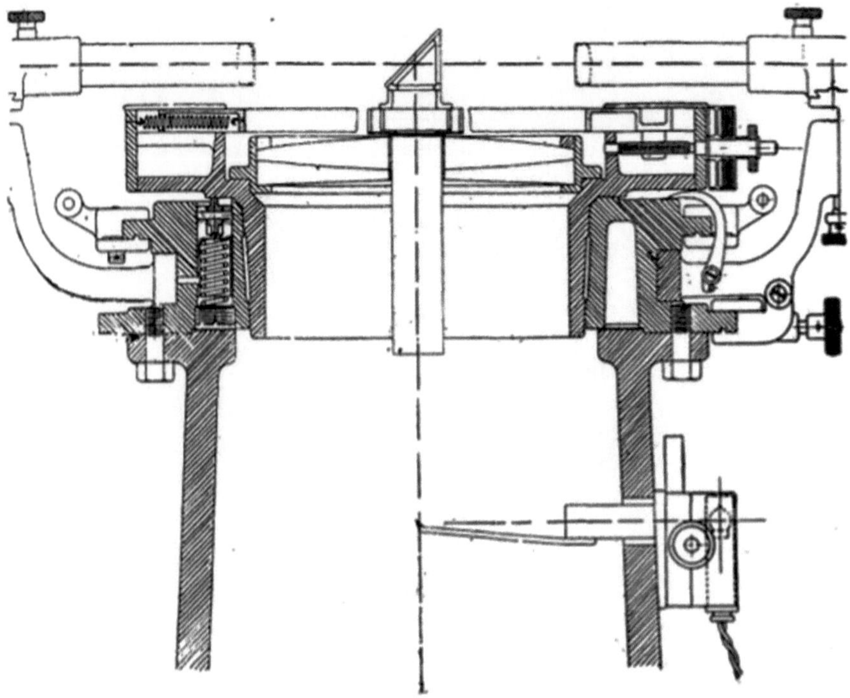

Fig. 12.7 The revolving dual eyepiece apparatus, with the diagonal held in the centre hole. This eliminates the arm holding the diagonal. An electric timing button is seen lower right

version of Doolittle's Wharton RZT was piqued.[15] Ross wrote, "In speaking of the Airy instrument Chandler[16] says: '* * * The theory of the instrument is perfect, its construction is of the last simplicity, and the quantity measured is obtained with a directness and thorough elimination of instrumental error as high as it is easy to conceive.'" Ross was familiar with earlier, similar instruments that employed photography, such as Marcuse's photographic zenith telescope[17] and Cookson's 1900 photographic floating zenith telescope at Cambridge, England.[18] There was also the aforementioned Photo-chronograph at Georgetown. But none of these embodied the advantages of the Airy and Wharton RZTs. Ross quickly concluded that the only remaining source of significant errors in the elegant design of the RZT was *human error*, that is, the problem of the Personal Equation. He wrote.

[15] Morgan (1967).

[16] Chandler (1901c, p. 57–60).

[17] Adolf Marcuse's development is described in *Science* 4(98), 13 November 1896, at p. 721. Simply, he allowed stars to trail across a photographic plate, eliminating the personal equation.

[18] Jones (1919, p. 542).

Fig. 12.8 The Wharton Reflex Zenith Tube, ca 1905. Credits: Results of Observations with the Wharton Reflex Zenith Tube and the Zenith Telescope of the Flower Astronomical Observatory from January 9, 1904, to December 3, 1906

> It appeared to the writer that an instrument of this type was preeminently fitted for the application of the photographic method; that the substitution of the impersonal photographic plate for the human eye would eliminate the last and only known source of error existing in this type of instrument. With a focal length sufficiently great, systematic errors in measuring the photographic images can be made negligible, or even entirely eliminated by suitable methods of measurement. It is well known that the personal equation in visual observations has amounted to 30% of the maximum variation of latitude. With a photographic zenith tube of a focal length not too great for successful manipulation, this error should not exceed 1%. There are weighty reasons why the focal length of visual instruments is limited to 4 or 5 ft. There seems to be no reason why a reflex zenith tube should not have a focal length of 30 ft or more.[19]

At the conference of the International Geodetic Association at Cambridge, England, in September 1909, Ross's plans were submitted for the construction and operation of a photographic reflex zenith tube to be used in determining the latitude variation. The contract for the construction of the photographic zenith tube and the

[19] Ross (1915, p. 11).

comparator was awarded to William Gaertner & Co., of Chicago and the completed instruments were received at Gaithersburg in May 1911. Observations commenced almost immediately in early June. In order to avoid the deleterious double pass problem, the Ross design incorporated placing the plate holder below the objective lens, which meant that the optician, Octave L. Petitdidier, didn't need to bore a centre hole as Brashear had. And there would be no double pass. After much consideration, it was concluded that a sufficiently accurate moving carriage could be produced, allowing for photographing stars as points, rather than using a stationary plate holder and photographing star trails.[20]

The Ross Photographic Zenith Tube (Fig. 12.9) proved to be a highly successful design. Ross summarized the programme results in his 1915 report and concluded with the following statement,

> The accuracy of the individual latitudes obtained with the photographic instrument appears to be considerably greater than that obtained with the visual instrument. For the study of the minute quantities here dealt with this is a matter of considerable importance. That there is also greater freedom from systematic errors has been proved for some of the classes (of observations), and is probably true for all.[21]

Frank E. Ross was nothing if not dogged and thorough in his zeal to perfect his PZT system and he identified one further potential source of errors—the emulsions on the photographic plates. So, he dived into a side project to study the possible movement of the emulsion layer during the development process, working with Kodak. Methods were developed to minimize any movement or distortion of the emulsion layer after exposure.

The International Geodetic Association had paid for the PZT and supported its use in Gaithersburg yet even after Ross's excellent report of its merits, *the PZT was not replicated for use at the other international stations.* It was an expensive 'one-off' compared to the Wanschaff instruments[22] bought for all the ILO locations. "In 1915 the instrument was purchased by the United States Naval Observatory and brought to Washington, D.C. Observations for the variation of latitude were begun at Washington in October 1915."[23]

While the value of studying the variation of latitude observations was clear, and the IGA extended and expanded its work, morphing into first, the International Polar Motion Service and then into the International Earth Rotation and Reference Systems Service, it continued to completely disregard the PZT as a suitable (and the most precise!) instrument. The programme continued the use of visual Zenith instruments until switching to the latest high technology resources, such as Very Long Baseline Interferometry.[24]

---

[20] Ross (1915, p. 8).

[21] Ross (1915, p. 127).

[22] Bell (2016, p. 15).

[23] Markowitz (1961, p. 90).

[24] Bell (2016, p. 17).

Fig. 12.9 The Ross PZT. Credits: US Gov Print Office, US Coast and Geodetic Survey 1915, Ross, Frank Elmore

A New Purpose; a New Lease on Life ... the Precise Keeping of Time

Latitude measurements continued at the U.S. Naval Observatory, using the original Ross Photographic Zenith Tube purchased from the ILS. Then, "About 1923 John E. Willis suggested that the PZT should be used to *determine time* as well as the variation of latitude. A timing contact was added to the PZT, and experiments were carried out by Littell and Willis from 1923 to 1933."[25] But even before that, by 1931, it was becoming clear from Littell's and Willis' preliminary results that they were on to something big. The PZT was not just the world's most accurate latitude telescope,

[25] Markowitz (1961, p. 90).

it was the world's most accurate time-keeping telescope. "Preliminary experiments indicate that the second [PZT] method will give good results for time determinations. The advantages which may be expected from the method as compared with the usual method depending on visual observations made with transit instruments are (1), the application of photography, (2), the comparatively great stability of the zenith tube which weighs about a ton, (3), the greater focal length, 203 inches, (4), the elimination of personal equation effects, (5), the elimination of the effects of instrumental errors, and of their variations."[26] After February 1934, the Ross PZT was used for the official determination of time. During World War II and later, "Several improvements in Ross' PZT No. 1 were made by Sollenberger.[27] In addition to utilizing a new Quartz crystal clock, a synchronous motor was installed to control the plate drive, and an automatic reversal system was added."[28]

The original Ross PZT and its several successors operated by the USNO had become the nation's best time keeping instrument.

> The instrument which has been the backbone of our observations from 1915 to the present is the Photographic Zenith Tube (PZT). It consists basically of an inverted telephoto lens that forms an image, after reflection in a mercury pool, of stars near the zenith. [...] It is now succeeded by a large 65-cm telescope (f = 13 m) [a PZT] with major new features.[29]

After WWII, Photographic Zenith Tubes became the global standard astrographs for keeping Universal Coordinated Time [UTC] until radio astronomy and the cesium atomic clock supplanted them. There were at least 13 or more stations in the UTC system utilizing PZT's of various designs and sizes, and built by several manufacturers (Fig. 12.10). However, they all worked on the same principles as the original.

The USNO, which had led the way for the international adoption of the method, built its final, giant 65-cm iteration of the PZT, conceived in 1972 and installed in 1974. It had a four-element objective—the expectation was to improve the precision and it was successful. It had a plate scale of 15.867″/mm, an increase by a factor of three [20.3 × 25.4 cm plate size], and a new photo-electric timing system with 0.1 ms precision, an improvement by a factor of 10.

Conclusion

As shown, the novel Reflex Zenith Tube and its successor, the Photographic Reflex Zenith Tube were successfully used from the mid-nineteenth century until the latter twentieth century. The Airy RZT results, utilized by Seth Carlo Chandler, were a critical part of the data enabling him to characterize the Variation of Latitude and

---

[26] Littell and Willis (1931, p. 266).

[27] Sollenberger (1942, p. 74); Sollenberger and Mikesell (1945, p. 124).

[28] Markowitz (1961, p. 90).

[29] Winkler (1981).

10-inch, Royal Greenwich Observatory, 1947 Grubb-Parsons	10-inch, Zvenigorod INASAN Observatory, 1986 Zeiss
10-inch, Neuchatel (Switzerland), 1954 Grubb-Parsons[35]	8-inch USNO (3) Washington 1915, etc., Gaertner
10-inch, Royal Greenwich Observatory, 1955 Grubb-Parsons	8-inch USNO [2] Richmond, Florida 1949
10-inch, Mt. Stromlo, Australia, 1956 Grubb-Parsons	8-inch Calgary (Canada), RASC 1951, Dominion Observatory shop (lens by Glenbrook)
?-inch, Deutsches Hydrographisches Institut (Hamburg), 1959 Zeiss	8-inch (2) National Astronomical Observatory of Japan, Tokyo & Mizusawa stations, 1959 (maker unknown)
10-inch, Pulkovo Observatory, 1957 GOMZ[36] – in 1979 moved to Kitab station (Uzbekistan)	8-inch Ottawa, RASC 1967, USNO Instrument Shop[37]
10-inch, Sternberg Observatory (Moscow State University), 1959 GOMZ	8-inch Punta Indio station (Argentina), 1968 USNO Instrument Shop
10-inch, Ondřejov Astronomical Observatory, Czech Republic, 1969 Zeiss	?-inch Tianjin, China, 1976 Nanjing Astronomical Instrument Factory

Fig. 12.10 List of some UTC Photographic Zenith Tubes, compiled by the author

prove its loose periodicities. As the precision of the design improved with the Wharton RZT, and then the Ross PZT, further refinements of the VOL measurements continued to be made. Then, in the early twentieth century, at the U.S. Naval Observatory, it was realized that the same increase in precision and elimination of errors found in visual instruments made the PZT an ideal candidate for attaining the highest precision for time-keeping. After WWII, the UTC system was developed, with PZTs furnishing the transit timings photographically and later, electronically. But what of the original problem, the Variation of Latitude? Characterizing it was one thing, but determining its cause was another and astronomers were still no closer to solving the problem than when Bradley first noticed it in 1720.

Afterword

Because it is essentially a problem of Geodesy, it took a Geodesist to finally identify the causes. An investigation was made in 2000 by Dr. Richard S. Gross [NASA-JPL and CalTech].[30] He showed the causes to be a combination of atmospheric and oceanic processes with the dominant cause being ocean-bottom pressure fluctuations. Gross found that two-thirds of the "wobble" were caused by fluctuating pressure on the seabed due to changes in the circulation of the oceans, caused by variations in temperature, salinity, and wind. The remaining third is due to atmospheric pressure fluctuations. Also, it appears that there may be some effects caused by strong solar storms reaching Earth. This needs further study.[31]

[30] Gross (2000, pp. 2329–2332).

[31] The author hypothesizes that global warming trends will cause an increase in the temperature component of these fluctuations, with a corresponding increase in the excitation of the Chandler wobble. He preemptively proposes to call this hypothetical result the Fried effect.

Acknowledgements Special thanks to Dr. Roger Ceragioli, for his analysis of the Wharton lens and many thanks to Vitaliy Kim, David Pantalony, Santiago Paolantonio, Taro Mimura, Zhao Ke, Gudrun Wolfschmidt for their help in retrieving information about PZTs for Fig. 12.10.

References

Airy, G. B. (1854). Appendix I. Description of the reflex zenith tube of the royal observatory, Greenwich. *Greenwich Observations in Astronomy, Magnetism and Meteorology made at the Royal Observatory, 2*(16), FI–FXV.

Bell, T. E. (2016). Wandering pole, Wobbling grid. *The Bent of Tau Beta Pi* (online). https://www.tbp.org/pubs/Features/Sp16Bell.pdf (accessed on 16 October 2024).

Brashear, J. A., & Hale, G. E. (1904). *California Institute of technology archives and special collections: George Ellery Hale Papers.* California Institute of Technology.

Carter, W. E., & Carter, M. S. (1995). Seth Carlo Chandler, Jr. In *Biographical Memoirs* (Vol. 66. pp. 51–52). National Academy of Sciences.

Chandler, S. C. (1881). *Altitude-instrument US Patent 239315.* US Patent Office.

Chandler, S. C. (1901b). Contribution to the history of the reflex Zenith-tube. *Monthly Notices of the Royal Astronomical Society*, 122

Chandler, S. C. (1891). On the variation of latitude. *Astronomical Journal, 11*, 59–61.

Chandler, S. C. (1901a). Variation of latitude from Molyneux's and Bradley's observations. *The Astronomical Journal, 513*, 71.

Chandler, S. C. (1901c). The Greenwich reflex Zenith-tube. *Astronomical Journal, 22*(511), 57–60.

Dick, S., McCarthy, D., Luzum, B. (2000). *Colloquium #178.* ASP.

Doolittle, E. (1906). The Wharton reflex Zenith tube. *Popular Astronomy*: 530.

Doolittle, C. L. (1912). Results of observation with the zenith telescope and the Wharton reflex zenith tube. *The Astronomical Journal, 641*, 133.

Eddington, A. S. (1911). Errors of instruments and observation: Discussion of the greenwich reflex Zenith-tube observations, 1906–1609. *Monthly Notices of the Royal Astronomical Society, 71*, 541.

Fargis, G. A. (1892). Georgetown college observatory, 1843–1893. *Publications of the Astronomical Society of the Pacific, 7*(41), 89.

Gressot, J. (2023). Automating time determination: The Photographic Zenith Tube (PZT) of the neuchatel observatory. *Bulletin of the Scientific Instrument Society, 156*, 16–23.

Gross, R. S. (2000). The excitation of the chandler wobble. *Geophysical Research Letters, 27*, 2329–2332.

Littell, F. B., & Willis, J. E. (1931). A new method of determining time. *Publications of the American Astronomical Society, 6*, 266.

Markowitz, W. (1961). The photographic Zenith tube and the dual-rate moon-position camera. In G.P. Kuiper and B.M. Middlehurst (Eds.), *Telescopes, vol. 1: Stars and stellar systems* (pp. 88–114). The University of Chicago Press.

Morgan, W. W. (1967). *Frank Elmore Ross: A biographical memoir*. National Academy of Sciences.

Polar Motion: Historical and scientific problems (Vol. 208). ASP Conference Series.

Ross, F. E. (1915). *Latitude observations with photographic zenith tube at Gaithersburg*. Md. Special Publications 27(11). U.S. Coast and Geodetic Survey.

Sollenberger, P. (1942). Time determination and time keeping. *Popular Astronomy, 50*, 74.

Sollenberger, P., & Mikesell, A. (1945). Quartz crystal astronomical clocks. *Astronomical Journal, 51*, 124.

Spencer Jones, H. (1919). Cookson floating zenith telescope. *Monthly Notices of the Royal Astronomical Society, 79*, 542.

Winkler, G. M. (1981). Developments in the time service division. In *Proceedings of the sesquicentennial symposia of the U.S. Naval Observatory* (Vol. 125). U.S. Naval Observatory.

Bart Fried is the founder and current President of the Antique Telescope Society. He is the recipient of the ATS Isaac Newton Medal for meritorious service to the Society. As a recognized authority on the history of the telescope, he has lectured for several decades around the U.S., Europe, the U.K., Ireland and Canada. He has written articles for Sky & Telescope magazine; the Journal of the Antique Telescope Society; Eyepiece newsletter of the Amateur Astronomers Association and other journals and publications. He is a member of the American Astronomical Society—Historical Astronomy Division, and the British Astronomical Association.

Part V
Astronomy and Chronometry in a Global Perspective

Chapter 13
Observatories, Marine Chronometers, and Terrestrial Applications: From Chronometric Longitude Expeditions to Observational Efficiency

Trudy E. Bell

Abstract International competitions and John Harrison's 18th-century invention of a marine chronometer accurate enough for a ship to determine its transitory longitude at sea inspired a host of technological improvements to the chronometer that enabled wider applications of chronometers to 19th-century astronomical problems: notably to cartography and to data recording. From the 1820s through the 1840s, European and American astronomers collaborated on major international chronometric expeditions, transporting dozens of marine chronometers between fixed and temporary astronomical observatories to determine accurate longitude differences between land masses and to improve the precision of interior maps. The widespread commercial adoption of the telegraph in the 1840s and 1850s enabled longitudes to be further refined over long overland or transoceanic distances. Both chronometric and telegraphic techniques became important through the 19th century in systematically mapping the American west. They serendipitously led to the discovery of the finite speed of an electromagnetic signal through telegraph wires. Moreover, they revolutionized the efficiency of transit instrument and meridian circle observations of stellar positions for star catalogues, becoming standard astronomical data-recording techniques within the walls of an individual observatory.

Key Words American method of longitudes · American method of transits · Chronometric expeditions · Telegraphic longitudes · Transit instruments

Background

Over the last century, several works have recounted the history of the development and use of the marine chronometer. Possibly most famous is the popular book *Longitude* by Dava Sobel, which focuses on the eighteenth century travails of John Harrison

T. E. Bell (✉)
Rocky River, OH, USA
e-mail: t.e.bell@ieee.org

G. Bernasconi et al. (eds.), *Astronomical Observatories and Chronometry, 18th-20th Century*, History of Physics, https://doi.org/10.1007/978-3-031-95874-8_13

in devising four workable prototypes of marine chronometers that were extensively tested.[1] A pioneering work that placed Harrison in historical context and detailed many essential eighteenth and nineteenth century technical developments by other artisans who made chronometers more precise, practical, available, and affordable is *The Marine Chronometer* by Rupert T. Gould.[2] More recent works carry the chronometer's history up into and through the twentieth century.[3]

Meantime, the revolutionary role of the telegraph in determining terrestrial longitudes beginning in the 1840s is described in *Longitude by Wire* by Richard Stachurski[4]; it is also discussed in sections of *Selling the True Time* by Ian R. Bartky[5] and in *The Coast Survey 1807–1867* by Capt. Albert E. Theberge.[6] James Lequeux devotes a chapter of his biography *Le Verrier: Magnificent and Detestable Astronomer* to summarizing the use of the telegraph in the 1840s and 1850s to determine longitudes in France and between Paris and observatories in other nations.[7]

Most histories of the marine chronometer, however, have primarily focused on its use for its original intended purpose in its intended environment, i.e., navigation at sea, rather than as a cartographic tool for fixing longitudes on land. One important exception is *From Celestial to Terrestrial Timekeeping: Clockmaking in the Bond Family* by Donald Saff,[8] which discusses innovations in both chronometric and telegraphic devices and methods for determining longitude. Another reference is the second part of a major two-part paper on the Liverpool Observatory by Richard Schmidt and Paul Dearden,[9] which analyzes the extensive multi-year transatlantic chronometric expeditions between Harvard College Observatory and the Liverpool Observatory in the 1840s and 1850s.

This paper aims to provide a brief sketch of two under-recognized applications of the marine chronometer to terrestrial applications: (1) dedicated chronometric expeditions from the 1820s into the 1850s culminating with measurements of longitude differences across the Atlantic Ocean, and (2) use of chronometers in conjunction with telegraph technology within the walls of individual fixed nineteenth century U.S. astronomical observatories beginning in the 1840s for improving the efficiency of recording regular nightly astrometric observations.

---

[1] Sobel (1995).

[2] Gould (1923).

[3] Whitney (1984); Bertele (1991); Cronin (2010).

[4] Stachurski (2009).

[5] Bartky (2000).

[6] Theberge (2007).

[7] Lequeux (2013).

[8] Saff (2019).

[9] Schmidt and Dearden (2020).

Introduction

As early as 1530, Dutch mathematician and cartographer Gemma Frisius wrote how differences in longitude at sea could be determined by transportation of sufficiently accurate timepieces.[10] The basic principle that Frisius doubtless realized was that, for a spherical earth, longitude bears a fixed relationship to the earth's rotation and thus to time: in essence, determining a difference in longitudes meant determining a difference in *local time* between the instant a celestial body transits the local meridian of an unknown point and the instant it transits the meridian of a reference location. But in the sixteenth century—indeed, until the 1840s—there was no way astronomers could simultaneously, in real time, compare the local time on a clock at an unknown distant location with the time on a clock at a reference meridian.

So, by the early nineteenth century, astronomers had developed or adapted several methods of determining approximate longitudes for terrestrial cartography. At least four involved the moon. One method compared the timings by various observers of the contacts of a total solar eclipse from various locations on the earth along the path of totality.[11] A second method was timing occultations (eclipses) by the moon of specific stars listed in the Nautical Almanac. A third method was timing transits of so-called moon-culminating stars, i.e., stars that transit the meridian at about the same time as the moon.[12] A fourth method was that of "lunar distances" or "lunars," which essentially treated the moon's orbit like the face of a clock centered on an observer with a sextant; the observer measured the ever-changing apparent angular distance between the sun and the moon, and calculated the observing location's longitude from tables of lunar distances tabulated in the time of some standard meridian.[13]

However, all the methods using the moon suffered from the same weaknesses. The moon's motion is nonuniform and mathematically complex, a full three-body problem since the moon feels the gravitational pull not only of Earth but also of the Sun, and the moon itself also pulls on the Earth. As a result, the various astronomical techniques using the moon gave inconsistent results.[14] For this and other reasons, in the early nineteenth century, where different terrestrial maps met, meridians of longitude sometimes differed from one another by large fractions of a mile or even whole miles.[15] That was not particularly a problem for mariners in the open ocean,

---

[10] Pogo (1935).

[11] First contact is the instant when the leading edge of the moon is first seen to start passing in front of the sun, marking the beginning of the eclipse; second contact is the instant when the moon first fully covers the sun, beginning totality; third contact is the end of totality; fourth contact is when the trailing edge of the moon is seen to depart the disc of the sun, marking the end of the solar eclipse.

[12] Epps (1839).

[13] Van der Werf (1997).

[14] Borden (1846, p. 55); Stachurski (2009, p. 49 ff).

[15] Borden (1846, p. 56, 58); Green (1875, p. 427); Schott (1898) noted that as late as 1869, when San Francisco, California, and Cambridge, Massachusetts were directly connected by the first transcontinental telegraph, discrepancies in the measured longitude differences between the Atlantic and Pacific coasts previously determined by moon culminations and other pre-telegraphic means were revealed to be "almost three-quarters of a mile" (pp. 201–202).

but it could be a significant obstacle for terrestrial map-making, especially for the accurate determination of state or national boundaries, or for the precise mapping of treacherous shoals and sunken rocks near coastlines.

In the late eighteenth century and well into the nineteenth century, various prizes offered to develop an accurate marine chronometer plus Harrison's demonstrations of his prototypes spurred international interest in improving the design and precision of escapements, balance springs and other mechanical components, as well as methods of temperature compensation.[16] Still, despite the growth of chronometer manufacturing businesses in various countries including the U.S., the demand for marine chronometers by navies, exploring expeditions, commercial whalers, merchant ships, passenger vessels, and others outstripped the supply until around 1840.[17]

Chronometric Expeditions

By the 1820s, marine chronometers had become reliable, compact, rugged, resistant to diurnal changes in ambient temperature, available, and affordable enough to serve as a portable time standard, keeping Greenwich time—or the time of any other desired reference meridian—for a few weeks at a stretch with an error as small as a second. Thus, it became clear to a number of European astronomers and cartographers that good marine timekeepers were finally accurate enough to be used to determine longitudes *of fixed locations on land* as well as transient locations of ships at sea.

in 1821, Danish-German astronomer and geodesist Heinrich Christian Schumacher (founding editor of *Astronomische Nachrichten* and director of the Altona Observatory in Hamburg, then under Danish administration) first used chronometers to determine the difference in longitude between two points—Copenhagen, Denmark, and Hamburg, Germany—nearly 200 miles apart to within a second of time.[18] It was a promising start, demonstrating the practical capability of the chronometric method of determining differences in terrestrial longitude.

From the 1820s through the 1840s, various European expeditionary astronomers sailed or traveled overland from one major observatory to another, using each observatory's transit instrument or meridian circle to determine accurate local time at each survey meridian. Between fixed observatories, they sequentially set up a portable astronomical transit on headlands or other prominent landmarks. From either type of expeditionary location, they timed meridian transits of stars to determine the accurate local time. Then, the observer immediately checked the expeditionary marine chronometer(s) to note the time that same instant at a reference meridian (Greenwich or elsewhere). From the time difference (after accounting for whether individual chronometers ran at slightly different rates, gaining or losing time), the observer could calculate the two locations' difference in longitude.

[16] Burton and Nicholas (2016).

[17] Moriarty (1888, pp. 263–264); Gould (1923, p. 131).

[18] Schumacher (1822); Struve (1844, p. 1).

A word about reference meridians and *differences* in longitude: as explained in one practitioner's manual, "Places are fixed absolutely by astronomical observations, or relatively by chronometers. This distinction must be clearly kept in view if ever we wish to arrive at final and conclusive results..."[19] Until the 1880s, longitude projects were hampered by lack of a universal worldwide reference prime meridian from which the absolute longitudes of all places could be determined. Although Greenwich, England, had been established as the British prime meridian in 1721, through much of the nineteenth century Paris, Berlin, Cadiz, Lisbon, Naples, Pulkova, Rio de Janiero, and a dozen more cities in other countries and continents also served as prime meridians for their own nations.[20] It was not until the International Meridian Conference in 1884 that the Royal Observatory at Greenwich, England, was voted to be longitude 0 degrees, the reference longitude from which every location around the planet should be determined as well as international time zones,[21] although some nations did not accept that for decades.[22] Even so, given the vastness of the Atlantic Ocean, into the mid-nineteenth century there was serious discussion whether there should be a separate American prime meridian.[23]

In 1822 and 1823, by order of the Board of Longitude, British astronomer John Lewis Tiarks (who had just achieved distinction for surveying the U.S.-Canada boundary) conducted on behalf of the Admiralty the first chronometric expeditions between Greenwich and the island of Madeira, southwest of Portugal some 1,350 miles away. He transported 15 chronometers from Greenwich to Falmouth by sea and then to Madeira, retracing the route on the return. Each way at Falmouth, he compared the resulting chronometric longitudes to those determined by the Trigonometrical Survey of the Board of Ordnance—a precision triangulation of angles and distances among 300 key landmarks across Great Britain (including Ireland and Wales) that had been underway since 1791.[24]

Tiarks was nonplussed to find that after all chronometric sources of error had been taken into account, "the means of all results nearly agreed in giving the longitude of Pendennis Castle (the station near Falmouth) 4″ (time) greater than it is laid down in the Account of the Trigonometrical Survey." Four seconds of time implied a longitude discrepancy between the Survey and Tiarks's chronometric expedition of a magnitude "little expected."[25] "As this question affected however the longitude of all places in England determined in the same manner...," Tiarks devoted another six weeks to accounting for all possible sources of error, repeatedly measuring the chronometric longitude difference between the Trigonometric Survey stations at Falmouth and Dover, two points accessible by sea and separated by nearly the greatest possible distance across England. When all the calculations were complete, they still revealed

[19] Shadwell (1861, p. 3).

[20] Borden (1846, p. 56).

[21] Howse (1997); Bartky (2000); Withers (2017).

[22] Black et al., (1984, p. 388).

[23] Dick (2003, pp. 126–127).

[24] Hewitt (2010).

[25] Tiarks (1824, p. 361).

longitude differences across Britain measured by the Trigonometrical Survey were proportionately (depending on distances between points) up to 3 s of time too small. Tiarks then examined the literature on measurements of the shape of the earth and concluded that the discrepancies arose from deviations in the earth's shape away from the perfect oblate spheroid that had been assumed by the Trigonometric Survey.[26]

Schumacher's and Tiarks's findings inspired several major national and international chronometric expeditions over the next two decades, notably ones underwritten by the British Admiralty and the Russian Emperor. In 1833, for example, Lt.-Gen. Fyodor Schubert in the Russian Corps of Military Topographers, in collaboration with the governments of Prussia, Denmark, and Sweden, commanded an expedition in which a steamboat transported 56 chronometers to some 70 points around the Baltic Sea to update their longitudes. These points included fixed astronomical observatories, lighthouses, cathedral steeples or other prominent landmarks, and (in remoter areas) temporary expeditionary observatories under tents, all east of Altona and referred to the Greenwich meridian. To complete three circuits of the Baltic and visit so many points in a single short northern summer, observers with transit instruments fanned out to the various points beforehand to establish each one's accurate local time before and after the ship's visits.[27] The Russian Emperor also underwrote chronometric expeditions between the Pulkovo and Altona observatories in 1843 (Fig. 13.1), and between the Altona and Greenwich observatories in 1844.[28]

Meantime, in 1830 in the U.S., the Massachusetts legislature passed a resolution authorizing the governor to appoint a chief surveyor to conduct a trigonometric survey of the entire state, incorporating astronomical methods as appropriate. The survey (specifying latitudes, longitudes, and elevations) of more than 8,400 square miles and 300 miles of coastline required most of the decade, somewhat hampered by inadequate equipment, funding, and staffing; the resulting triangulation measurements were estimated to be accurate to within 5 or 6 feet.[29] The longitudes of 27 points (23 within Massachusetts and the others just outside its boundaries) were also determined by multiple exchanges of chronometers, crosschecked by astronomical observations (lunar occultations, solar eclipses) to ascertain the shape of the underlying spheroid of the earth. From this effort—essentially a rigorous pilot comparative analysis of the chronometric and trigonometric methods—Borden concluded in 1841, "[C]hronometric comparisons do afford the means of determining…the value of a degree perpendicular to the meridian, and that, too, with a degree of uniformity quite too great to be the result of a happy accident."[30]

Not all astronomers or geographers were enthusiastic, however, about using chronometers to determine longitudes on land. One prominent practitioner who strenuously opposed Borden's conclusion was Ferdinand Rudolph Hassler, the first

---

[26] Tiarks (1824, p. 371).

[27] Kupffer (1836, p. 413).

[28] Struve (1844); Struve and Struve (1846).

[29] Borden (1846, pp. 44–45).

[30] Borden (1846, p. 91). Although Borden presented his paper to the American Philosophical Society in 1841, it was not published in the society's proceedings until 1846.

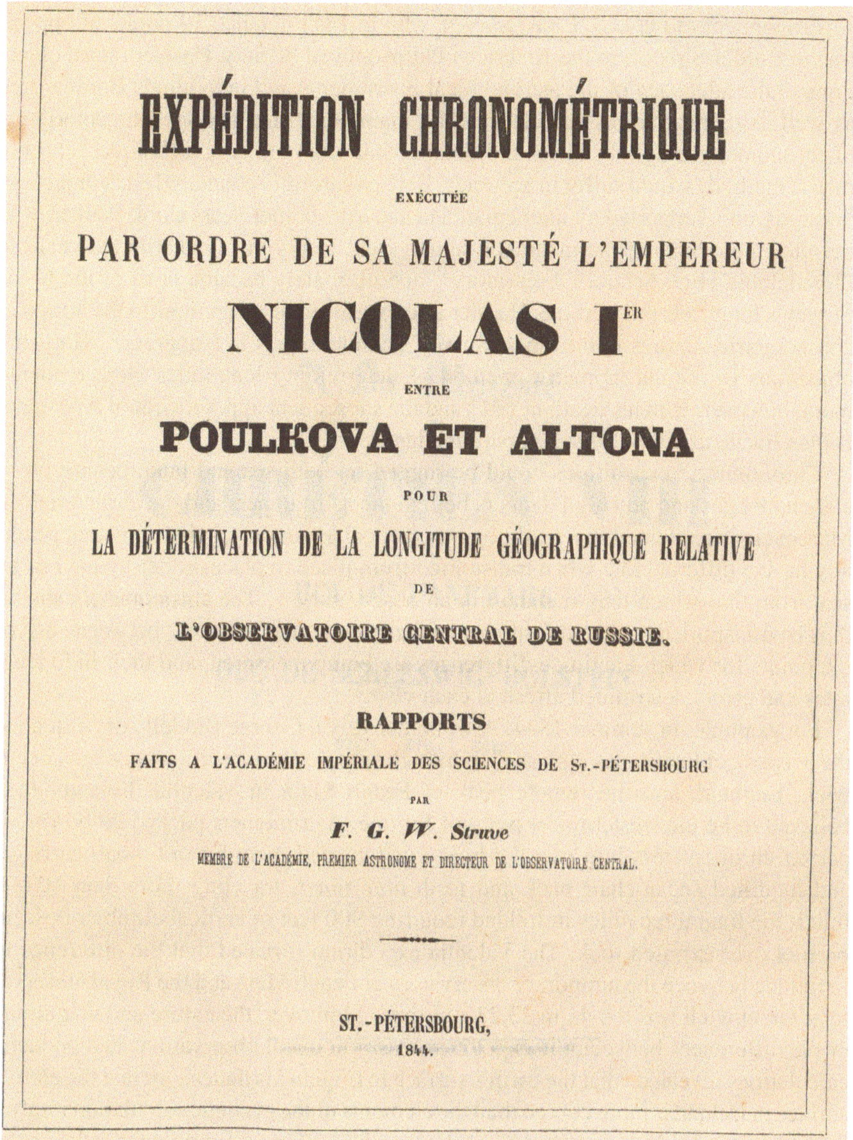

EXPÉDITION CHRONOMÉTRIQUE

EXÉCUTÉE

PAR ORDRE DE SA MAJESTÉ L'EMPEREUR

NICOLAS IER

ENTRE

POULKOVA ET ALTONA

POUR

LA DÉTERMINATION DE LA LONGITUDE GÉOGRAPHIQUE RELATIVE

DE

L'OBSERVATOIRE CENTRAL DE RUSSIE.

RAPPORTS

FAITS A L'ACADÉMIE IMPÉRIALE DES SCIENCES DE St.-PÉTERSBOURG

PAR

F. G. W. Struve

MEMBRE DE L'ACADÉMIE, PREMIER ASTRONOME ET DIRECTEUR DE L'OBSERVATOIRE CENTRAL.

ST.-PÉTERSBOURG,

1844.

Fig. 13.1 Title page of the report of the Russian chronometric expedition between the Pulkova and Altona observatories in 1843 (Struve, 1844)

superintendent of the U.S. Coast Survey, who in 1841 detailed 19 points of opposition to Borden's report to the American Philosophical Society. Hassler raised objections to the adequacy of the astronomical instruments and individuals Borden used as well as of his various calculations; he maintained that although transportation of chronometers might be faster and cheaper than trigonometric techniques, chronometric methods would suffer in accuracy. Borden carefully countered each objection, his arguments supported by mathematician and astronomer Sears Cook Walker, who despite employment as an insurance actuary was deeply involved with the year-old Philadelphia High School Observatory[31] and ultimately became head of the Coast Survey's longitude programme. Hassler also expressed his objections to the notion of chronometric surveys in the Coast Survey's annual report to Congress.[32] Hassler's objections to the chronometric method of determining longitudes were rendered moot, however, with his death in 1843 and the subsequent appointment of Alexander Dallas Bache as Coast Survey superintendent.

Chronometric expeditions could be rugged on both sea and land, posing physical challenges and potential dangers both to the astronomers and the chronometers. Moreover, as explained in a textbook on practical astronomy, "chronometers generally have a different rate when transported from place to place, either by land or by sea, from that which they maintain in an observatory. ...The chronometers should thus be transported back and forth a considerable number of times" between each of the points for which longitude differences are being measured, and their individual rates and errors determined afresh at each visit.[33]

For example, in summer 1844, Astronomer Royal George Biddell Airy extended the network of longitudes with an elaborate chronometric expedition between Greenwich, England, and the remote peak of Feagh Main in Valentia, Ireland—then believed to be the westernmost point of Europe. Astronomers packed 30 borrowed pocket chronometers into two flat boxes and carried them by rail, boat, carriage, and modified sedan-chair back and forth *nine times*, traveling more than 10,000 miles, the final three miles in Ireland requiring 900 feet of vertical climb or descent on foot over exposed rock. The Valentia expedition revealed that the difference in longitude between the temporary observatory at Feagh Main and the Royal Observatory, Greenwich was 0 h 41 m 23.23 s of time. Moreover, the nature and magnitude of the differences between the expedition's astronomical observations and geodetic calculations revealed "that the earth's surface in England is flatter...or that the earth's surface in Ireland is more curved than the elements of the geodetic calculation imply." Last, the expedition ascertained that "the length of 1$''$ in an arc perpendicular to the meridian in latitude 51° 40' is 101.6499 feet."[34] Most importantly, European chronometric expeditions culminating with Airy's had revised the longitudes of key points from Moscow to Valentia—an east–west distance exceeding 2,000 miles.[35]

---

[31] Stated Meeting (1842); Stachurski (2009, pp. 14–15).

[32] Hassler (1842, p. 12).

[33] Loomis (1855, p. 300).

[34] Airy (1847, p. 289).

[35] Chronometer-Expeditionen (1846); Bond (1856, p. cxxxix).

Spanning the Atlantic

Meantime, the 1840s brought several important, independent technological developments that together pointed a way for chronometric expeditions to measure a major unknown: the longitude difference across the Atlantic Ocean. A couple of the new developments were astronomical. In 1843, the new Liverpool Observatory at Waterloo Dock was completed with a 4-inch transit instrument for daily astronomical determinations of time and accurately and officially rating marine chronometers to Greenwich time.[36] Late that same year (in November 1843), the Harvard College Observatory installed a new 4 1/4-inch (focal length 5 feet) Troughton & Simms transit circle.[37]

But one major non-astronomical development was equally important. Fast steamships were replacing sailing vessels, reducing the time for a transatlantic crossing from more than a month down to a couple of weeks. In 1845, the English government awarded a 13-year contract to the British and North American Royal Mail Steam Packet Co., headed by Samuel Cunard, to dispatch a steamship every two weeks carrying mail west from Liverpool to Boston.[38] In 1848, Cunard introduced the large, powerful steamship the *America* (249 feet long and over 1,800 tons), which immediately set transatlantic crossing speed records to as short as 10 days.[39] From 1845 through 1848, Harvard College Observatory director William Cranch Bond—who also owned the Boston clock-making shop Wm. Bond & Son, which rated the "going" of shipboard marine chronometers to Cambridge time as determined from the Harvard observatory—kept meticulous lists of Greenwich time as shown on chronometers of the Cunard mail steamers, and calculated the resulting longitude differences.[40]

In 1849, with all the essential developments in place, Bond wrote to the Harvard Overseers, "It therefore has now become *our* duty to complete the chain which shall unite us more intimately with the Observatories of the Old World."[41] He conceived of using Harvard's new transit circle as the western end point for a grand chronometric expedition with the Liverpool Observatory as the eastern end point, spanning more than 3,000 miles across the Atlantic Ocean. Instead of a dedicated costly U.S. government transatlantic expedition, Bond reasoned that fast mail steamers could provide an economical alternative. In early 1849, Bond interested John Hartnup, founding superintendent of the Liverpool Observatory, in his transatlantic proposal. Hartnup readily obtained promises of support from Astronomer Royal Airy and other key British observatory directors and chronometer makers. Bond presented his proposal

[36] Schmidt and Dearden (2019a, b); Schmidt and Dearden (2020). In 1866, the Liverpool Observatory was relocated and renamed the Bidston Observatory.

[37] Bond (1856).

[38] Saff (2019, p. 6). At that time, some 40 percent of the world's trade passed through the two port cities of Boston and Liverpool.

[39] North American Mail (1851).

[40] Stephens (1987); Baker (1890, p. 7, 15); Knox (1957, pp. 557–559).

[41] W. C. Bond (1856, Appendix, p. cxxxix).

to Coast Survey superintendent Bache in a letter of March 12, 1849.[42] Bache immediately allocated funds for renting and insuring three to five dozen chronometers and paying their cost of passage multiple times in both directions, to begin that current (1849) sailing season.

Although originally planned for just one sailing season, the Boston-Liverpool transatlantic grand chronometric expeditions ran for three summers: 1849, 1850, and 1851. Each year, Bond refined protocols for conducting the astronomical observations in Cambridge and Liverpool as well as for winding and reading the chronometers during a voyage, seeking to ensure as much uniformity as possible throughout. During the winters between sailing seasons, he tested the sensitivity of the chronometers' rates to variations in vibration, barometric pressure, temperature, and other factors. Over the three years, Bond and his adult sons George and Richard (partners in his clock-making shop) sent a total of 92 individual chronometers on 19 one-way voyages both eastward and westward. Altogether, the Harvard and Liverpool observatories recorded 1,065 exchanges of chronometers—greater than any previous or later chronometric expedition.[43] In 1855, another six voyages (three round trips) were made to measure diurnal temperature variations even more stringently, seeking to understand and eliminate a longitude discrepancy between the eastward and westward voyages.[44]

"Cambridge is now the central geographical point of this Continent," wrote George P. Bond. "... Our longitude has, undoubtedly, been investigated with more care than that of any other spot on the globe."[45] Because its position was so minutely determined, the Harvard College Observatory served as a de facto American prime meridian until the first successful direct measurement of the longitude between North America and Europe by undersea transatlantic telegraph cable in 1866.[46] The Harvard meridian was referenced even as late as the 1883–1884 adoption of standard time zones and Greenwich as the world's prime meridian.[47]

Telegraphic Longitude Expeditions

Meanwhile, Samuel F. B. Morse's improvement of the telegraph in the late 1830s bore the promise of adding a powerful new tool for determining longitude differences. The telegraphic longitudes concept appears to have originated in a conversation between Morse and Paris Observatory director François Arago in 1839, when Morse

---

[42] This unpublished letter resides in Harvard University Archives (in UA V 630.349.1 Box One, in folders mostly categorized by year).

[43] Knox (1957, p. 559).

[44] Schmidt and Dearden (2020).

[45] Quoted in Stephens (1990, p. 30); Saff (2019, p. 55).

[46] B. A. Gould, (1870, p. 2).

[47] Bartky (2000).

was in France seeking to secure patent rights for his new technology.[48] In brief, the seemingly-instantaneous transmission speed of electromagnetic signals through telegraph wires promised to allow chronometers hundreds of miles apart to be compared in what today would be called real time; in other words, as Walker later asserted, "the comparison of the clocks together is effected with the same degree of precision as if they were placed side by side."[49] Thus, observers from the Coast Survey arranged to make longitude observations as soon as commercial telegraph wires were strung to any new location.

The telegraphic method of determining differences in longitudes was actually several methods, which evolved within the telegraph's first five years. Three were techniques using chronometers to compare local times with varying rigor, which subsequently became standard to use in tandem. The first technique, used on Morse's demonstration experimental line between Baltimore and Washington, D.C.—just 2 1/2 weeks after Morse's 24 May 1844 inaugural message "What hath God wrought?"—was an "exchange of clock signals," on 9 June 1844. On three days, chronometric expert Commodore (later Admiral) Charles Wilkes of the U.S. Navy[50] and an associate took turns comparing two solar chronometers set to the respective local times of the two cities. One observer tapped the telegraph's battery key every 10 s in time to the beats of his chronometer. Forty miles away, the other noted on the face of his own chronometer the instant he heard the corresponding click of the armature of his receiving electromagnet, estimating fractions of a second. After several minutes of 10-s signals, the first observer transmitted the exact local hour, minute, and second at which he had begun tapping with the chronometer's beats. Assuming an instantaneous transmission time, subtracting the two local times would give the minutes and seconds difference of longitude between the two telegraph offices.[51] The mean of all the observations at both ends was taken to be the longitude difference between the cities. Although some later writers dismissed Wilkes's experiment as crude, an exchange of clock signals became standard at the outset of a determination for quickly narrowing longitude differences to within a second of time.

After Wilkes's pioneering experiment for the U.S. Navy, the U.S. Coast Survey took the lead. As soon as commercial telegraph lines linked offices in major cities, the Coast Survey paid to have additional wires extended to any nearby astronomical

---

[48] Which one of the men suggested it to the other is unclear as claims are made in both directions. For example, in 1849, Sears Cook Walker told the AAAS, "The first mention of the electro-magnetic telegraph, in connection with longitude operations, as far as I know, was made, in 1837 [sic], by M. Arago to Dr. Morse" (Walker 1850, p. 183). In 1856, however, Elias Loomis (professor of mathematics and astronomy at the University of the City of New York, now New York University) wrote, "In 1839, Professor Morse suggested to M. Arago, that the electric telegraph would afford the means of determining the difference of longitude between distant places with an accuracy hitherto unattainable" (Loomis 1856, p. 304).

[49] Bache (1848b, pp. 125–126).

[50] Bryan (1939); Borthwick (1965). Wilkes had been superintendent of the U.S. Depot of Charts and Instruments—forerunner of the U.S. Naval Observatory—for three years before leading a four-year, six-ship worldwide exploring expedition 1838–1842, for which he had selected 29 chronometers and various astronomical instruments.

[51] Walker (1850, p. 183).

observatory and contracted with local astronomers or sent the agency's own observers equipped with its own transit telescopes. To remove the estimation error inherent in an exchange of clock signals, a "method of coincidences" technique was developed in July and August of 1847,[52] when longitude differences were being measured between the U.S. Naval Observatory in Washington, D.C., the Philadelphia Central High School Observatory, and a temporary observatory in Jersey City, N.J.[53] The method of coincidences took advantage of the fact that solar and sidereal chronometers beat audibly at slightly different rates. An astronomer at one end of the line tapped a telegraph key each second in time to his *sidereal* clock; at the other end of the line a second astronomer listened to the clicks of his receiving armature magnet along with the beats of his own *solar* chronometer. Since a sidereal clock gains upon a mean solar clock one second in about six minutes, whenever the two ticked in exact coincidence, the listener noted his local solar time. After several such coincidences, the second astronomer began beating and the first astronomer began listening. Within half an hour, the longitude difference could be calculated—its value accurate to within the clock errors and the observers' reaction times.[54]

In 1848, a third technique called the "exchange of star signals" was refined when a telegraph line between Boston and New York City was extended to the Harvard College Observatory in Cambridge, Mass., and also to the grounds of the private observatory of Lewis M. Rutherfurd in New York City. The line was used on seven nights to determine the longitude difference between the two observatories. Each observer tapped on a telegraph key each time a pre-selected star near the zenith was seen to cross each of the seven wires in the eyepiece of a transit instrument. The local solar time of the signals from each observatory was noted at both observatories, the time difference yielding the longitude difference. Pairs of stars separated by a small angular distance were used so the differences in local times were independent of any uncertainties in the known celestial positions of the stars.[55]

All three techniques—exchange of clock signals, method of coincidences, and exchange of star signals—were used in October 1848 to determine the difference in longitude (some 37 min of time) between Philadelphia and the Cincinnati Observatory nearly 600 miles west, a new distance record.[56] By the end of that year, the Coast Survey's superintendent Bache declared in his annual report to Congress that the telegraphic method of determining longitude "may be considered to have passed into one of the regular methods of geodesy."[57]

---

[52] Airy (1847, p. 70). However, Astronomer Royal George B. Airy attributed the innovation of the "coincidence of beats" method to British astronomer Richard Sheepshanks for Airy's 1844 chronometric expedition between Greenwich and Valentia, noting that "a good ear, with a little practice, can certainly fix on the time when the beats coincide to within less than the hundredth part of a second."

[53] Until submarine cables were perfected, the Hudson River posed a barrier to extending reliable telegraph lines directly into New York City from the south and west.

[54] Walker (1850, p. 184).

[55] Loomis (1856, p. 310–312).

[56] Mitchel (1848); Loomis (1856, pp. 312–313),

[57] Bache (1848a, p. 19).

But key technical issues troubled Bond and several other astronomers and instrument makers. Among them were mathematician Walker, Cincinnati Observatory director Ormsby MacKnight Mitchel, Coast Survey instrument-maker James Saxton, and Cincinnati instrument maker John Locke. One technical problem was each observer's unknown individual human reaction time or "personal equation" in tapping a telegraph key simultaneously with a clock beat or star transit.[58] Another was the lack of a recording device to provide a permanent written record of a longitude determination. A third was the absence of a method for getting a clock or chronometer to record its own beats on such a record without degrading the precise timekeeping of the clock.

As Mitchel recounted much later, "On the 26th of October, 1848, Professor Walker…first presented to me the mechanical problem of causing the clock to send its own beats by telegraph from one station to the other." If clock beats "could be received on a uniformly flowing time scale, the star transit could also be sent by telegraph and received on the same scale; and thus a new method of transits would at once spring from the resolution of the first mechanical problem." Walker mentioned to Mitchel that he had also presented the problem to others, "but so far as he knew, [it] had never been solved."[59]

The next month, on Nov. 17, 1848, Locke experimented with connecting a clock in Cincinnati with the telegraph wire to Pittsburgh, so that the standard Morse register (electromagnetic mechanism for receiving and printing telegraph messages on a paper tape "fillet") in Pittsburgh and all the telegraph offices in between recorded the beats of the Cincinnati clock.[60] Two months later, Locke was the first to try using the Morse register for recording star signals as well as clock signals at a distance. In a seminal experiment on the night of January 23, 1849, a clock of Locke's design was set up in the Philadelphia observatory and connected to the respective observatories in Cambridge, New York City, and Washington, D.C. The star signals telegraphed from each observatory were recorded on Morse fillets at all four observatories along with the time ticks from the Philadelphia clock. Fractions of a second could be measured at leisure by scale and dividers. Locke was awarded $10,000 by U.S. Congress for the invention and was directed to build such a clock for the U.S. Naval Observatory, which was installed in 1850.[61]

Although the Morse register was used in several early longitude determinations, astronomers soon rejected it as the ultimate solution for two major reasons. The fillet ran irregularly depending on whether the pen was writing. Moreover, it ran out of the register at an inch a second, so an hour of longitude determinations produced

[58] The quantification of personal equation in astronomical observations and its resultant effect on observatory practice and culture, primarily in Europe, has been examined by Schaffer (1988); Canales (2001); Rieznik (2013).

[59] Mitchel (1860, p. 235).

[60] Loomis (1856, p. 330).

[61] Loomis (1856, pp. 330–333); Dick (2003, pp. 88–92).

300 feet of paper tape, and "for a single night's work of an observatory, a length of nearly half a mile would be required"[62]—impractical for either analysis or storage.

In 1849, Mitchel developed a flat disc 22 inches in diameter by pasting a sheet of paper over a circular wooden hoop, which dried to become as taut as a drumhead. The disc revolved horizontally once per minute, making concentric circles (Fig. 13.2). A make-circuit clock marked every other second with a tiny dot. At the end of every revolution, a new concentric circumference was started with a slightly different radius. Two hours of observations were recorded on each flat circular sheet, on which alternate seconds appeared as radial dotted lines and observations as dots irregularly in between. At least two of Mitchel's revolving-disc chronographs were placed in actual operation, one at the Cincinnati Observatory itself and the other at the Dudley Observatory in Albany, N.Y.[63]

In 1850, Bond and his sons unveiled a new cylindrical chronograph whose precision and uniformity of movement was revolutionary. The Bonds' cylinder, covered with a large rectangular sheet of paper, revolved once per minute while a pen drew a continuous line; the pen was offset momentarily by each tick of an astronomical clock, marking seconds of time. The cylinder revolved and the paper advanced with uniform motions controlled by a pioneering mechanism the Bonds called a "spring governor," consisting of a train of clockwork connected with the axis of a flywheel (Fig. 13.3). The pen was also offset whenever an observer pressed a telegraph key to signal the transit of a star behind a vertical wire in the eyepiece. Each sheet of the Bonds' spring-governor chronograph could record two hours of observations.[64] The Bonds unveiled their design at the Great Exhibition of the Works of Industry of All Nations, in London's Crystal Palace in 1851 (Fig. 13.4). Ultimately, the Bonds' spring governor design became standard for chronographs used in most nineteenth century observatories.[65]

With the addition of a method of permanently recording observations, the telegraphic method of determining longitudes was essentially complete. One significant modification by the early 1850s was the practice of having observers switch sites midway through a series of observing nights to cancel most effects of their respective personal equations.[66] The telegraphic method won immediate accolades and adoption in Europe, and rapidly became known as the "American method of longitudes" in contemporary literature; it was one of the first major contributions of the American scientific community to worldwide astronomical practice[67] and was acknowledged as such at the time.[68]

---

[62] Loomis (1856, p. 327).

[63] Mitchel (1849); Mitchel (1860, pp. 243–244); Description of the Observatory (1866, pp. 26–43).

[64] Loomis (1856, p. 327–328); Saff (2019, pp. 125–130).

[65] The controversial history of priority and invention of the printing chronograph and the bitterness between Locke and the Bonds has been discussed elsewhere Evans (2002); Dick (2003, pp. 88–92); Saff (2019).

[66] Loomis (1856, pp. 340–41).

[67] Rothenberg (1999, p. 514).

[68] Airy (1849, p. 26).

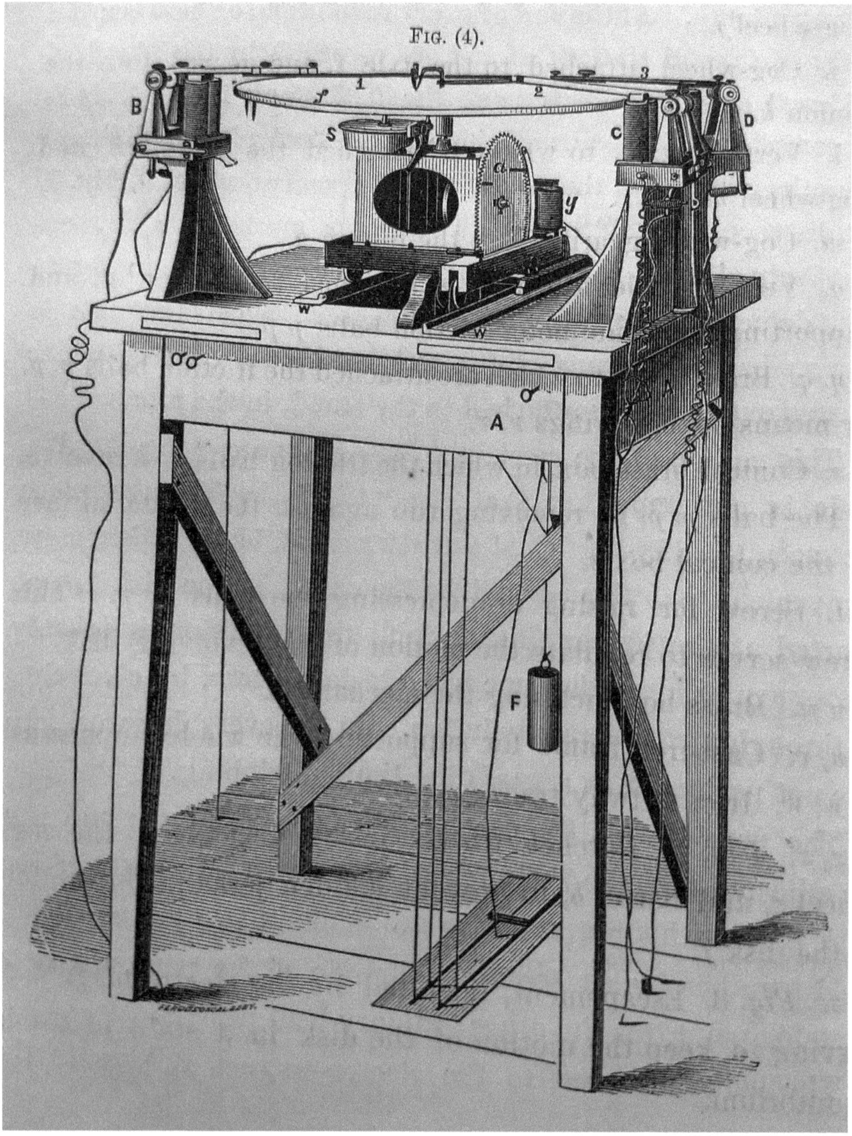

Fig. 13.2 Ormsby MacKnight Mitchel's rotating disc chronograph (Description of the Observatory, 1866, 33)

By the mid-1850s, the U.S. Coast Survey had crisscrossed the eastern half of the United States with baselines representing differences in longitude measured telegraphically between reference points of geodetic significance. Eventually, the American system of longitudes was linked with key British points in Canada from

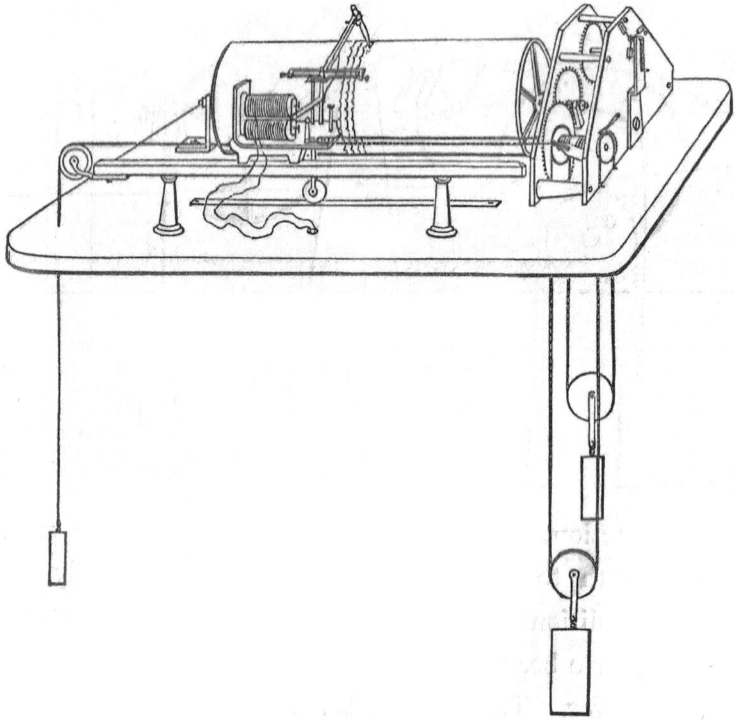

Fig. 13.3 The Bonds' spring governor drum chronograph (Bond, 1856, lii)

New Orleans, Louisiana to New Brunswick, Canada.[69] For their fundamental reference meridian, most observations referred either the Harvard College Observatory in Cambridge or the U.S. Naval Observatory in Washington, D.C. As the telegraph spread throughout Europe, British and Continental astronomers used it to check baselines between Liverpool, Greenwich, Paris, and other major observatories.

Serendipity and Spinoffs

Applying the telegraph to geodesy unwittingly furthered basic knowledge of the fledgling science of electromagnetism. In a classic case of scientific serendipity, the four Morse fillets from the seminal January 1849 experiment of simultaneously registering clock and star signals at the four observatories at Cambridge, New York

[69] Details of the progress and end points of telegraphic longitude determinations and baselines were documented each year in the annual reports of the progress of the U.S. Coast Survey (later renamed the U.S. Coast and Geodetic Survey), which are scanned and available from the U.S. National Oceanic and Atmospheric Administration at https://library.noaa.gov/Collections/Coast-Geodetic-Survey-Heritage (accessed 15 April 2023).

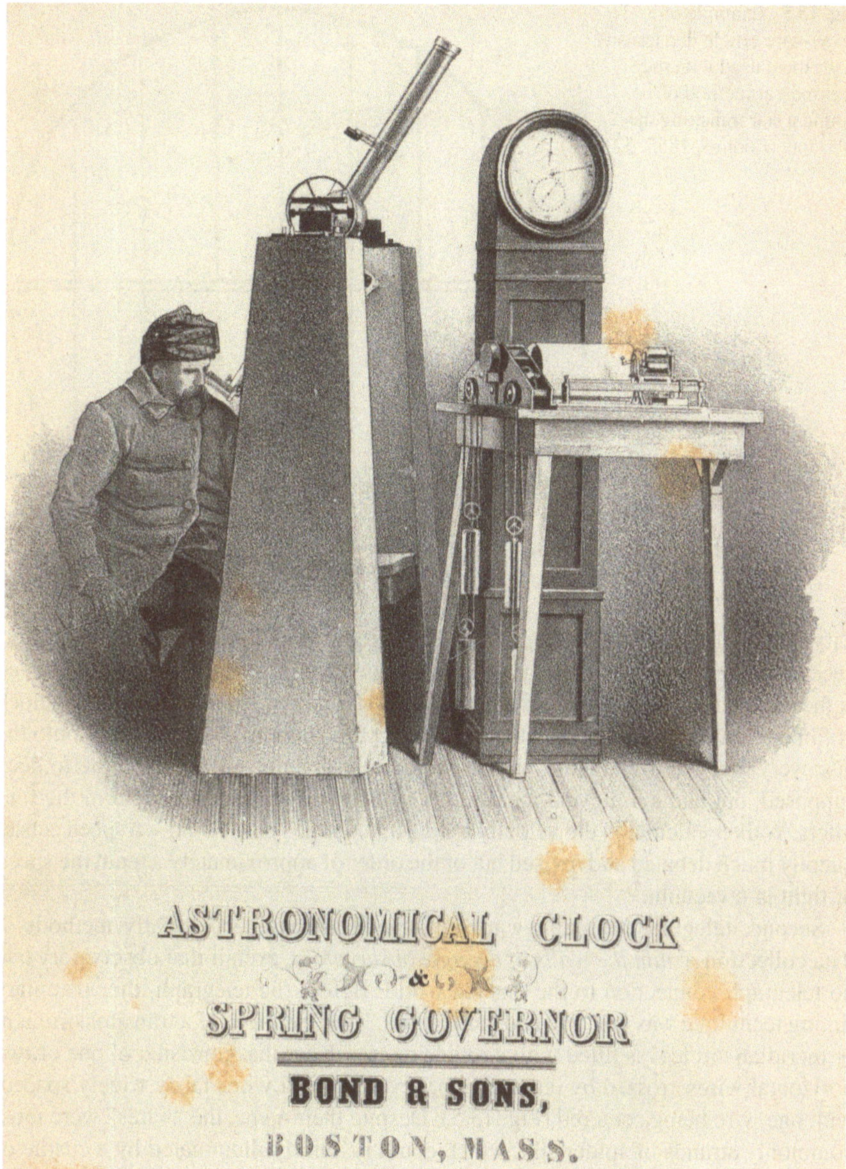

Fig. 13.4 The Bonds' exhibit at the 1851 great exhibition of the works of industry of all nations, in London's Crystal Palace (Public Domain/The New York Public Library)

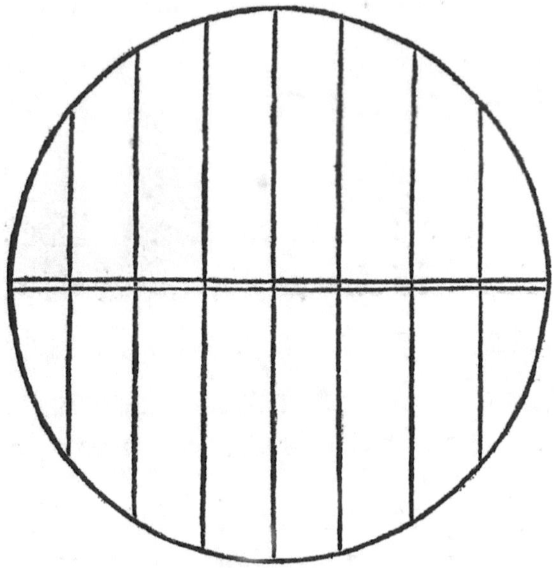

Fig. 13.5 Example of a seven-wire reticle in a transit instrument used with the eye-and-ear method of timing a star transiting the meridian (Loomis, 1855, 52)

City, Philadelphia, and Washington, D.C., revealed "small, but appreciable, differences...in the respective readings of the apparent date of the same event as recorded at the different stations."[70] The farther an observatory was from the graduating clock at Philadelphia, the greater was the discrepancy. In short, Walker stumbled onto the discovery that electromagnetic signals were not instantaneous, as had hitherto been supposed, but had a finite and measurable velocity through a circuit. From the four fillets, Walker calculated the velocity to be 18,800 miles per second—a speed subsequently much debated and revised but of the order of approximately a tenth the speed of light in a vacuum.[71]

Second, telegraph technology revolutionized astronomers' nightly methods of data collection *within the walls of a single observatory*, even if that observatory had no telegraph connection to the outside world. Before the telegraph, the customary timing technique was the "eye and ear method." The eyepiece of a transit instrument or meridian circle was fitted with a reticle or crosshairs that consisted of one or two horizontal wires crossed by either five or seven vertical wires fairly widely spaced, with one wire being centered (Fig. 13.5). Despite their name, the "wires" were most commonly strands of spider silk, which could be faintly illuminated by a candle or oil lamp in the telescope so as to be just visible against the night sky.

The rough timing of an expected transit was calculated in advance. The observer directed the telescope of a transit instrument or meridian circle to the anticipated

[70] Bache (1849, p. 75).

[71] Many of the issues hypothesized to potentially affect the speed of transmission (e.g., composition and gauge of the telegraph wires, ambient temperature, etc.) were discussed at length by Benjamin Apthorp Gould before the AAAS in 1850, Gould (1851). Modern values are higher, depending on the medium and other circumstances. Bache (1849); Walker (1850, p. 189).

altitude. Then either he or an assistant sat holding a quill pen at the ready above ruled paper on an adjacent table, listening intently to the ticking clock, and peering through the eyepiece. Soon, the star would enter one side of the eyepiece's field of view, drift across the field of view, and disappear off the opposite side. As the star crossed, it was momentarily bisected by each vertical wire. Timing each momentary bisection was the astronomer's task—one demanding keen attention, long training, and exacting skill, especially if the observer had no assistant. As described by Airy,

> In ordinary observations the observer listens to the beat of a clock while he views the heavenly bodies passing across the wires; and he combines the two senses of hearing and sight (usually by noticing the place of the body at each beat of the clock) in such a manner as to be enabled to compute mentally the fraction of the second when the object passes each wire, and he then writes down the time in an observing book.[72]

Possible sources of error were numerous. Some outside noise could drown out the sound of the beating clock. The astronomer could miscount clock beats and be off by half a second or more. The astronomer could misestimate the fraction of a second between clock beats or star bisections. The astronomer could momentarily lose concentration. If more than one astronomer were making timings with different telescopes in different rooms of the observatory, each might be using different clocks, which may or may not be beating the seconds exactly simultaneously, putting the time standard into question. Worst of all, there was no independent written record of exactly what did happen and when, which could be cross-checked if calculations revealed some error.

Telegraph technology diminished many of the sources of error and enabled new capabilities. It was more accurate, as timings recorded graphically on paper sometimes could be measured to within hundredths of a second instead of to the nearest quarter or tenth. The paper records could be stored for later reference.[73] The telegraphic technique was simple enough that inexperienced observers could readily master tapping a break-circuit key at the moment of stellar bisection faster than mentally interpolating the audible beats of a clock with the passing of the star behind a wire and writing down the result. Thus, it was comparatively easy to train novices to make acceptable observations.

Additionally, tapping a key was so fast that observers could record more observations during a star's passage across the field of view, eliminating the need for multiple nights of observation to attain the same statistical precision.[74] Thus, the vertical wires in the reticle could be much more closely spaced: instead of being 15 s apart (meaning that 15 s elapsed as a star drifted from one wire to the next) to allow a moment for the astronomer or assistant to write down the time, wires could be spaced a mere 2 s apart, or even just 1.5 s (Fig. 13.6). In Walker's words, "*fifty* wires may take the place of *seven*, and one *month's* work may take the place of a year's work" [his emphasis].[75] Indeed, some astronomers estimated that the potential

---

[72] Airy (1849, p. 27).

[73] Walker (1850, p. 191–192; Schaffer (1988).

[74] Saff (2019, p. 130).

[75] Walker (1849, p. 213).

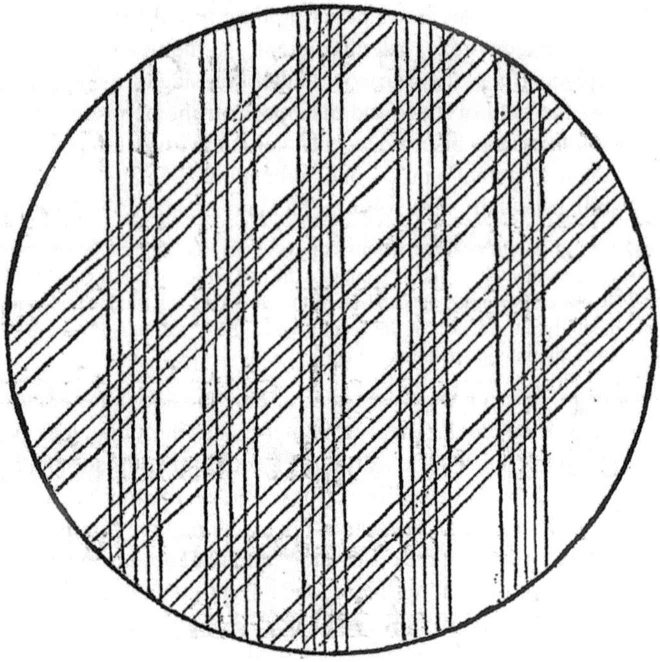

Fig. 13.6 US Naval Observatory reticle for a transit instrument used with the American method of transits for measuring north polar distance while timing the meridian transit of a star (Loomis, 1856, 366)[78]

gain in observational productivity—a boon especially for creating star catalogues or measuring stellar proper motions—could range up to factors of 10, to 36, to even as high as 70.[76] In short, the telegraph directly influenced the design of eyepieces. By 1850, the telegraphic data-recording technique became adopted and known on both sides of the Atlantic as the "American method of transits."[77]

Conclusion

The advent of the American method of transits inspired the development of completely new techniques, such as the simultaneous measurement of differences in declination, not just right ascension. For example, both the U.S. Naval Observatory and the Cincinnati Observatory invented variants on recording differences in

[76] Gould (1853).

[77] Airy (1849, p. 26).

[78] Walker credited the initial design of a reticle or "diaphragm" with seven or nine tallies of wires to U.S. Coast Survey instrument maker William Wurdemann. Walker (1850, p. 191).

declination or north polar distance during a single transit.[79] The Naval Observatory's technique required timing the passage of a star, or a pair of stars, across two sets of wires, one set inclined at 45 degrees to the other, see Fig. 13.6. In contrast, Mitchel experimented with devising a long electromechanical arm that could be connected to an ordinary transit instrument and a chronograph to record north polar distance simultaneously with the time of a star transit. By his mechanism, he asserted, "the transit instrument is converted, (at trifling expense,) into a declinometer" and the observer is "released from the necessity of reading a graduated circle."[80]

Telegraph technology was widely adopted in both U.S. and European observatories in the nineteenth century for longitude determinations in places where telegraph wires were installed, increasingly superseding chronometric methods. It remained important well into the twentieth century, evolving as continued technological innovations provided new opportunities.[81] Nonetheless, transportation of chronometers remained a standard Coast Survey technique for determining longitudes for astronomical, cartographic, and exploring expeditions to remote regions not yet reached by telegraph.[82] In the twentieth century both chronometric and telegraphic methods were gradually supplemented and eventually replaced by radio positioning techniques.[83] And well into the twentieth century, the American method of transits remained standard for observatories' routine nightly recording of star positions.[84]

Acknowledgements The author expresses gratitude to Sara Schechner, Richard E. Schmidt, and Philip Shoemaker for helpful suggestions or useful references.

References

Airy, G. B. (1847). Account of the measurement of the astronomical difference of longitude from Greenwich to the island of Valentia. *Memoirs of the Royal Astronomical Society, 16*, 55–174.

Airy, G. B. (1849). On the method of observing and recording transits, lately introduced in America. *Monthly Notices of the Royal Astronomical Society, 10*(2), 26–34.

Bache, A. D. (1848a). *A report of the superintendent of the coast survey during the year ending Nov 1848*. Rpt Sec Treas 30th Congress, 2nd Session. Senate. Executive No. 1.

Bache, A. D. (1848b). Schreiben des Herrn A. D. Bache an den Herausgeber. *Astronomische Nachrichten, 632*, 119/120-125/126.

Bache, A. D. (1849). Letter of March 1, 1849 to R. M. Patterson. *Proceedings of the American Philosophical Society, 5*(42), 74–78.

79 Loomis (1856, pp. 364–367).

80 Mitchel (1852, p. 2).

81 Bowie (1941).

82 Hayford (1899).

83 Jayne (1913); Littell and Hill (1918); Littell et al. (1929).

84 For example, incidental mentions of using the telegraph key for timing meridian transits appear in the unpublished correspondence in the Dudley Observatory archives of the astronomers at Dudley's southern observing station in San Luís, Argentina, for observations 1909–1912.

Baker, D. W. (1890). *History of the Harvard college observatory during the period 1840–1890.* Boston Evening Traveler.
Bartky, I. R. (2000). *Selling the true time: Nineteenth-century timekeeping in America.* Stanford University Press.
Bertele, H. von (1991). *Marine and pocket chronometers: History and development.* Schiffer.
Black, H. D., et al. (1984). The centenary of the prime meridian and of international standard time. *Johns Hopkins APL Technical Digest, 5*(4), 381–389.
Bond, W. C. (1856). Report of the director to the visiting committee of the board of overseers of Harvard University. *Annals of Harvard College Observatory, 1*(1): Appendix.
Borden, S. (1846). Account of a trigonometrical survey of Massachusetts with a comparison of its results with those obtained from astronomical observations by Robert treat Paine. *Transactions of the American Philosophical Society, 9*(1), 33–91.
Borthwick, D. E. (1965). Outfitting the United States exploring expedition: Lieutenant Charles Wilkes' European assignment, August-November, 1836. *Proceedings of the American Philosophical Society, 109*(3), 159–172.
Bowie, W. (1941). The determination of the difference of longitude of two stations. In: *Determination of time, longitude, latitude, and azimuth, part II,* pp. 78–102. U.S. Coast and Geodetic Survey, Department of Commerce.
Bryan, G. S. (1939). The Wilkes exploring expedition. *Proceedings of the U.S. Naval Institute, 65,* 1452–1464.
Burton, M. D., & Nicholas, T. (2016). Prizes, patents, and the search for longitude. In: *Harvard Business school working paper,* pp. 16–46.
Canales, J. (2001). Exit the frog, enter the human: Physiology and experimental psychology in nineteenth-century astronomy. *British Journal for the History of Science, 34,* 175–197.
Chronometer–Expeditionen zwischen Pulkowo, Moskau und Warschau (1846). *Astronomische Nachrichten, 558,* 87–90.
Cronin, J. (2010). *Marine chronometers.* Crowood Press.
Description of the Observatory. (1866). *Annals of the Dudley Observatory, 1,* 1–126.
Dick, S. J. (2003). *Sky and ocean joined: The U.S. naval observatory 1830–2000.* Cambridge University Press.
Epps, M. (1839). On the method of determining the longitude by moon-culminating stars. *Monthly Notices of the Royal Astronomical Society, 4*(21), 187–188.
Evans, R. B. (2002). Morse's register and the American method. *Rittenhouse, 16*(2), 65–83.
Gould, B. A. (1851). On the velocity of the galvanic current in the telegraph wires. In: *Proceedings of the American association for the advancement of science, 4th meeting,* August 1850, pp. 79–100.
Gould, B. A. (1853). On the comparative precision of the electro-chronographic, or American method of observation. *Annals of Science, 1*(22), 257.
Gould, B. A. (1870). *The transatlantic longitude, as determined by the coast survey expedition of 1866.* Smithsonian contributions to knowledge, Vol. 16(223): article 7.
Gould, R. T. (1923). *The marine chronometer: Its history and development.* J.D. Potter.
Green, F. M. (1875). Telegraphic determination of longitude. *Popular Science Monthly, 7,* 426–433.
Hayford, J. F. (1899). The determination of longitude of two stations. Report of the Superintendent of the U.S. Coast Geodetic Survey for July 1, 1897 to June 30, 1898, Appendix 7(2), pp. 320–340.
Hassler, F. R. (1842). Questions of the resolution of congress of June 24, 1841, relating to the survey of the coast of the United States, with the answers to the same. Coast survey. Letter from the Secretary of the Treasury. 27th Congress, 2nd Session. House of Representatives. Document No. 28.
Hewitt, R. (2010). *Map of a nation: A biography of the ordnance survey.* Granta.
Howse, D. (1997). *Greenwich time and the longitude.* Philip Wilson.
Jayne, J. L. (1913). Radio-determination of longitude. *Popular Astronomy, 21*(10), 594–596.
Knox, R. W. (1957). Precise determination of longitude in the United States. *Geographical Review, 47*(4), 555–563.

Kupffer, M. (1836). Notice of the chronometric expedition of Lieut-Genl Schubert, executed in 1833, to determine the longitude of the most important points on the coast of the Baltic. *Journal of the Royal Geographical Society of London, 6*, 413–416.

Lequeux, J. (2013). Telegraphic longitudes. In *Le Verrier: Magnificent and detestable astronomer*. Springer.

Littell, F. B., & Hill, G. A. (1918). *Determination of the difference of longitude between Washington and Paris*. Publications of the U.S. Naval Observatory (2nd series) 9: Appendix.

Littell, F. B., et al. (1929). *World longitude operation of 1926*. Publications of the U.S. Naval Observatory (2nd series) 12: Two parts and appendix.

Loomis, E. (1855). *An introduction to practical astronomy*. Harper.

Loomis, E. (1856). *The recent progress of astronomy especially in the United States*. Harper.

Mitchel, O. M. (1848). Differences of longitude by magnetic telegraph—details of work for obtaining the longitude of the Cincinnati observatory—U.S. Coast Survey. *Sid Mes, 3*(1), 6–8.

Mitchel, O. M. (1849). Report of Professor O. M. Mitchel of cincinnati on the mechanical record of astronomical observations. Report of the superintendent of the U.S. Coast Survey. Letter from the secretary of the treasury. 31st Congress, 1st Session, Senate Executive Document 5, December 27, Appendix 5, pp. 72–78.

Mitchel, O. M. (1852). On a new method of recording differences of North polar distance or declination by electro-magnetism. *American Journal of Science, 13*(27), 2–9.

Mitchel, O. M. (1860). *Popular astronomy*. Phinney, Blakeman & Mason.

Moriarty, H. A. (1888). Navigation. In: *Encyclopedia British* (9th edn, vol. 17, pp. 250–277). Henry G. Allen & Co.

North American Mail. Ocean steamers—foreign mails. In: *The companion to the Almanac: Or year-book of general information for 1851* (pp. 61–65). Charles Knight.

Pogo, A. (1935). Gemma Frisius, his method of determining differences of longitude by transporting timepieces (1530) and his treatise on triangulation (1533). *Isis, 22*(2), 469–506.

Rieznik, M. (2013). The Cordoba observatory and the history of the 'personal equation' (1871–1886). *Journal for the History of Astronomy, 44*(3), 277–301.

Rothenberg, M. (1999). Sears cook walker. In *American national biography*, Vol. 22. Oxford University Press.

Saff, D. (2019). *From celestial to terrestrial timekeeping: Clockmaking in the bond family*. Antiquarian Horological Society.

Schaffer, S. (1988). Astronomers mark time: Discipline and the personal equation. *Science in Context, 2*(1), 115–145.

Schmidt, R. E., & Dearden, P. (2019a). The Liverpool observatory at waterloo dock: Part I. *Antiquarian Astronomer, 13*, 2–22.

Schmidt, R. E., & Dearden, P. (2019b, 2020). The Liverpool observatory at waterloo dock: Part II. *Antiquarian Astronomer, 14*, 77–108.

Schott, C. A. (1898). The telegraphic longitude net of the United States and its connection with that of Europe, 1866-1896. Report of the Superintendent of the U.S. Coast Geodetic Survey for June 30, 1897. Part II, Appendix 2, pp. 201–261.

Schumacher, H. C. (1822). Chronometrische bestimmung des längenunterschiedes zwischen Copenhagen und Hamburg. *Astronomische Nachrichten, 19*, 303–304 and *20*,305–306.

Shadwell, F. A. (1861). *Notes on the management of chronometers and measurement of meridian distances*. J. D. Potter.

Sobel, D. (1995). *Longitude: The true story of a lone genius who solved the greatest scientific problem of his time*. Walker & Co.

Stachurski, R. (2009). *Longitude by wire: Finding North America*. University of South Carolina Press.

Stated Meeting, April 15 (1842). *Proceedings of the American Philosophical Society, 2*(21), 164–170.

Stephens, C. E. (1987). Partners in time: William bond & son of Boston and the Harvard college observatory. *Harvard Library Bulletin, 35*(4), 351–384.

Stephens, C. E. (1990). Astronomy as a public utility. *Journal for the History of Astronomy, 21*(1), 21–35.
Struve, F. G. W. (1844). *Expédition chronométrique entre Poulkova et Altona.* Imperial Academy of Sciences.
Struve, F. G. W., & Struve, O. W. (1846). *Expédition chronométrique entre Altona et Greenwich.* Imperial Academy of Sciences.
Theberge, A. E. (2007). *The coast survey 1807–1867.* National Oceanic and Atmospheric Administration.
Tiarks, J. L. (1824). A short account of some observations made with chronometers, in two expeditions sent out by the admiralty. *Philosophical Transactions, 114*, 360–371.
Van der Werf, S. Y. (1997). The lunar distance method in the nineteenth century. *Navigation, 44*, 1–13.
Walker, S. C. (1849). Application of the galvanic circuit to an astronomical clock and telegraph register in determining local differences of longitude, and in astronomical observations generally. *American Journal of Science, 7*(20), 206–217.
Walker, S. C. (1850). Fourth day, August 17, 1849. In: *Proceedings of the American association for the advancement of science*, 2nd meeting, August 1849, pp. 182–192.
Whitney, M. E. (1984). *The ship's chronometer*. American Watchmakers-Clockmakers Institute.
Withers, C. W. J. (2017). *Zero degrees: Geographies of the prime meridian.* Harvard University Press.

Trudy E. Bell (M.A., New York University, 1978) is a contributing editor for *Sky & Telescope* and member of the editorial advisory board for Springer's *Historical & Cultural Astronomy* series of books. She is co-editor (along with William Sheehan, Robert W. Smith, and Carolyn Kennett) of *Neptune: From Grand Discovery to A World Revealed* (Springer, 2021). She has been senior writer for the University of California High-Performance AstroComputing Center (UC-HiPACC), senior editor for *IEEE Spectrum,* and member of the editorial board for *Scientific American.* She has been a recipient of the David N. Schramm Award from the American Astronomical Society and the Herbert C. Pollock Award of the Dudley Observatory. Her research interests are nineteenth-century U.S. observatories and their instruments, instrument-makers, and philanthropists.

Chapter 14
Astronomy and Chronometry: Experimental Practices to Improve Hydrography during the Early Nineteenth Century

Emily Akkermans

Abstract Developments in astronomical and chronometric instruments and methods in the late-eighteenth century transformed the practice of navigation at sea. Improvements in maritime navigation went hand in hand with improvements in surveying and the determination of longitude on land. Astronomical observations determined longitude on land; chronometrical longitudes linked these to other places. In the early nineteenth century, however, there was still little consensus on the exact longitudes of many maritime ports. This article examines the first British expedition to measure the difference in longitude using chronometric differences. An examination of the correspondence between the Admiralty and the Board of Longitude reveals issues of authority and trust in methods, instruments and their users. This one-sided correspondence, from John Croker to Thomas Young, reveals many of the uncertainties surrounding the issuing of chronometers, Pond's authority as superintendent, the relationship between the Board and the Admiralty, and the accuracy of the methods to be used.

> The chronometric longitudes of places visited are not required, but the "meridian distance," or difference of longitude in time, as shown by each chronometer between those places from the observations, is to be carefully recorded.[1]

[1] Shadwell (1861, p. 216).

E. Akkermans (✉)
National Maritime Museum (trading as Royal Museums Greenwich), Greenwich, London, England
e-mail: EAkkermans@rmg.co.uk

G. Bernasconi et al. (eds.), *Astronomical Observatories and Chronometry, 18th-20th Century*, History of Physics, https://doi.org/10.1007/978-3-031-95874-8_14

So wrote Captain Charles Frederick Alexander Shadwell (1814–1886) in his *Notes on the Management of Chronometers and the Measurement of Meridian Distances*, first published in 1861. As chronometrical measurements of meridian distances increased, so too did the need for standardised instruction and record keeping. Much of what Shadwell wrote was founded on the chronometric expeditions of the nineteenth century. In his words "many scattered hints, and some valuable information relative to the application of chronometers to the accurate deduction of differences of longitude, exist in many detached works, but the subject seems to require condensation on some points, amplification on others, and systematic arrangement as a whole."[2] The scattered hints Shadwell referred to consisted of previous expeditions,[3] papers and publications on navigation,[4] and various French publications on the subject.[5] The principle of chronometric meridian distances relied on transporting a larger number of chronometers between two places and using the measurements obtained to determine the relative distance between the two places. From the 1820s, the Admiralty in collaboration with the Board of Longitude launched several expeditions to put this method to the test.

However, this method was experimental in nature, as at the time opinions on the appropriateness of different methods for determining longitude differed, and not all users trusted chronometers implicitly.[6] In addition, instruction on the use and management of chronometers was sparse. This meant that officers measuring chronometric differences took different approaches when dealing with instruments and managing the data they produced.

Through collaborations between the Admiralty, the Board of Longitude, and the Royal Society, instrumental precision and accuracy transferred from the domain of astronomers and observatories to an emerging group of scientific naval officers.[7] These naval officers commanded scientific expeditions during the nineteenth century. Many of these officers were trained under the supervision of Henry Kater (1777–1835), a Royal Navy officer who worked closely with Thomas Young, Secretary to the Board of Longitude, (1773–1829) on pendulum experiments to measure the figure of the earth and trigonometrical surveys in the 1820s. As such, "naval officers were taught the operation of Kater's invariable pendulum as well as the most effective way to record observations".[8] A result of this was that when instructions on the use of chronometers did appear they were based on the practices adopted by naval officers who had considerable experience in astronomical observations conducted on land.

[2] Shadwell (1855, p. viii).

[3] Foster, Adventure and Beagle, Owen, Belcher.

[4] Henry Raper's *Practice of Navigation*, various papers in the Nautical Magazine (Raper, Fisher, Bayfield, Bedford) and from the Memoirs of the Astronomical Society.

[5] Pierre Daussy's *Connaissance des Temps*, Aristide Lieussou's *Sur la Marche Des Chronometres* and *Recherches sur les Variations de la Marche des Chronometres*.

[6] The methods listed for example by Henry Raper's papers in the Nautical Magazine were chronometers, lunar distances, observations of Jupiter's satellites, occultations of the stars and moon, and Eclipses of the Sun and Moon.

[7] Cock (2005, p. 95-112).

[8] Waring (2014, p. 109).

This was because shore-based observations remained crucial in determinations of meridian distances.

Operations of an Experimental Nature: Tiarks's Expedition to Madeira

The first expedition to formally take place in Britain with the aim of measuring meridian distances was a collaboration between the Admiralty and the Board of Longitude and it considered the difference of longitude between Falmouth and Madeira. Madeira was an important stop for Royal Navy vessels, but the longitude of its port in Funchal was disputed. In 1818, by orders of the Admiralty, Captain Bartholomew (1767–1821) received notice to "ascertain that point, and that, generally, all HMS [His Majesty's Ships] proceeding in the direction of Madeira, with time keepers on board, will have directions to attend to the same subject."[9] But by 1822, there was still no agreement on the matter, and Astronomer Royal John Pond (1767–1836) determined that the longitude of Funchal was 5" in error. Following this, that same year, John Lewis Tiarks (1789–1837) was appointed as astronomer to a voyage solely to determine the meridian distance between Falmouth and Madeira.[10] In a letter to Young, First Secretary to the Admiralty John Croker (1780–1857) asserted that "their Lordships … perfectly agree with the Board of longitude as to the expediency of ascertaining with the greatest possible accuracy the longitude of Funchal" and to facilitate the matter, they would "furnish a proper vessel for this purpose" whilst the Board should "point out the time, and have prepared the instruments and the instructions for the observers."[11] This one-sided correspondence, from Croker to Young, reveals a great deal of the uncertainties surrounding the issuing of chronometers, Pond's authority as superintendent, the relationship between the Board and the Admiralty and the accuracy of the methods to be employed.

After receiving the order described above, Young received another letter on 22 June 1822. Here, Croker stated that "the Board of Admiralty cannot give any opinion as to the person who may be fit to make the observations at Madeira. It will be the duty of the Board of Longitude to select a proper person for that purpose and to notify the same to the Admiralty" and that the "Resident Committee should prepare the instructions for the person and should determine as to the number of chronometers & tc to be taken out and the mode of superintending them."[12] Ten days later, Croker wrote again to Young asking him to act immediately, as "the season [was] proper" and if he saw "no prospect of getting a competent observer to undertake the observations

---

9 Barrow to Young (1818, p. 157r).

10 Dr John Ludwig Tiarks was a German astronomer (1789–1837). He served as assistant librarian to Joseph Banks and was later appointed as British Astronomer to determine the American Boundary line.

11 Croker to Young, Admiralty Office (1822). CUL RGO 14/49 (p. 161r).

12 CUL RGO 14/49 (p. 164r).

ashore, not a moment should be lost in selecting ten or a dozen of the best going of the chronometers in order to their being immediately embarked in a frigate who should be directed to convey them to Madeira and back, departing from and returning to Falmouth."[13] This appears to have resulted in a miscommunication between the two, as Croker wrote again three days later asking "when will the doctor and his instruments be ready to embark? I presume that you mean, tho' it is not clearly expressed, that the frigate and the chronometers should run out to Madeira, land the Doctor and return immediately to Falmouth and thence sail back to fetch the Doctor and the instruments. To save time I have written to the astronomer royal to select 6 8 or 10 of the best chronometers and to give their makers notice of the intended trial."[14]

The Admiralty organised the endeavour and selected a suitable vessel. In turn, they trusted the Resident Committee and the Astronomer Royal to select the appropriate individuals for making celestial observations and another for the care of the instruments. However, the letters reveal that Young believed a chronometric measurement alone would be sufficient, whilst it was the Admiralty's intention that the method should be tested against astronomical observations, to establish the accuracy of chronometric measurements. Croker made that very clear the following day when he again corresponded with Young:

> Of course we must be satisfied with what shall be thought satisfactory by the Board of Longitude, but it occurs to us that the observations on shore should be made with great care, nicety, and deliberation and although it may be expected that the chronometers will give a nearer approximation to the truth, than the observations, this must be mere expectation and one advantage of the experiment will be the bringing careful observations to bear upon the best chronometrical calculations. It seems therefore that the observations by Dr. Tiarks should be made in the most complete and scientific manner, else, they had better, for all our sakes, not be made at all.[15]

The Admiralty wanted two determinations of longitude; one made on shore by observations, the other by "chronometers afloat."[16] Tiarks was to be the on shore observer, and a second person, selected by Pond and the Resident Committee, was to have care of the chronometers. Croker had made clear in a conservation with Young that he wished "to keep the result of the chronometer time a secret from the observer."[17] In addition to this, Young believed that one trip to Funchal should be sufficient, whilst Croker wished for the chronometers to take two trips. Young also proposed that "little stress is to be laid on the observations and a great deal on the chronometers" leading Croker to reply that this would "diminish the chronometer experiment by one half without at all improving the other."[18] Croker then requested Young and Tiarks to call on him to discuss the matter. Judging by the manner in which

[13] CUL RGO 14/49 (p. 167r).

[14] CUL RGO 14/49 (p. 169r).

[15] Croker to Young, Admiralty Office (6 July 1822). CUL RGO 14/49 (p. 171r–171v).

[16] Croker to Young, Admiralty Office (8 July 1822). CUL RGO 14/49 (p. 173r).

[17] CUL RGO 14/49 (p. 173r).

[18] CUL RGO 14/49 (p. 173r).

the voyage proceeded, it seemed that Young was able to convince the Admiralty that the method of chronometric distances would be sufficient for the purpose. Tiarks subsequently sailed with his chronometers to Madeira and back between 24 July and 29 August 1822. At Falmouth and Madeira, he determined the rates of the chronometers by equal altitudes of the sun. No other astronomical observations were made.

Meanwhile, Pond was ordered to "select a sufficient number of the best going chronometers."[19] Though he was trusted with selecting these instruments, Pond was seen to overstep his authority by authorising an additional four instruments to the total of twelve issued by Admiralty order, which they did not approve. As with the discussion between Croker and Young, we can only follow the comments made from the Admiralty in the digest held at The National Archives in Kew and so Pond's reason for adding the chronometers remains unclear. An additional pocket chronometer was approved, but their Lordships disapproved of the other instruments and ordered "Captain Spencer and Dr Tiarks not to take any account of their rates of going."[20] Eventually, Pond managed to convince the Admiralty of his reasons for supplying these additional instruments, who permitted "all the chronometers to proceed, tho' there are some points which they do not yet perfectly understand and could have wished that the Private Watches had not been sent."[21] It is surprising that the Admiralty were so opposed to Pond supplying these additional instruments, considering a greater number of instruments could improve the calculations. However, from this example we can see that the Admiralty had to approve each additional chronometer.

Pond reported back to the Admiralty on return from the voyage. He found that the "discordances [were] not greater than I expected from previous experiments and I have little doubt but that the difference of longitude of these two places is now as correctly known as that of Greenwich and the principal observatories in Europe." He added that though "future revision may improve this result a very small quantity" the "performance of the chronometers will remain exactly the same." "Upon the whole" Pond was "persuaded the experiment has been as successful as it was important."[22] In 1823, on recommendation from the Board of Longitude and agreed by the Admiralty, Tiarks embarked on a similar voyage to determine the difference of longitude between Dover, Falmouth, and Portsmouth. Following this, Young again suggested that Tiarks measure the distance of Heligoland in relation to the coast of Germany and Greenwich: "employing a steam boat in this service, and adopting the same method that was adopted in regard to the Island of Madeira."[23] Consequently, the Admiralty ordered similar experiments to be performed upon a variety of Royal Navy expeditions, notable ones including the voyages of HMS *Adventure,* under command of Captain King (1826–1830); HMS *Chanticleer,* under command of Captain Foster (1828–1831) and HMS *Beagle*, under command of Captain Fitzroy (1831–1836). All

---

[19] Minute of the Board of Admiralty (1822). TNA ADM12/210.

[20] Admiralty Digest (1822). ADM 12/210, tab 98a Mathematical Instruments.

[21] Admiralty Digest (1822). ADM 12/210, tab 98a Mathematical Instruments.

[22] Pond to the Admiralty (1822). TNA ADM 1/3462.

[23] Barrow to Young, Admiralty Office (1822). CUL RGO 14/49 (p. 186r).

three expeditions sailed with a high number of chronometers to measure meridian distances in the same manner as Tiarks had done before them.

Managing Instruments: Collecting and Collating Data

Overall, the Admrialty, astronomers and naval officers involved in measuring chronometric meridian distances agreed that the method could improve hydrographic knowledge. However, there were still many smaller disagreements regarding the instrumental practices adopted and the method for calculating rates and errors. The expedition of HMS *Chanticleer* demonstrates these ongoing issues.

Henry Foster (1797–1831) was one of these officers trained by Kater. Appointed to command the *Chanticleer* in 1828, Foster set out on "an enterprise destined solely and simply for the promotion of scientific research, and the extension of the bounds of human knowledge."[24] The voyage was a continuation of the pendulum experiments to "establish the true figure of the earth and the law of the variation of gravity in different points of its surface."[25] Foster had previously accompanied Captain Basil Hall (1788–1844) to South America in 1820 to assist in pendulum measurements and in 1827, he received the Copley Medal for his pendulum experiments on Parry's third voyage. An ideal candidate, Foster was described by a midshipman on the *Fury*, as "a distinguished navigator; he was an excellent officer, the best nautical scholar I ever knew, and a good astronomer."[26] He was also a "most intimate friend" of Hydrographer Francis Beaufort (1774–1857) and enjoyed the support of Young.[27] In collaboration with hydrographers from the Navy and the East India Company, the Royal Society selected fourteen pendulum stations that were critical for the expedition, but also beneficial for hydrography and navigation in general.[28] The Royal Society considered all other objects of the expedition secondary to the pendulum experiments. A secondary aim of the expedition was to measure relative meridian distances with chronometers. The chronometric objective of the expedition was adopted primarily to connect points of longitudes determined astronomically at each of the pendulum stations to determine differences of longitude. By employing many chronometers, and by determining the time at principal ports and commercial stations, the committee hoped to judge the efficacy of the method and to approximately measure "a great number of smaller intervals of high importance to navigation" without interfering

[24] Minutes RS committee (1828). Royal Society Archive CM1.

[25] Minutes RS committee (1828). Royal Society Archive CM1.

[26] SPRI MS 1562;D. An account of Parry's third voyage, probably written by Berkley Westrope, midshipman HMS Fury, Published in *The Times* (Halifax Nova Scotia) as follows: *17 March 1835, 21 April 1835, 21 April 1835, 6 October 1835*. photocopied typescript.

[27] Webb (2010, p. 63); Waring (2014).

[28] The stations were: Para, Maranham, Fernando de Noronha, Ascension Island, Porto Bello, Trinidad, St Helena, Cape of Good Hope, Monte Video, Greenwich, Staten Island, Cape Horn, Deception Island, Port Bowen.

with the primary aim.[29] For these determinations, the Admiralty ordered Pond to supply Foster with twelve of the best chronometers available.[30] However, similar to the expedition to Madeira, extra instruments were added and the *Chanticleer* sailed with 15 chronometers.

Fortunately, we can gain more insight into the practices from the *Chanticleer* as not only the chronometric data and calculations remain, but also Tiarks comments on the practices adopted by Foster. They allow us to understand the importance of managing all the observational data, the comparison and rate books, and the calculations and corrections made during and after the voyage. As Foster drowned after falling from a canoe in the river Chagres in 1831, the *Chanticleer* returned to England under command of first lieutenant Horatio Thomas Austin (1800–1865). On its return, the Admiralty appointed Tiarks to examine the measurements and data that Foster had produced.

Foster primarily followed the same instructions as Tiarks had been given. On arrival and pre-departure at each station, he established the error and the rate of each chronometer. The run between two stations was kept as short as possible as the accuracy of the method was believed to decrease over longer interval.[31] Pre-departure, the error of the chronometer on the Mean Time at the place was determined by either Equal or Absolute Altitudes of the Sun. When the error was determined over consecutive days, a rate was calculated and applied to the next run of the journey. During the run, the daily noon comparison was kept to check the function of each chronometer. Compiled into a table, Foster used this data to select which chronometers he thought reliable to use for the meridian distance. The function of this table was two-fold: it allowed Foster to check the function of each chronometer over time, but it also noted the exact difference between each chronometer and the standard at noon each day.[32] Because of this, the noon tables of comparison were one of the most important parts of this practice. Without these, each chronometer would need to be rated separately. In addition, relevant information could be easily extracted from this table, and finally, the data could also be readily recalculated if they found the longitude used to establish one meridian distance changed (Fig. 14.1).

The data that Foster sorted and organised in this way was essential for two reasons. It was "data in comparable, numerical form," which followed the growing trend of scientific pursuit of the nineteenth century.[33] This data was useful to others. It could be analysed, compared, adjusted, and transferred to support other scientific pursuits. It also proved that the chronometric measurements had been performed adequately by the officers of the voyage. Astronomical observations could not be redone, and so trust in the abilities of the officers performing these observations

---

[29] Minutes RS committee (1828). Royal Society Archive CM1.

[30] Admiralty minutes (1828). TNA ADM3/216.

[31] Basil Hall to the Admiralty, Edinburgh, 7 April 1820, TNA, ADM1/1956. See Bennett (1987).

[32] The standard chronometer was checked against astronomical observations, and officers then compared all other chronometers against it. Foster selected the chronometer that had performed the best during trial at the Royal Observatory.

[33] Godlewska (1999, p. 245).

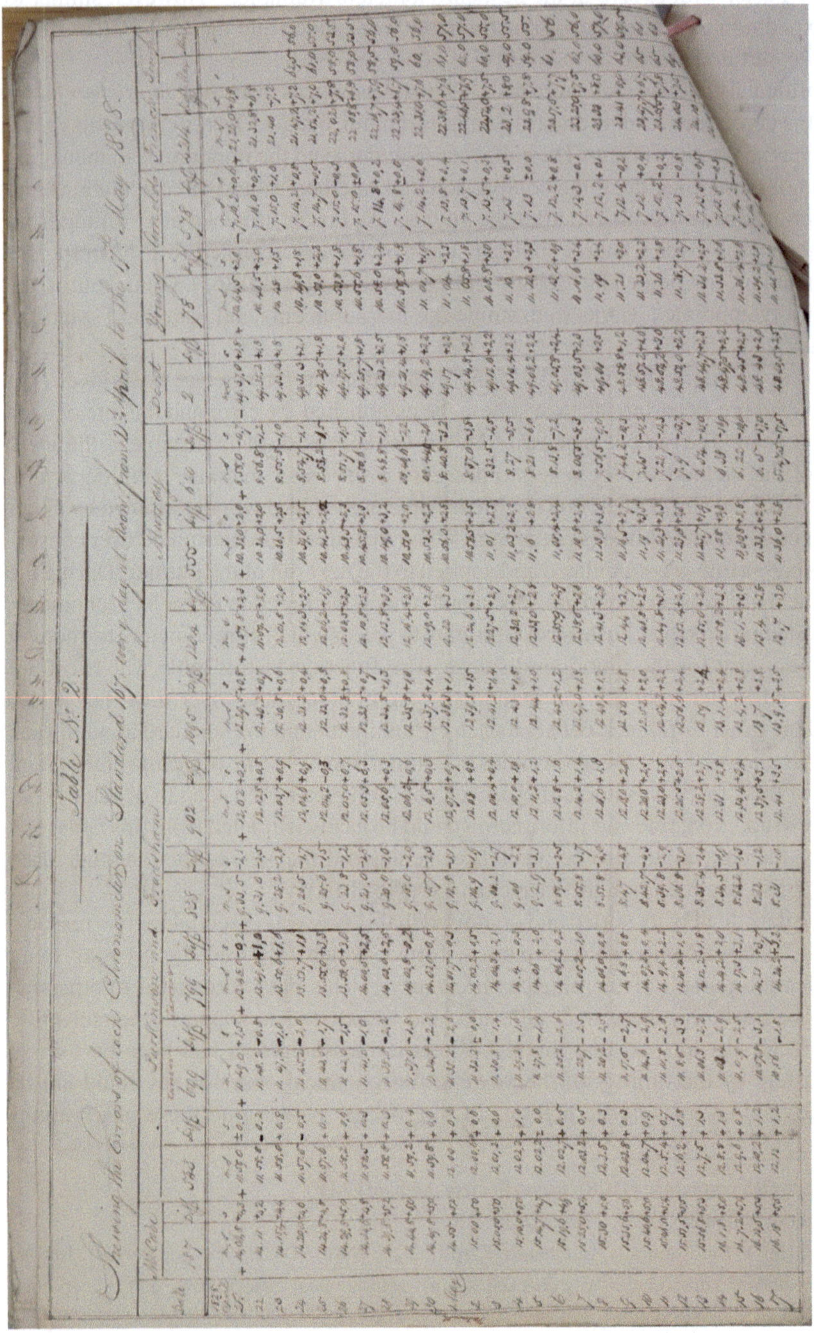

Fig. 14.1 “Table N. 2 shewing the errors of each chronometer on standard, every day at noon, from 21st April to the 17th May 1828.” UKHO, AO32: SFD7/7/1

remained essential. By organising and tabulating the calculations and corrections resulting from these observations, they could be examined and corrected if necessary. It was common for the officer of the voyage to perform this duty himself. But in Foster's case this was done by Tiarks, who corrected some errors that Foster had made in his calculations, errors that Foster would have detected himself. But Tiarks also modified the data by using different rates from Foster. He found fault with the rates Foster had determined at Falmouth, and therefore discarded those completely. Instead, rather than calculating the meridian distance between Falmouth and Madeira using the mean rates as had Foster, Tiarks used only the rates determined at Madeira. This data thus mattered: it could be moulded to support an outcome. As Tiarks illustrated: "with a view to examine the correctness of Captain Foster's calculations, and at the same time to see whether more accordant results could be adduced, I have tried other rates, generally those nearest to the observations on which the longitudes depend."[34] Tiarks applied his own judgement when selecting which rates to apply to a meridian distance. He "endeavoured to find the most probable results for the meridian distances chronometrically determined during the expedition." According to Tiarks, the "near agreement of Captain Foster's equal altitudes shows his great skill in observations of this kind, and the advanced state and correctness of the calculations afford unequivocal proof of his indefatigable industry." Tiarks did not doubt the reliability of the instruments, "as all were considered very excellent ones," but rather questioned Foster's method of storing them during the voyage.[35] Tiarks had issues with another practice followed by Foster: namely the suspension of the chronometers from the deck in his cabin. Due to the alteration of their rates over very short intervals, Tiarks was "inclined to believe that Captain Owen is right in finding fault with the manner in which the chronometers were suspended from the deck of the Chanticleer." He added that his own experience confirmed that "suspension from the upper deck of a vessel is not favourable to the going of timepieces."[36] He corrected Foster's longitude for Madeira, using the rates Foster had determined at Funchal without including the Falmouth ones. Tiarks thought the run had been too long and that the rates had altered after the ship's departure.[37]

Collecting and Correcting Longitudes: The Importance of Place

[34] Tiarks (1834, p. 228).
[35] Tiarks (1834, p. 225).
[36] Tiarks (1834, p. 226).
[37] Tiarks (1834, p. 229).

> In fact hydrography has now arrived at a phase corresponding to that of Astronomy, at which the positions of the smaller stars, instead of being made the object of direct observations, are inferred with more convenience and precision from comparison with a few fundamental stars.[38]

With this quote, Henry Raper sums up in essence what he saw as the ideal way to use chronometers to improve hydrography. Whether making *absolute* astronomical observations of longitude on shore or measuring *relative* chronometrical differences of longitude by ship, no two measurements would ever agree. For Raper this meant that "the whole mass of positions is kept in a state of perpetual fluctuation, from which it is impossible that universal precision can ever be obtained."[39] Every data set that the Hydrographic Office received was slightly different from those previously gathered. Absolute universal precision may not be obtained, but by analysing the accumulated data, agreement could be achieved. Raper defined *absolute* positions as those determined astronomically and *relative* positions as those determined chronometrically. He believed that these two aspects would lead to the "ultimate perfection of hydrography," which would be attained "solely by the chronometric measurement of small distances, finally depending upon certain points determined by unimpugnable astronomical observations."[40] These certain points he termed "secondary meridians" (as according to him only Greenwich or Paris could be considered the zero, or prime meridian) which would serve as a "regulating meridian of a survey."[41] Raper thus set out to establish fixed longitudes for a total of eighteen secondary meridians, based on information accumulated at the Hydrographic Office. These were "determined nearly enough for present purposes" but would require a "long series of astronomical observations" to settle their positions.[42] This brings back into focus the discussion between Croker and Young about trust in the different methods: only extensive astronomical observations could settle the longitude of Raper's secondary meridians.

Place was thus key to coordinating the efforts to standardise longitude. Navigational manuals commonly included lists containing the latitude and longitude of places around the globe useful to navigation; these were not only listed as a destination, but also for navigators to use as a departure.[43] Many of these positions were inaccurate. This added to the difficulty in establishing where one was and where one was going. It was not until progress in charting had been made that a more accurate determination of these positions became beneficial. Both factors, the accuracy of charts and the positions of latitude and longitude, were imperative for chronometric methods to succeed. Astronomer Royal John Flamsteed had made this point in 1697: "Tis in vain to talk of the Use of finding the Longitude at Sea, except you know the

---

[38] Raper (1839a, b, c, d, e, f, p. 402).

[39] Raper (1842, p. 379).

[40] Raper (1839a, b, c, d, e, f, p. 320).

[41] Raper (1839a, b, c, d, e, f, p. 399).

[42] Raper (1840, p. xi).

[43] Moore (1804); Norie (1805); Bowditch and Kirby (1809); Riddle (1824).

true Longitude and Latitude of the Port for which you are designed" (May, 1973, 29).[44]

This is also apparent from Foster's material. Foster indicated the exact spot where he took his observations for the latitude and longitude in a detailed survey, since it was important that other observers could locate it for use in their own observations, or at least, refer to this precise spot (Fig. 14.2). This would also help those working in the Hydrographic Office coordinate the charts and surveys the Office received. Astronomer Edward Sabine also noted this in 1825:

> The revised tables should contain an additional column to those in the tables at present esteemed as of the best authority, for the purpose of specifying the spot to which the geographical position refers; without such specification, it is quite superfluous to insert the data, as is now done; to seconds of space. The spots should also be selected, as far as might be possible, with reference to their conveniency of access, with instruments, from vessels in the harbour.[45]

Pond raised this point when discussing the longitude determinations made by Tiarks, as he had

> Great objections against the plan proposed by Major Capt. Kater and thinks that the vessel must go to a spot as near as possible to the meridian of Greenwich which is connected with the Great Trig. Survey and will admit of the vessel being safe at anchor and my going daily on shore for making the observations for time. The captain of the vessel must, therefore by consulted in selecting a proper spot.[46]

This did not always work as planned. Robert Fitzroy, when arriving at Fernando Noronha to take observations for the chronometers, found it "difficult to ascertain the house in which his pendulum observations were made. Not even the Governor could tell me, for he had arrived since Captain Foster's departure."[47] The same difficulties were experienced during the "Trigonometrical Operations" which intended to measure the difference of longitude between Greenwich and Paris in 1821 where the exact spots used for previous measurements could not be determined.[48]

In discussing the validity of a point of longitude, Raper made clear that secondary meridians should be determined by astronomical observations which could then be connected chronometrically with other places. These astronomically determined positions could then be connected chronometrically to surveys across the globe. Corrections in relation to one of the secondary meridians could be transferred through these connections to any other places affected. Raper also elaborated on the difficulty of judging the results: "when two or three successive determinations agree nearly together we are apt to assume in general that they are right. This is however by no means certain," he continued citing the example of the difference of longitude between Greenwich and Paris, "in which two or three equally bad results will often

[44] Flamsteed, writing to Pepys on 21st April 1697.

[45] Sabine (1825, pp. 401–402).

[46] Tiarks to Young (1823). CUL RGO 14/49 (p. 183r).

[47] Fitzroy (1839, p. 59).

[48] Withers (2017, p. 110).

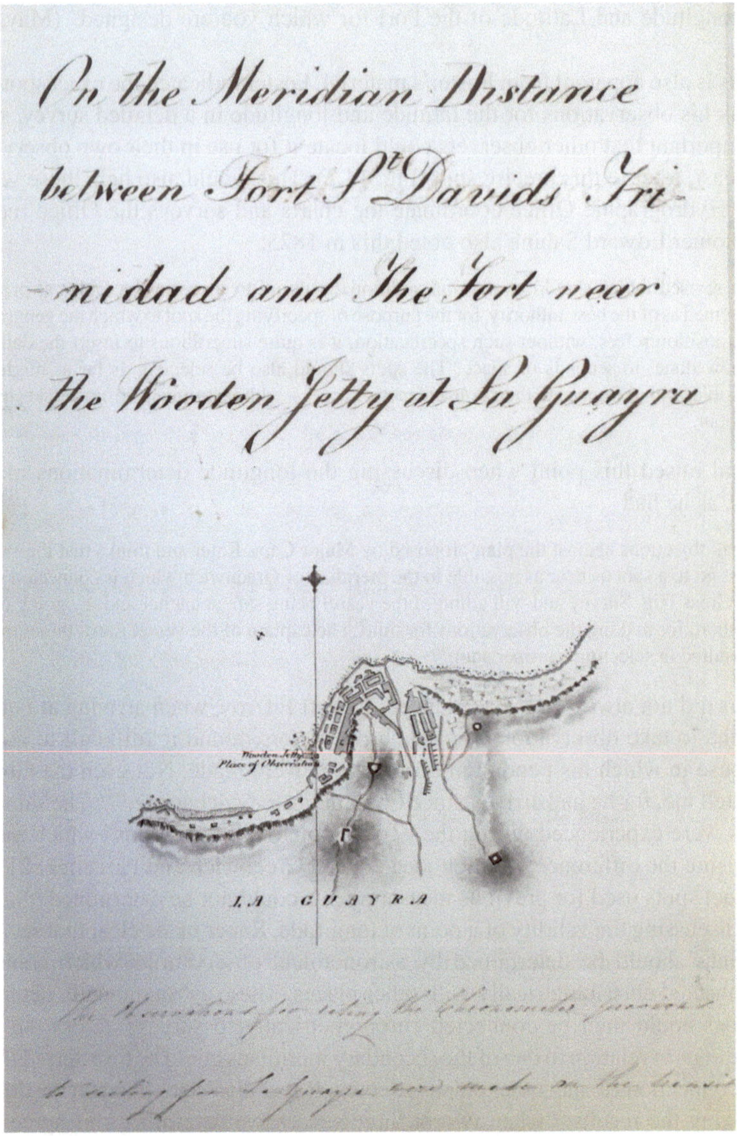

Fig. 14.2 Foster's report on the meridian distance between Trinidad and La Guayra. UKHO, AO32: SFD7/7/1/2

be found following each other. It is therefore open to discussion whether the first determinations of a place, even though they agree, are right or wrong." Raper did believe however, that if determinations differed widely from one another, then the truth was likely to be found somewhere "between them, because such great discrepancy most probably arises from the presence of errors of opposite kinds." In some cases, however, Raper suggested that "the evidence divides itself in support of two distinct determinations … in such cases the probability seems to be that one of the determinations is true and the other false."[49]

Raper's lists of maritime positions published in 1839 included descriptions of how these determinations were made. These details included whether they were made astronomically (by Jupiter's satellites, lunar observations, occultations, eclipses of the Sun or Moon or moon-culminating stars) or chronometrically, when and by whom. Raper compared these to form a determination, which although not itself always precisely accurate, was at least a fixed point of reference that all navigators and surveyors should refer to.

Conclusion

During the early nineteenth century, the Admiralty increasingly sought to improve their charts by refining the accuracy of longitudes of important ports. Experimental practices were carried out by a small group of astronomers and naval officers under the oversight of the Board of Longitude and the Admiralty.

These earlier voyages demonstrate some of the uncertainties that occurred in the measurement of relative longitudes by several chronometers. Due to these uncertainties and the necessity of undertaking astronomical observations that underpinned the method, only naval officers with exceptional astronomical abilities were trusted to undertake these experiments. They were funded and supported by collaborations between the Board of Longitude, the Admiralty, and the Royal Society who oversaw the issuing of instruments and the evaluation of the results.

There were varying views on how much trust could be placed on chronometric versus shore-based astronomical measurements. On the expedition under Tiarks to Madeira, the Board of Longitude advised an astronomer should be appointed to the task, and the Astronomer Royal John Pond was tasked with supplying the best chronometers held by the Royal Observatory. This was certainly a question of trust that extended beyond the method employed, to the trust placed in the instruments and the abilities of the chosen observer.

The officers performing these roles were highly skilled in astronomical observations and often trained by members of the Royal Society to follow instruction and record instrumental practices. Despite this, there were still many variations in how each officer approached both the management of the instruments, the method for determining error and rate, the process of recording the data, and the reliance placed on individual instruments. As a result, the determinations of longitude measurements varied. Tiarks examination of Foster's calculations shows how even with the same instruments and observer, different outcomes could be achieved.

Therefore, an important aim of the naval officer Henry Raper was to evaluate all these varying determinations and select what he considered the best approximation

[49] Raper (1839a, b, c, d, e, f, p. 476).

of the true value, to be printed in lists of longitudes that would then serve as the base points for surveying. As we saw with Raper's evaluation of methods, astronomical measurements on shore were again seen to be more trustworthy than chronometric measurements. However, combining the two would lead to huge improvements in hydrography.

Twenty years later in 1861, Shadwell combined all of this in his "Notes on the Management of Chronometers and the Measurement of Meridian Distances". His aim was to establish guidelines for those wishing to pursue this method at sea. It brought together the concerns that the officers faced in managing chronometers and explored different ways of calculating the rate and error of the instruments. Shadwell's publication shows how the experimental practices examined here contributed to guidance written in the latter half of the nineteenth century.

References

Admiralty Digest. (1822). *ADM 12/210, tab 98a mathematical instruments.*

Admiralty minutes. (1828). *TNA ADM3/216.*

Barrow, J., to Young, A. (1818). *CUL RGO 14/49* (p. 157r).

Barrow, J., to Young, A. (1822). *CUL RGO 14/49* (p. 186r).

Bennett, J. (1987). *The divided circle: A history of instruments for astronomy, navigation and surveying*. Phaidon.

Bowditch, N., & Kirby, T. (1809). *The improved practical navigator* (3rd edn.). James and John Hardy.

Cock, R. (2005). Scientific servicemen in the royal navy and the professionalisation of science 1816–1855. In: D. Knight, & M. D. Eddy (Eds.), *Science and beliefs: From natural philosophy to natural science 1700–1900* (pp. 95–112). Ashgate.

Croker, J., to Young, A. (1822). *CUL RGO 14/49* (p. 161r).

Fitzroy, R. (1839). *Narrative of the surveying voyages of his majesty's ships adventure and beagle, proceedings of the second expedition 1831–36*, vol. II. Henry Colburn.

Flamsteed, J. (1697). *Letter to Pepys, 21 April 1697.*

Godlewska, A. M. C. (1999). From enlightenment vision to modern science? Humboldt's visual thinking. In: D. N. Livingstone, & C. W. J. Withers (Eds.), *Geography and enlightenment* (pp. 236–275). The University of Chicago Press.

Hall, B. (1820). *To the admiralty, Edinburgh. TNA ADM1/1956.*

May, W. E. (1973). *A history of marine navigation.* G. T. Foulis & Co Ltd.

Minute of the Board of Admiralty. (1822). *TNA ADM12/210.*

Minutes RS committee. (1828). *Royal Society Archive CM1.*

Moore, J. H. (1804). *The new practical navigator* (16th edn.). G. and J. Robinson.

Norie, J. W. (1805). *A new and complete epitome of practical navigation.* Printed for the Author and William Heather.

Pond, J. (1822). *To the admiralty. TNA ADM 1/3462.*

Raper, H. (1840). *The practice of navigation and nautical astronomy* (1st edn.). R. B. Bate.

Raper, H. (1842). *The practice of navigation and nautical astronomy* (6th edn.). J. D. Potter.

Raper, H. (1839a). On the longitudes of the principal maritime points of the globe. *Nautical Magazine,* 259–270.

Raper, H. (1839b). On the longitudes of the principal maritime points of the globe. *Nautical Magazine,* 543–549.

Raper, H. (1839c). Discussion of the longitudes of the principal maritime points of the globe. *Nautical Magazine,* 475–483.
Raper, H. (1839d). On the necessity of adopting secondary meridians. *Nautical Magazine,* 399–402.
Raper, H. (1839e). Remarks on the mode of determining longitude. *Nautical Magazine,* 319–326.
Raper, H. (1839f). On the longitudes of the principal maritime points of the globe. *Nautical Magazine,* 697–704.
Riddle, E. (1824). *Treatise on navigation and nautical astronomy.* Baldwin, Cradock, and Joy.
Sabine, E. (1825). *An account of experiments to determine the figure of the earth by means of the pendulum vibrating seconds in different latitudes.* John Murray.
Shadwell, C. F. A. (1855). *Notes on the management of chronometers and the measurement of meridian distances.* J. D. Potter.
Shadwell, C. F. A. (1861). *Notes on the management of chronometers and the measurement of meridian distances.* J. D. Potter.
SPRI MS 1562;D. (1835). Published in *The Times* (Halifax Nova Scotia) as follows: 17 March 1835, 21 April 1835, 6 October 1835. Photocopied typescript.
Tiarks, J. L. (1823). *To young. CUL RGO 14/49*, p. 183r.
Tiarks, J. L. (1834). Report on captain Foster's chronometrical observations. In: W. H. B. Webster (Ed.), *Narrative of a voyage to the Southern Atlantic Ocean, in the years 1828, 29, 30, performed in H. M. Sloop Chanticleer*, vol. 2. Richard Bentley.
Waring, S. (2014). *Thomas young, the board of longitude and the age of reform.* Unpublished PhD Dissertation, University of Cambridge.
Webb, A. (2010). The hydrographer, science and international relations: Captain Parry's contribution to the cruise of HMS chanticleer. *The Mariners Mirror, 96*(1), 62–71.
Withers, C. W. J. (2017). *Zero degrees: Geographies of the prime meridian.* Harvard University Press.

Emily Akkermans is Curator of Time at Royal Museums Greenwich. She obtained her PhD at the University of Edinburgh and has a background in clock restoration and metals conservation. Her research interests include the time keeping in the Royal Navy and the history of timekeeping at the Royal Observatory, with a particular interest in nineteenth-century time distribution.

Chapter 15
Colonial Time Machines: Chronometry and the Personal Equation Between Europe and South Asia

S. Prashant Kumar

Abstract The chronometric observatories of colonial India were, in a sense, both literal and figurative time machines. In the eighteenth century it became possible to think that the local structure of time provided access to its global structure – that clocks provided privileged access to universal history, both morally and practically. As an example of this process, I discuss the history of the Madras Observatory, a chronometric station built in 1791 under the rule of the East India Company in India to provide ephemerides and time signals to sailors and revenue surveyors throughout the Madras Presidency. Both revenue collection and the empire's extractive trade required ephemerides and chronometric data, which in turn depended on the astronomical regulation of chronometers. But measuring the variation in the measuring instruments themselves was a laborious process. To meet the demand, the Observatory employed specialised scribes, experts in jyotiśāstra [Sanskrit astral science], as observers and computers. The Brahmins carry and maintain modern instruments, but are fully identified with an ancient past by the Company's astronomers, who styled themselves as continuing the astral sciences in the subcontinent. Colonial chronometry meant giving the modern observatory sciences an autochthonous ancient prestige, while at the same time tying the Brahmins to the instruments they were hired to operate.

It has become a commonplace to say that assistants in nineteenth century observatories were trained and calibrated like machinery. Indeed, one could argue that observatories brought about a curious process of reversal. Humans became instrumental parts in a larger machine, while instruments themselves became objects of care, attention, and even affection.[1] In this paper, I ask how much of this picture was

---

[1] Tresch (2010).

S. Prashant Kumar (✉)
Department of History, University of Chicago, Chicago, USA
e-mail: spkumar@uchicago.edu

Archives at NCBS, Tata Institute of Fundamental Research, Bengaluru, India

G. Bernasconi et al. (eds.), *Astronomical Observatories and Chronometry, 18th-20th Century*, History of Physics, https://doi.org/10.1007/978-3-031-95874-8_15

specific to certain aspects of chronometric work, such as the maintenance of theodolites or the taking of meridional observations for ephemeris catalogues. Moreover, how does the situation change when the regular work of observatory assistants begins to encompass more than the rote work of chronometric observation? As I show, certain peculiarities of colonial observatories make it possible to ask and answer such questions in detail. My argument is that the routine work of data collection using the transit circle required a totally different kind of labour order than that required for discovery, which usually made use of an equatorial or altazimuth mount. I proceed via a comparative history of the use of the personal equation in organizing chronometric labour at Madras and Greenwich Observatories.

The personal equation's appearance in astronomy, where it was used to correct for individual errors of observation, has been understood by historians of science as a consequence of the intensification and increasing stratification of scientific labour schemes towards the middle of the nineteenth century.[2] The personal equation has also been used by historians to understand the epistemic traffic between astronomy, at one end of the Comtean hierarchy of disciplines, and those at the other, such as psychology and physiology.[3]

But historians have only begun to study the personal equation in the everyday practice of observational science worldwide, and the global career of quantified personality remains to be worked out.[4] By 1874, the personal equation was employed as a labour management technique by hydrological engineers in India. Indian observers at Roorkee, stationed along the banks of the Ganges river, used transit methods to measure the speed at which a barge passed through a canal. Drilled in the use of the personal equation by a team of observers from Greenwich, their work was part of a "global experimental programme, stretching from Mississippi to Paraguay," to study the performance of canals constructed by imperial powers. In Roorkee, as Michael Lewis shows, the personal equation appears as a technique of social control amongst British "hydraulicians" because observers' accuracy was seen to depend upon their "social, moral, and racial pedigree."[5]

Where Madras differed from Greenwich was in the ultimate purpose of chronometric observation. The ephemeris tables produced by Madras were used not only by mariners, as in Greenwich, but also by the colonial revenue and survey administration. Increasingly accurate mapping allowed the East India Company to collect revenue from cultivated land which might otherwise escape notice. So whereas in Greenwich the division of labour meant the everyday life of the observatory came to resemble a factory or an accounts office, in Madras, astronomy and accounting look similar because they were both part of the Company's revenue apparatus from the observatory's very founding. Indeed, the observatory's "native assistants" were hired from the same Brahmin communities as were employees of the district tax offices

---

[2] Schaffer (1988, pp. 115–145); Hoffmann (2007, pp. 333–365).

[3] Boring (1961, pp. 238–257); Benschop and Draisma (2001, pp. 1–25); Canales (2001, pp. 173–197).

[4] Rieznik (2013, pp. 277–301); Brinkmann et al., (2019, pp. 41–71).

[5] Lewis (2007, pp. 967–994).

in the revenue administration. The recruitment of Brahmins with scribal knowledge for observatory work had to do with the textual and computational nature of chronometric observation and maintenance.[6] As I show here, the racialization of the personal equation shows up earlier, in Madras observatory's caste-based labour order.

Inheriting Algebra at Greenwich

The story of the personal equation's use in organizing astronomical labour is by now well known. In 1835, George Biddell Airy became the seventh Astronomer Royal. Airy recognized soon after taking office that most observations needed for navigation and chronometry were amenable to the division of labour, and could be reduced to a number of specialized rote tasks.[7] So he hired a small army of assistants, and rigidly arranged them by education and social prestige. Cambridge graduates performed observational tasks, and supervised reductions carried out by youths of fifteen or sixteen, on so-called "computerships," hired young enough to be "tractable." Where observation was reformed along factory lines, computation was made to resemble an accountant's office, full of teenaged clerks examining not ledgers of sale or income tax schedules, but registers of stars and planets.[8] Most of the observational work was done by Airy, his first assistant, Robert Main (who had been sixth wrangler in the Cambridge Mathematical Tripos, and supervised most of the computation) and other Cambridge-educated observers.[9] Airy regarded the computers mere obedient drudges, trained for a specific task, and to be kept from learning much else. Most burned out by their mid-twenties, and cycled into other clerical work. For Airy, solutions to the problem of social order were ipso facto solutions to the problem of navigation.

But the massive expansion of Greenwich's astronomical labour pool brought about new problems. Before electromagnetic registration of transit times was introduced in the 1850s, positional astronomers used what was called the eye-and-ear method. Listening to the beat of a clock, the observer would register the passage of a star across reticule lines on the eyepiece of the transit instrument, recording time down to tenths of a second. But by the first decade of the nineteenth century, astronomers began to realize that experienced observers seemed to differ as to the actual time of transit by a "constant difference," measurable for any given pair of observers, but apparently impossible to eliminate entirely (Fig. 15.1).

A generation earlier, the Astronomer Royal Nevil Maskelyne's assistant, David Kinnebrook, had been fired for having "unfortunately commenced a vicious way

[6] Madras (2022).

[7] Chapman (1985, pp. 321–328); Dewhirst (1985, pp. 147–158); Schaffer (1988, pp. 115–145); Smith (1991, pp. 5–20); Ashworth (1998, pp. 151–178); Donnelly (2014, pp. 479–503).

[8] Belteki (2023, pp. 65–81).

[9] Hollis (1925, pp. 388–398); Belteki (2022, pp. 575–593).

1833.	Observed Right Ascension. Feb. 9.		Feb. 15.		Sum of Equations. Feb. 9.	Feb. 15.	Sum of Equations. Feb. 9.	Feb. 15.	Clock Rate in 6 Days.
	h m s		s		s	s	h m s	s	s
Persei - - -	2 53 48, 22	Taylor	43, 00	Ellis	+0, 80	+1, 03	2 53 49, 02	44, 03	− 4, 99 − c
τ^2 Arietis - - -	3 13 55, 06	,,	49, 88	,,	+0, 60	+0, 70	3 13 55, 66	50, 58	5, 08
Camelopardalis,	3 16 24, 04	,,	18, 64	,,	+0, 61	+0, 82	3 16 24, 65	19, 46	5, 19
δ Eridani - - -	3 36 1, 00	,,	55, 73	,,	+0, 43	+0, 51	3 36 1, 43	56, 24	5, 19
32....... - - -	3 46 40, 48	,,	35, 22	,,	+0, 38	+0, 48	3 46 40, 86	35, 70	5, 16
35....... - - -	3 53 50, 56	,,	45, 14	,,	+0, 15	+0, 44	3 53 50, 71	45, 58	5, 13
40....... - - -	4 8 21, 20	,,	16, 00	,,	+0, 22	+0, 33	4 8 21, 42	16, 33	5, 09
47....... - - -	4 26 55, 36	,,	50, 10	,,	+0, 11	+0, 24	4 26 55, 47	50, 34	5, 13
μ....... - - -	4 37 55, 47	,,	50, 40	,,	+0, 10	+0, 19	4 37 55, 57	50, 59	4, 98
w^1 Orionis - - -	4 42 17, 04	Ellis	11, 90	Taylor	+0, 14	+0, 19	4 42 17, 18	12, 09	5, 09 + c
ω Eridani - - -	4 45 27, 68	,,	22, 37	,,	+0, 05	+0, 15	4 45 27, 73	22, 52	5, 21
ψ....... - - -	4 54 6, 82	,,	1, 70	,,	0, 00	+0, 10	4 54 6, 82	1, 80	5, 02
ψ^2 Orionis - - -	5 18 51, 80	,,	46, 58	,,	−0, 09	+0, 01	5 18 51, 71	46, 59	5, 12
31...... - - -	5 22 1, 58	,,	56, 20	,,	−0, 13	−0, 04	5 22 1, 45	56, 16	5, 29
δ...... - - -	5 24 15, 02	,,	9, 72	,,	−0, 12	−0, 01	5 24 14, 90	9, 71	5, 19
42...... - - -	5 27 55, 20	,,	50, 04	,,	−0, 16	−0, 07	5 27 55, 04	49, 97	5, 07
49...... - - -	5 31 34, 70	,,	29, 50	,,	−0, 20	−0, 11	5 31 34, 50	29, 39	5, 11
ζ...... - - -	5 33 6, 27	,,	1, 23	,,	−0, 17	−0, 08	5 33 6, 10	1, 15	4, 95
κ...... - - -	5 40 36, 60	,,	31, 54	,,	−0, 25	−0, 16	5 40 36, 35	31, 38	4, 97
α...... - - -	5 46 54, 16	,,	49, 02	,,	−0, 20	−0, 12	5 46 53, 96	48, 90	5, 06

Hence from the first series, clock rate in 6 days, = − 5, 104 − c

And from the second........................ = − 5, 098 + c

Hence 2 c = − 0, 006 and c = − 0, 003

which may be neglected.

Fig. 15.1 Greenwich observatory's personal equation scheme under John Pond (1832a, b). From John Pond, *Astronomical observations made at the royal observatory at Greenwich in the months of January*, February and March 1832 (London: T. Bensley, 1832), p. iv

of observing the times of transit too late."[10] To Maskelyne, that the error was constant indicated intransigence, if not outright insubordination. Kinnebrook would have remained entombed as a cautionary tale but for the bibliographic excavations of Friedrich Bessel, astronomer at Königsberg, who was then investigating the phenomenon of "constant differences" between observers. Bessel suggested in an 1823 report that Kinnebrook's attempts and subsequent failure to correct for the effect meant that it was involuntary. However, as Hoffman has shown, even though it proved influential on the continent, Bessel's report failed to receive immediate notice at Greenwich. By the time Greenwich details attempts to control for the effect, it is clear that the quantity Pond calls the "personal equation" is not the same as Bessel's "constant difference" between two observers, if only for the reason that Bessel never actually measured such a quantity.[11]

Pond introduced the term "personal equation" in an 1833 supplement to the *Observations* for 1832, together with details of his methods of calculation, mainly to dispel

[10] Maskelyne (1945, p. 63).

[11] Hoffmann (2007, pp. 333–365).

doubts that had lately arisen regarding the rigour of his superintendence at Greenwich.[12] The term referred to a quantity, denoted *c*, "which added to Ellis's observations will give the same result as Taylor's."[13] In order to determine *c*, Pond had two assistants observe 20 stars. On February 9th, Thomas Taylor—the father of Thomas Glanville Taylor, who was to become Company Astronomer at Madras in 1830—observed the transit of 9 stars, and Thomas Ellis another 11. On February 15th, they repeated the sequence, with Ellis observing the first 9, and Taylor the final 11 stars. After making standard corrections (listed in the first column marked "sum of equations") for aberration, refraction, and other effects, the observed times of transit on the 9th were subtracted from those on the 15th, to give the amount of time the clock lost in six days, called the clock rate. Pond averages the clock rate for each of the two series of stars, and in doing so finds he must quote the averaged rates to an additional decimal place, so that *c* will be nonzero. He next assumes the two rates must be equal but for the quantity *c*, which is to assume that "the difference in the times of noting the same phenomenon by these observers" is the only source of error.[14] He equates the expressions, shows that *c* is half of the difference between the two clock errors, and calculates a personal equation between Taylor and Ellis so small as to be conveniently negligible. Observers' personal equations were calculated every year until Pond's retirement in October of 1835, but his labour regime was not so exacting as to require much extension of the 1833 procedure.[15]

As observational programmes grew increasingly ambitious, the social order of the observatory became increasingly instrumentalized. The complex protocols to account for observers' personal equations, which Greenwich and other observatories instituted in the middle of the nineteenth century, were an attempt to adapt to the consequences of a labour regime requiring multiple observers to share the work of a single set of observations.[16] So long as the same person carried out all the observations for any one calculation, the error due to personal equation could be expected to cancel out. Very few observatories in the 1820s employed a large pool of assistants or attempted observational programmes demanding enough for the effect to

[12] Francis Baily, stockbroker, amateur astronomer, and soon-to-be founder of the Royal Astronomical Society, was a persistent opponent of what he saw as the outmoded order obtaining at Greenwich. In a letter to Friedrich Bessel, Baily complained at length about Pond's work: "The business of the observatory at Greenwich is conducted by 4 persons, of very inferior talent. Two watch by day, and two by night. Mr Pond himself seldom handles the instruments, but merely superintends … The King's warrant requires that each person should put the initials of his name to his observations: but this is never done. They are all presumed to be made by Mr Pond, although he has very little hand in the business. This communication however is *private*." F. Baily to F. W. Bessel, 30 October 1823, BP, No. 162, f.4., as cited in Hoffman (2007, pp. 357–358).

[13] Pond (1832a, b, p. iv).

[14] Pond's "personal equation" differs markedly in this respect from Airy's. For one, during Pond's tenure, "no notice [was] taken of any instrumental errors: they are so minute and fluctuating that it has been considered safer and less troublesome to eliminate their influence by multiplying observations." Pond (1832a, b, p. iv).

[15] A difference of one second of arc in Right Ascension typically translated to a difference in longitude of one nautical mile, at the equator.

[16] Schaffer (1988).

OBSERVATIONS for the PERSONAL EQUATION between Mr. DOWNS and Mr. W. ELLIS.

Day, 1846.		Interval of nearest Stars.	Order of Observations.	Clock Slow at 0h by T D.	Clock Slow at 0h by W E.	W E–T D.
September	8	2^{h}	WE TD	$57^{s}.94$	$58^{s}.61$	$+ 0^{s}.67$

From the above differences, if we form equations wherein each observer's initial shall successively stand first, we obtain the following groups of equations.

For M.

$M - E = -0^{s}.18$
$2(M - R) = -0.42$
$3(M - D) = +0.26$
$3(M - HB) = -0.17$
$M - JB = -0.44$

For E.

$E - G = -0^{s}.14$
$E - TD = +0.03$
$E - H = +0.15$
$E - M = +0.18$

For H.

$H - E = -0.15$
$14(H - R) = -4.31$
$19(H - D) = -0.18$
$2(H - HB) = -0.16$
$H - L = +0.27$
$H - JB = -0.36$
$H - T = -0.23$

For R.

$2(R - M) = +0.42$
$2(R - HB) = +0.19$
$14(R - H) = +4.31$
$2(R - WE) = -0.19$
$R - D = +0.27$
$R - JB = +0.20$

Greenwich Astronomical Observations, 1846. f

Fig. 15.2 Greenwich Observatory's personal equation scheme under George Biddell Airy, 1846. From George Biddell Airy, *Astronomical Observations made at the Royal Observatory, Greenwich, in the Year 1846* (London: Palmer and Clayton, 1848), p. xxxiii

require extensive accommodations.[17] John Pond, for example, had in 1833 at most two assistants taking transit observations for the purpose of regulating the clock (Fig. 15.2).

A decade into his administration at Greenwich, Airy was regularly calculating the personal equation for—that is, the pairwise differences between—12 observers, including himself, so that the effect of any one observer standing in for another could be corrected for. By the time Airy makes algebraically explicit his method of calculating the personal equation, in 1846, it is clear that his scheme, although vastly more complicated, is algebraically a straightforward generalization of the procedure detailed by Pond in 1833.[18]

Nevertheless, the details and consequences of Airy's generalization of Pond's algebra have gone unnoticed in the literature on the personal equation. Where Pond

[17] Göttingen, Königsberg, and Dorpat were the only German observatories to employ two or more assistants during the 1830s. See Hoffman (2007, p. 355).

[18] Biddell Airy (1848, pp. xxiii–xxxv).

wrote the difference between Ellis's and Taylor's observations as a constant, *c*, Airy writes a difference between two variables denoted by each observer's initial. The tabulated differences for each observer are summed together, producing a linear equation. Taken together, these expressions compose a system of linear equations for the personal equations of all twelve observers. The system has solutions when the expression

$$\frac{n!}{2(n-2)!} - n \tag{15.1}$$

is equal to zero, meaning the number of pairwise equations is equal to the number of unknowns. But in practice the first term was always greater than n, meaning that Airy's system of equations was always indeterminate. Equation 15.1 has solutions (i.e. is equal to 0) for $n = 3$, but is always indeterminate for $n > 3$. It was always possible for Pond to calculate personal equations for his small establishment without need for an observational reference. (Although there is no evidence that Pond considered the n = 3 case, i.e. measured personal equations between himself, Ellis, and Taylor together). But Airy's vastly larger staff meant that the scheme required modification. Airy realized that one observer's personal equation had to be set to zero, in order to recover pairwise differences that could be ranked. In Greenwich, the assistant who had charge of the transit observations was always used as this reference, with John Henry's equation set to zero in the *Observations* for 1846. Airy's scheme generalized the pairwise differences which had originated in Pond's reply to Bessel. From an algebraic standpoint, then, the personal equation at Greenwich reveals the larger contradictions of bureaucratic inheritance and inertia. But it also shows that the fungibility of observational labour relies on standards and references. At Greenwich, the choice of a reference was largely arbitrary, and did not encode elements of the observatory's social order. At Madras, as we will see, it did.

A Solar Ritual at Madras

How did this same personal equation scheme play out elsewhere? When transplanted to Madras, the algebraic structure of Airy's personal equation scheme became the basis for an entirely different mode of organizing observational labour. The personal equation was first used in Madras by Thomas Glanville Taylor, the son of Thomas Taylor, first assistant to the Astronomer Royal, John Pond. T. G. Taylor had started working at Greenwich as a supernumerary in 1820, aged sixteen. Two years later he was hired full time. He remained at Greenwich until 1830, when he was selected to succeed John Goldingham as Company Astronomer, on Pond's own recommendation. Taylor resolved to begin an expansive programme of astronomical observation, which would later result in the Madras General Catalogue, comprising observations of some 11,015 stars, taken from 1830 to 1843.[19]

Because he was attached to the Great Trigonometrical Survey of India, which required his absence from the observatory for several months at a time, Taylor's labour regime aimed to carry out transit observations autonomously, without direct supervision. For this, the personal equation was essential. Like Airy's *Greenwich Observations,* Taylor's Madras catalogue also required the division and ritualization of labour, but he had to effect it from locally-sourced forms of domination. Like his predecessors, Taylor found caste easily adaptable to his end. Where Airy had used class, Taylor used caste to divide labour between instrumentally-disciplined and hierarchically-ordered hands. With the aid of several assistants, all "natives of high *caste*" (and therefore trustworthy collaborators) both observation and reduction could be accomplished in his absence. Taylor is the first Company astronomer to expand the observational labour pool beyond Brahmins, recruiting Mudaliars, a title denoting a Vellalar or land-holding caste. To allow for the rostering of observers, all high-caste assistants' personal equations were regularly measured. It was here that Indian assistants could be verified as almost, but not quite equal, to European observers. Yet, improvement was still possible, something German observers had failed to remark upon.[20]

Leaving the chronometric work to Indian assistants also freed the Company astronomers to publish on topical astronomical matters more likely to win notice abroad. Jacob, for example, began to examine the double star system 70 Ophiuchi, noting that systematic errors in the orbital parameters showed a periodic regularity. Either the regularity resulted from "an erroneous assumption of the universal application of the law of gravitation," or, more simply, that the observational anomalies were inconsistent with a binary model, first suggested by Herschel, and were in fact evidence for a planet (a so-called "dark body") in orbit around the system.[21] Treating the existence of extrasolar planets as distinctly possible, in 1855 Jacob published a rejoinder to William Whewell's (anonymously published) essay *Of the Plurality of Worlds*, itself published in 1853.[22] Whewell's argument hinged on the premise that

---

[19] Airy acknowledged Madras' special status (albeit in a somewhat backhanded fashion) on the occasion of an award presentation: "In the number of observations and the number and distribution of the stars, and in the circumstance that the observations were made, reduced, combined, and printed, at the same place and under the same superintendance, it bears the palm from all others. But this was the fruit of an endowed observatory, *the work of an astronomer and competent assistants, whose strength was not exhausted by any other employment*" Airy (February 1854, p. 154 [italics mine]). This of course is not true, Taylor having more often than not been out on survey, and the assistants having been tasked with chronometric and meteorological observation on top of transit work. Airy's assertion is intended to elevate the stock of colonial observatories (the majority of which globally were in British possession), and establish their capacity to produce reliable navigational data.

[20] "I generally find that unpractised observers note the time in excess from three to seven tenths of a second when compared with those long accustomed to this species of observation." Glanville Taylor (1833, p. iii). See Schumacher's account of the "constant difference" between himself and his assistant: Hansen (1825, p. v ff).

[21] Jacob (1855a, 1855b, pp. 228–230).

[22] Anonymous [William Whewell] (1853).

the only stars whose mass can be measured are binary stars, and that it wasn't possible for binary stars to have planets in orbit. The basis for Jacob's objection to Whewell is the possibility of other, extrasolar, worlds.[23] Although Jacob doesn't cite the case of 70 Ophiuchi explicitly, his essay was published the same year as the paper on Ophiuchi's anomalies, and it appears the two were in preparation simultaneously. Though now known to be an erroneous claim, this episode is sometimes regarded as the first claimed exoplanet detection.[24] The first attempt to detect other worlds, then, was made possible by the autonomous, rostered labour order of Madras observatory.

The fact that Taylor left most of the actual observatory work to his Indian assistants produced a curious contradiction. An Indian assistant's work had to be equivalent to their European superintendent's, but if it were, the European's position would be undermined. It is in the context of calibration, or accounting for individuality, that the high-caste assistants are named, and their observations labelled, "in order that a confidence proportional to their merit may be established."[25] The first two observers, "M for Mootoosawmy Moodeliar [and] A for Anuntacharyer Braminy" were replaced when Taylor found that their personal equations "did not become as small as I had anticipated." Ragavacharyer Braminy and Soontharum Moodeliar, "from whom [Taylor] had reasons for expecting an increased degree of accuracy and attention," replaced them as observers. But his experiments with Tamil Mudaliars and Brahmins suggested that improvement was possible, and that the involuntary difference may not after all be constant, as Heinrich Christian Schumacher had suggested in the preface to his *Helgoland Observations*.[26]

If what Taylor was measuring was indeed Bessel's "involuntary constant difference," improvement should not be possible. But improvement, to be effected by proper discipline and error correction, was fundamental to the logic that governed the observatory. Trustworthy natives could be trained in proper methods, corrected relative to a British standard, and thereby become practised observers, elevated in the hierarchy produced by instrumental (including algebraic) means. So unlike in Greenwich, ready means of characterizing observational accuracy did not render assistants entirely fungible. Caste still underwrote the social order of the observatory. Despite performing poorly as an observer, Anuntacharyer Braminy, for example, had by 1842 become head assistant.[27] Mudaliars, by contrast, could become skilled observers, but they did not climb in the hierarchy of assistants.

It is a fact of no little significance that in 1859, a year after the Sepoy Mutiny of 1857–58, an assistant at Madras Observatory, Chintamani Ragoonathachary, published a scheme, of his own invention, for calculating the personal equation.

---

[23] "It is certainly *possible*…that our Earth may be the only body brought to the degree of perfection in which we see it—the only one as yet inhabited by intelligent creatures capable of knowing their Creator, or even the only one inhabited at all; but it is in the highest degree *improbable* that such should be the actual state of things." Jacob (1855a, 1855b, p. 43 ff [emphasis in original]).

[24] Kipping (2011, p. 4 ff).

[25] Taylor (1832a, b, p. iii).

[26] Taylor (1831, p. 36).

[27] Taylor (1848, p. 69).

Thomas Metcalf has described the aftereffects of the Mutiny as "an enduring legacy of fear," which intensified patterns of anxiety and information panic.[28] The threat of uprising, and mistrust of potentially turncoat native subordinates, became an imperial preoccupation, in India as well as in the broader empire. It is amid this climate of suspicion that Ragoonathachary published his first paper, a note in the *Monthly Notices of the Royal Astronomical Society*. The personal equation served to mitigate this uneasy tension in the relative status of Indian observers. What in Greenwich was an ex post facto method, intended to correct for the unintended consequences of rostered labour, became in Madras a scheme for producing observers whose work was indistinguishable from their European superintendent's.

Ragoonathachary was born on March 17th, 1828, the son of an assistant to Goldingham, and joined the observatory himself in 1847 as an assistant to Taylor.[29] Twelve years later, he set out a method of his own invention, for determining the relative personal equations of multiple observers simultaneously. Ragoonathachary's method involved projecting the Sun's image onto on oiled paper sheet, upon which were drawn grid lines. Three observers (Ragoonathachary himself, Sashoo Iyengar, and T. Mootoosamy Pillay) were arranged in a circle, and instructed to watch for the transit of the Sun's upper and lower limbs. Following Airy's algebraic scheme in the *Greenwich Observations* for 1846, Ragoonathachary's experimental method produced a system of linear equations, which could be solved by setting one observer's equation to zero.

As both observers and computers, observatory assistants had to be characterised, measured against some standard, if the information they processed was to be committed to print, under a British name. In practice, that standard was the imperial astronomer-superintendent. In Madras, the procedure involved setting the European superintendent, W. S. Jacob's equation to zero, and using his observational style as a reference for assistants. The Indian assistants were trained repeatedly, using the projected image of the Sun, until—one year after the initial set of observations were taken—setting the lowest ranked high-caste assistant's equation to zero produced pairwise differences similar to those produced if Jacob's own equation were set to zero. High-caste observers had to measure up to Jacob himself, to adopt his particular observational habits.

It is tempting to read the episode in terms of colonial mimicry, as a scheme for producing trustworthy Indian assistants. Jacob's equation was measured before his departure from India in late 1858, at the height of the Sepoy Mutiny. Ragoonathachary's personal equation method, sent to Jacob while Tennant was Acting Astronomer, was dated April 25th, 1859—a week after the last rebel leader, Tantia Tope, was executed.[30] It is as much an argument for the loyalty and reliability of Indian observers as it is a method for producing them: Ragoonathachary makes sure to mention that the accuracy of the procedure increases with additional observers.

---

[28] Metcalf (1997, p. 245).

[29] Dikshit and Sastra (1969, pp. 181–182); Briers et al. (2022); Kameswara et al. (2009, pp. 201–210).

[30] Metcalf (1964, pp. 289–328).

When his paper was published, Ragoonathachary was fourth assistant under Sashoo Iyer. Jacob appended a (somewhat contradictory) note to Ragoonathachary's paper, carefully limiting the credit due to Ragoonathachary.[31]

But who is mimicking whom is much more opaque than it would seem. For four months, at the height of the single greatest threat to colonial rule, Jacob, Ragoonathachary, and two junior assistants sat in a semicircle every day at noon and attempted to observe simultaneously. The procedure resembles a kind of seance, a ouija or spirit board reading, inasmuch as it distributes intentionality over the entire collective of observers. So the question of colonial difference becomes quantified only by the inseparable mixture of the agency of both colonizer and colonized.

The purpose of Ragoonathachary's method was to produce standardized observers suitable for the rigour and tedium of chronometric work. But, as later astronomers who took charge of Madras reported, this chronometric labour order was ill-suited to the work of discovery. After the India Act of 1858 and the dissolution of the East India Company, the command structures surrounding the observatory changed drastically. Topping and Goldingham understood the observatory as a military institution: Madras was "an Establishment by itself, immediately under the Government," to whom the director needed only submit reports and observations.

Half a century later, much had changed. Though the Madras Presidency itself was barely affected by the events of the Rebellion of 1857, the observatory saw significant administrative restructuring, and was placed under the supervision of the Astronomer Royal, George Airy. The first Government Astronomer, Norman Pogson, was hired in 1861, by Airy himself. In contrast to his predecessors, he had no connection either with the military or surveying in any capacity. Pogson, who was without a university degree, and had come up working as an assistant in the Radcliffe Observatory at Oxford, sought passage to Madras partly as an act of self-reinvention, as had his predecessors. In a letter to his former employer, John Lee, whose Hartwell House observatory he'd quit for Madras, Pogson described himself as the new "Astronomer Royal of India."[32]

But upon arrival he found he had inherited an institution whose observations—magnetic, meteorological, and stellar—were decades in arrears, "raw material wholly unreduced and awaiting attention."[33] Problems with the supply and race of available labour ("Natives superstitious and devious. Government schools are preparing better qualifications, but natives opt for Revenue Board or Public Works, not the observatory."[34]) appear to have slowed Pogson's attempts to process both his own and his predecessor's data, deepening the backlog of unreduced observations.

[31] "I am aware that the method does not possess any novelty, having been occasionally employed by myself more than 20 years ago, and probably by many others […] but it is not, I believe, used in any other observatory besides that of Madras as the mode of observing the sun at noon" Jacob (July 1859, p. 339).

[32] Reddy et al. (2007, p. 240).

[33] Reddy et al. (2007, p. 240).

[34] Reddy et al. (2007, p. 240).

To the new civilian administration, Madras and Pogson were increasingly seen as superfluous. In 1866, acting on Airy's recommendations, the Secretary of State for India, the Conservative Viscount Cranborne, recommended that since "systematic observations may be more advantageously taken at other observatories in the Northern and Southern Hemispheres," principally those at Greenwich and Sydney, "the best ultimate arrangement would be to close the Observatory at Madras, and to restrict that at Bombay to magnetic and meteorological observations, maintaining only so much of an astronomical establishment as may be necessary for the determination of true local time."[35] Bombay, not Madras, was to become the temporal centre of British India, flashing time to subordinate nodes at Madras and Calcutta.

Pogson immediately wrote to Airy (who was in fact the architect of the observatory's proposed abolition) "respectfully requesting you to aid me by your influence to avert a scientific catastrophe." Pogson wrote that although he had worked "steadily and laboriously" he had had "such fearful interruptions and delays to overcome…as you and Englishmen in general would never dream of…"[36] What eventually saved Pogson was Madras's connection with the Great Trigonometrical Survey of India. In 1867, J. T. Walker, Superintendent of the Survey, told Airy that Madras was still producing data of use to the Survey. Moreover, if Madras were closed, all of the data it had collected since 1790 would for practical purposes become useless.

During Pogson's tenure, a single Indian is named: "C. Ragoonatha Chary is a good clever self taught assistant; particularly ready at calculating eclipses and occultations. The rest are mere dolts—machines without the certainty of machinery."[37] Ragoonatha Chary alone "possessed either knowledge or intellect."[38] Pogson considered the others "at best mere machines requiring perpetual control, more tedious than the actual personal labour."[39] It is perhaps unsurprising that Pogson should see Indian agency as prosthetic to the instrument, and Indians themselves simple extensions of the machines they were hired to operate. The labour and moral order by which numerate Brahmins could be rendered subordinate and productive, made useful instruments, was by this point well established, and understood as a condition of the observatory's normal operation. As Pogson put it, each observer had to find and apply their own corrections, since "natives have greater and more constant personalities of observation than generally prevail among Europeans."[40] Yet these "mere machines," products of his predecessor's labour regime, irked Pogson. In his annual reports, Pogson complained at length that his requests for a European assistant had been denied:

> …for the refined researches of modern Astronomy, Natives are, with rare exceptions, perfectly useless. The attempt to find any suitable has been utterly futile and, except in the fundamental observations, viz. those made with the Transit Circle, the labours of the

---

[35] Airy (1864–1869, p. 32).

[36] Pogson (1864–1869, p. 36).

[37] Reddy et al. (2007, p. 242).

[38] Pogson (1871a, b).

[39] Pogson (1886, p. 194).

[40] Pogson (1890, p. i).

> Observatory are, therefore, those of a single individual, and necessarily of smaller amount than the same person could have accomplished more easily in a less enervating climate.[41]

His petitions for European aid often reached histrionic pitch: "The refusal of the long expected and urgently needed aid of an English Assistant is a misfortune, alike to the Astronomer, to the sciences he pursues, to the reputation of the Observatory, and to its efficiency as a Public Department."[42] All were ignored. The production of instrumental Brahmins was not the object of Pogson's intended labour regime, but a hindrance to it.

Conclusion: Beyond the Meridion

Unlike Taylor, Pogson was not content with an observational programme consisting solely of "fundamental observations." He sought to occupy himself with extra-meridional observations: those of minor planets, comets, and—his principal occupation whilst in England—the variable stars. This required an altazimuth mount, and the improvisatory use, and on-the-fly interpolation, of tables of refraction. Airy meanwhile maintained that the uncertain work of extra-meridional observation should not be conducted at the Government's expense, and by 1870 had drawn up a code of instructions limiting Pogson to meridional work. Although Airy was willing to allow that astronomers should "sometimes balance the formal and dry labour of meridional observations against the less regular, but more interesting, work of extra-meridional and micrometrical investigation," he nevertheless maintained that the uncertain work of the lone and heroic observer should not be conducted at the Government's expense.[43] Without sufficient regard for his own, rather insecure, position in the moral order of the science, Pogson consigned Airy's regime to the discipline's past: "in former times meridional observations constituted the sole duty of the Director of a Public Observatory, but such is no longer the case; and, in the present state of the science, the extra-meridional pursuits are by far the most arduous portion of his labours."[44]

Pogson thought most Indians could be trusted only with the fundamental drudgery for which they had been calibrated. If improvement and instrumentality were elements of a meridional regime, it was because this work could proceed best by the

---

[41] Pogson (1865, p. 39).

[42] Pogson (1864–1869, p. 140).

[43] Royal Astronomical Society (1856, p. 243). As Airy observed at a meeting of the Royal Astronomical Society on April 12th, 1872, "no observatory can ever sustain its position unless it is connected with some secular object." Greenwich's charter had tasked it with the improvement of astronomical means of navigation and chronometry, and these twin objects motivated the observatory's every activity, "not only to prevent astronomers going mad, as they have been apt to do since the time of Flamsteed... but because Governments want to have something to show for their money." Airy (1872, p. 115).

[44] Letter from the Government of India (1869–1870, p. 3).

division and strict scrutiny of labour. Discovery was far harder to make so programmatic: at a minimum, one had to be able to search the sky, interpolate tabulated values, as well as a host of other improvisatory measures, not amenable to the factory order which had obtained under the previous Madras astronomers. Methods like the personal equation could produce and regulate observers well enough for transit catalogues, but the kinds of Indian observers who were valuable to "the present state of the science," and therefore also Pogson's career, could not be produced quite so reliably (Fig. 15.3).

If the transit circle required drudges, the alt-az could produce exceptions beyond the colonial meridion. Under Pogson we see the emergence of a new labour regime, which even as it reproduces a meridional norm is at the same time obsessed with extra-meridional exceptions. In 1861, Ragoonathachary had been crucial to (and not just instrumental in) Pogson's discovery of a minor planet. Half of the data used to detect the planet, which Pogson named Asia, was collected and reduced by Ragoonathachary. In 1867, Ragoonathachary discovered the variable star R Reticuli, and was duly elected a Fellow of the Royal Astronomical Society in 1872.[45] Earlier technologies of power, like the personal equation, were necessary but insufficient means of obtaining control over observers who had to be creative in order to pursue "the refined researches of Modern Astronomy."

Under Pogson, Ragoonathachary is, by invidious comparison with the other assistants, described as a different kind of machine, one whose ability to self-calibrate underwrites their value and social position. When in 1874, Ragoonathachary published a pamphlet about the Transit of Venus, on the eve of the event, Pogson lauded his efforts as intended to show "the vast inferiority and great inaccuracy of the calculations of Hindu astronomers." His pamphlet "urge[d] them to abandon their worthless methods and antiquated tables and avail themselves of modern improvements."[46] To Pogson and the European establishment, Ragoonathachary was a time machine: a well-regulated chronometric assistant and also the means by which modernity would be conveyed to the colony.

But Ragoonathachary's real project was reform. In fact, the Kannada language version of the pamphlet is written in Sanskritized prose, and—far from disparaging Sanskrit astronomy—is concerned with citing precedent for carrying out ceremonial rites only upon celestial events directly observable with a telescope.[47] In April of 1874, Ragoonathachary delivered an address to an audience of Madras notables. The predictions of the makers of Hindu almanacks had come to disagree with observation because necessary cycles of correction could not be carried out. Ragoonathachary was requesting funds for a purpose-built, chronometric observatory. Inaccuracy in ritual calendars was "for the most part the result of the political changes which have swept

[45] Kameswara Rao et al. (2009, pp. 201–210).

[46] Letter from Pogson to R.S. Ellis, August 16, 1871. *Madras Observatory Papers*, Tamil Nadu State Archives, as cited in Venkateswaran (2019, pp. 263–296).

[47] See Shylaja (2012).

Fig. 15.3 Portrait and autograph of C. Ragoonatha Chary at Madras Observatory, from the collection of Cherry Armstrong, Pogson's great, great, granddaughter. Image reproduced from, N Kameswara Rao et al., "C. Ragoonatha Chary and his Variable Stars," arXiv 0908.3081 (2009): p. 9

over the country, and hitherto deprived modern native computers of...support."[48] Chronometry was to be a time machine of a different sort for the "native computer"—one that generates counterfactuals.

[48] See the Address delivered by C. Ragoonatha Chary, F. R. A. S., on Pacheappah's Hall, Madras, on 13th April 1874, to a large meeting of native gentlemen, in Chary (1874, p. 1).

Acknowledgements This paper is drawn from a larger project on Madras observatory, and in its early stages benefitted much from conversations with Projit Mukharji, James Poskett, Claire Sabel, Simon Schaffer, Charu Singh, Sujit Sivasundaram, Koyna Tomar, John Tresch, and Adelheid Voskuhl. I also owe much to participants of the workshop *Astronomical Observatories and Chronometry: Time, Science And Instruments (18th–20th Century).*

References

Airy, G. B. (1848). *Astronomical observations made at the royal observatory, Greenwich, in the year 1846*. Palmer and Clayton.

Airy, G. B. (1864–1869). Letters from observatories. *RGO, 6*(149), 32.

Airy, G. B. (1872). Response to Col. Strange. *Astronomical Register 10,* 115.

Airy, G. B. (1854). Address delivered by the President, G. B. Airy, Esq. F. R. S., astronomer royal, on presenting the gold medal of the society to Mr. Charles Rümker. *Monthly Notices of the Royal Astronomical Society, 14,* 145.

Anonymous [William Whewell] (1853). *Of the plurality of worlds: An essay*. Parker and Son.

Ashworth, W. J. (1998). John Herschel, George Airy, and the roaming eye of the state. *History of Science, 36*(2), 151–178.

Baily, F., to Bessel, F. W. (1823). *BP, No. 162, f. 4*, as cited in Hoffmann (2007, pp. 357–358).

Belteki, D. (2022). The spring of order: Robert main's management of astronomical labour at the royal observatory, Greenwich. *History of Science, 60*(4), 575–593.

Belteki, D. (2023). The winter of raw computers: The history of the lunar and planetary reductions of the royal observatory, Greenwich. *The British Journal for the History of Science, 56*(1), 65–81.

Benschop, R., & Draisma, D. (2001). In pursuit of precision: The calibration of minds and machines in late 19th-century psychology. *Annals of Science, 57*(1), 1–25.

Boring, E. G. (1961). The beginning and growth of measurement in psychology. *Isis, 52*(2), 238–257.

Briers, M., et al. (2022). Chasing change: The lasting legacy of India's 1871 eclipse. *Astronomy & Geophysics, 63*(1), 24–29.

Brinkmann, R., et al. (2019). The rise and fall of the "personal equation" in American and British medicine, 1855–1952. *Perspectives in Biology and Medicine, 62*(1), 41–71.

Canales, J. (2001). Exit the frog, enter the human: Physiology and experimental psychology in nineteenth-century astronomy. *The British Journal for the History of Science, 34*(2), 173–197.

Chary, C. R. (1874). Address delivered by C. Ragoonatha Chary, F. R. A. S., on Pacheappah's Hall, Madras, on 13th April 1874, to a large meeting of native gentlemen. *Chintamani Ragoonathachary, Transit of Venus*, p. 1.

Dewhirst, D. W. (1985). Meridian astronomy in the private and university observatories of the United Kingdom: Rise and fall. *Vistas in Astronomy, 28*, 147–158.

Dikshit, S. B., & Sastra, B. J. (1969). *History of Indian astronomy*. Government of India Press.

Donnelly, K. (2014). On the boredom of science: Positional astronomy in the nineteenth century. *The British Journal for the History of Science, 47*(3), 479–503.

Fellmann, E. A. (2007). *Leonhard euler*. Birkhäuser Verlag.

Government of India, Home Department. (1869–1870). *BL IOR/L/PJ/3/1108* No. 61.

Hansen, P. (1825). *Journal of observations made for ascertaining the time of the place, in the observatory, which was erected at Helgoland for that purpose*. Langhoff's Printing Office.

Hoffmann, C. (2007). Constant differences: Friedrich Wilhelm Bessel, the concept of the observer in early nineteenth-century practical astronomy and the history of the personal equation. *The British Journal for the History of Science, 40*(3), 333–365.

Hollis, H. P. (1925). The Greenwich assistants during 250 years. *The Observatory, 48*, 388–398.

Jacob, W. S. (1855a). *A few more words on the plurality of worlds*. Bell and Daldy.

Jacob, W. S. (1855b). On the theory of the binary star 70 Ophiuchi. *Monthly Notices of the Royal Astronomical Society, 15*(9), 228–230.

Jacob, W. S. (1859). Note in reference to the preceding communication. *Monthly Notices of the Royal Astronomical Society, 19*, 339.

Kameswara, R. N., et al. (2009). C. Ragoonatha Charry and variable star astronomy. *Journal of Astronomical History and Heritage, 12*(3), 201–210.

Kipping, D. M. (2011). *The transits of extrasolar planets with moons*. Springer.

Lewis, M. (2007). The personal equation: Political economy and social technology on India's canals, 1850–1930. *Modern Asian Studies, 41*(5), 967–994.

Maskelyne, N. (1945). *Astronomical observations made at the royal observatory at Greenwich*, vol. 3 (originally published in 1799). Buckton and Bensley.

Metcalf, T. R. (1964). *Aftermath of revolt*. Princeton University Press.

Metcalf, T. R. (1997). *Ideologies of the raj*. Cambridge University Press.

Pogson, N. R. (1864–1869). *Letter from N.R. Pogson, esquire, government astronomer, to A.J. Arbuthnot, esquire, chief secretary to the government of Fort St. George—dated Madras, the 25th February 1867*. Royal Greenwich Observatory (RGO) 6/149.

Pogson, N. R. (1864–1869). *Letter to George Airy, December 22nd, 1866*. RGO 6/149.

Pogson, N. R. (1865). *Madras observatory report*. RGO 6/149.

Pogson, N. R. (1871). *Norman Robert Pogson papers*. British Library, Mss Eur B251.

Pogson, N. R. (1871b). *Letter from Pogson to R.S. Ellis*. Madras Observatory Papers, Tamil Nadu State Archives.

Pogson, N. R. (1890). *Results of observations of the fixed stars made with the Meridian circle at the government observatory, Madras in the years 1868, 1869, and 1870*. Lawrence Asylum Press.

Pogson, N. R. (1886). *Papers of George Airy*. RGO 6/154.

Pond, J. (1832a). *Astronomical observations made at the royal observatory at Greenwich in the months of January, February and March 1832*. T. Bensley.

Pond, J. (1832b). *Greenwich observations*. Bensley and Murray.

Rao, N. K., et al. (2009). C. Ragoonatha Charry and variable star astronomy. *Journal of Astronomical History and Heritage, 12*, 201–210.

Reddy, V., et al. (2007). Scaling the magnitude: The fall and rise of N.R. Pogson. *Journal of the British Astronomical Association, 117*, 240.

Rieznik, M. (2013). The Córdoba observatory and the history of the "personal equation" (1871–1886). *Journal for the History of Astronomy, 44*(3), 277–301.

Royal Astronomical Society. (1856). *Memoirs*, vol. 24. The Society.

Schaffer, S. (1988). Astronomers mark time: Discipline and the personal equation. *Science in Context, 2*(1), 115–145.

Shylaja, B. S. (2012). *Chintamani Ragoonatha Charry and contemporary Indian astronomy*. Bangalore Association for Science Education.

Smith, R. W. (1991). A national observatory transformed: Greenwich in the nineteenth century. *Journal for the History of Astronomy, 22*(1), 5–20.

Taylor. (1831). *Astronomical observations*, vol. 1. Printed at the American Mission and C. K. S. Presses.

Taylor, T. (1832a). *Result of astronomical observations made at the honorable east India company's observatory at madras* (vol. 2). Orphan Asylum Press.

Taylor. (1832b). *Astronomical observations*, vol. 2. Printed at the American Mission and C. K. S. Presses.

Taylor. (1848). *Astronomical observations*, vol. 6. Printed at the American Mission and C. K. S. Presses.

Tresch, J. (2010). Even the tools will be free: Humboldt's romantic technologies. In: D. Aubin et al. (Eds.), *The heavens on earth: Observatories and astronomy in nineteenth-century science and culture* (pp. 253–284). Duke University Press.

Venkateswaran, T. V. (2019). Ragoonatha Charry and his 'scientific' Pañcāṅga. In: M. Vahia et al. (Eds.), *The growth and development of astronomy and astrophysics in India and the Asia-pacific region* (pp. 263–296). Springer.

S. Prashant Kumar is a Postdoctoral researcher at the Rank of Instructor, in the Department of History, at the University of Chicago. He is a social and intellectual historian of astronomy and mathematics, whose work examines the role of race and caste in constructing infrastructures of time measurement in South Asia, from the late eighteenth to the mid twentieth century. He is also an Honorary Fellow at the Archives at NCBS, Tata Institute of Fundamental Research, Bengaluru, a collecting centre for the history of science in modern India, where he works on acquiring and describing archival collections relating to the space sciences in India.

Part VI
In Memoriam—Remembering Paolo Brenni

Chapter 16
Paolo Brenni Scholar, Educator and Artisan

Paolo Galluzzi

Abstract This short *in memoriam* piece honors Paolo Brenni, tracing his life and his significant contributions to historical scholarship and conservation practice.

In the late 1970s, Maria Luisa Righini Bonelli invited me to take part in the planned renovation of the scientific instrument collections at the Istituto Tecnico per Geometri Gaetano Salvemini in Florence.[1] This training institute for surveyors had inherited the collections from the ambitious but short-lived Istituto Tecnico Toscano founded by Leopold II of Lorraine in the last years of the Grand Duchy of Tuscany.

So deeply was I impressed by the discovery of this forgotten but miraculously intact treasure trove that I committed myself to a far-reaching plan to recover and promote the Institute's rich heritage—a vast complex that offered a detailed picture of technical education in nineteenth-century Europe.

Besides the challenge of raising the substantial funding required, there was the immediate difficulty of finding specialists with the technical and historical skills needed to restore and catalogue the countless measurement, observation, and experimentation apparatuses. At the suggestion of the late Gerard Turner, I approached the Ticinese physicist Paolo Brenni, who, despite his young age, had already acquired considerable experience in restoring and developing such collections.

Convincing Paolo to join in the project proved extremely easy. His participation was anything but ephemeral, for it was to last—with an ever more intense personal commitment—until a few months before his untimely death.

To realise the scale and quality of Paolo's work in reorganising, cataloguing, historical contextualisation, and exhibition design for over three decades of unstinting efforts, we need only compare the splendid displays in the Gabinetto di Fisica (Physics Cabinet) at what is today the Fondazione Scienza e Tecnica (FST) on Via Giusti with photographs showing the collection's neglected state as late as the

[1] This text was first published in 2022 in the *Bulletin of the Scientific Instrument Society* (152): 3–4.

P. Galluzzi (✉)
Florence, Italy
e-mail: p.galluzzi@museogalileo.it

G. Bernasconi et al. (eds.), *Astronomical Observatories and Chronometry, 18th-20th Century*, History of Physics, https://doi.org/10.1007/978-3-031-95874-8_16

mid-1970s. In addition, we should not forget Paolo's crucial role in promoting the institution's impressive resources within the international community of specialists in old instruments and historians of science. Nor should we forget the publication of his fundamental studies on makers of instruments and experimental apparatus, on the major International Exhibitions at which they were triumphantly displayed, on their use in training schools for technical occupations, and on methods for restoring such artifacts and making them function again properly.

As is well known, Paolo did not make his expertise available to Florentine institutions alone. There is practically no major collection of historically valuable scientific instruments—whether in Europe or elsewhere—that he did not closely examine in situ, framing them in their specific context, and offering penetrating insights on how to promote them.

The exemplary importance of Brenni's work soon drew the attention of Italian national authorities in charge of preserving cultural heritage; it also appealed to those who believed in the strategic importance of disseminating at the grass-roots level an image of scientific research to counter the absurd but widely believed opposition between science and culture. One way to achieve this would be to showcase physical objects bearing witness to the historical evolution of science. It was in this context that the Italian National Research Council (CNR) signed a four-year cooperation contract with Paolo, which, at the expiry date, was turned into an open-ended agreement. This arrangement enabled Paolo to play a role in raising public awareness of the importance of old instruments for understanding the history of scientific research, innovation in industrial processes, technical education, and the institutions tasked with that essential mission.

South of the Alps, in Italy, Paolo found himself working within an academic tradition that has always paid little or no attention to the instruments and apparatus used over the centuries to understand and measure natural phenomena. Thanks to his interdisciplinary training, however, he succeeded in generating "openings" and expressions of interest among humanities scholars and militant scientists with regard to this heritage. Complementing his deep scientific and technical knowledge, Paolo possessed great expertise in the philological and historical analysis of sources. These included not only printed and manuscript texts, as well as instruments and experimental apparatus, but also other equally valuable evidence such as sales catalogues of technical and scientific instruments, promotional leaflets, and instruction sheets for assembling devices. Paolo combined these competencies with his talent as a technician who subjected the material artifacts of past science to careful anatomical scrutiny—and who successfully reconstructed their original configuration through expert restoration work. Lastly, he excelled in another field, essential for promoting this type of cultural assets: the production of videos and audio recordings in which his personal participation stimulated curiosity and interest for such relics of the past even among non-specialists.

Paolo adored his work. His unique and universally appreciated skills had led him to collaborate with famed institutions around the world. But his heart beat most strongly in Florence, at the Museum that now bears Galileo's name, and on the premises of the former Istituto Tecnico Toscano on Via Giusti (now FST). At the latter, the instruments and experimental apparatuses of the Lorraine collection had been gradually restored to their original splendor, installed in a technically adequate and secure infrastructure, and expertly displayed—all this with Anna Giatti's valuable collaboration.

Paolo's love for the Via Giusti institution—his second home for so many years—was great indeed. It was long reciprocated by an affectionate recognition of his priceless service, and of his efforts to promote the FST's precious collections internationally.

No one could ever have imagined that, one day, this warm mutual appreciation would be abruptly terminated in such an incomprehensible and appalling manner. A few months before his passing, Paolo received from the FST Presidency a letter informing him that his services were no longer required, and requesting him to clear out his personal effects and return his key to the building. As I recall, no sooner had he opened the letter—truly a lightning bolt out of a clear sky!—than he ran into my office at the Museo Galileo to announce the terrible news in a trembling voice. It was terrible both because of Paolo's deep sense of belonging to the institution and because of the shocking realisation that decades of productive and selfless commitment had been so callously rewarded.

In the following months, none of his many friends, eminent colleagues, and profound admirers was able to ease the pain of that wound. Paolo did continue to receive prestigious invitations and stimulating proposals for scientific collaboration from Italy and the rest of the world, which went some way to fill the void created by that brutal "cancellation." But the dramatic severing of his long-standing ties to the FST continued to haunt him like a black cloud until the very end—a feeling exacerbated by his concern about the future of the magnificent Gabinetto di Fisica, which he rightly regarded as his own creation.

For me, Paolo was more than an extraordinarily gifted colleague and an utterly trustworthy partner in critically important research projects. He was a friend with whom one could converse on matters of common interest, even outside the scope of our scholarly pursuits.

The international community of historians of science and technology is deeply indebted to Paolo for his lifelong commitment and achievements. I hope that the valuable documentation gathered in the course of an incredibly active career will be acquired by an institution that will preserve it, catalogue it, and make it accessible to the public. That would be the most effective way to make sure the path marked by Paolo Brenni will live on in our collective memory, and continue to inspire and educate future generations.

Paolo Galluzzi is an Academician of the Lincei and a member of the Royal Academy of Sciences in Stockholm. He has taught at the universities of Siena and Florence, and has held visiting professorships at several prestigious American universities. From 1982 to 2021, he directed the Museo Galileo in Florence. He has authored more than 300 publications on a variety of subjects, such as Leonardo da Vinci, Renaissance engineers, Galileo and his school, philosophical-scientific lexicography, the historiography of science, early modern academies, and contemporary Italian debates on science, religion and society. He has designed innovative digital library models and created numerous exhibitions on important figures and events in the history of scientific thought, as well as on the relationship between art, science, culture and society.

Chapter 17
Beyond the Written Word. Paolo Brenni's Foundational Contribution to the Conservation of Scientific Instruments

Jean-François Gauvin

Abstract This short *in memoriam* piece honors Paolo Brenni, tracing his life and his significant contributions to historical scholarship and conservation practice.

The last time I saw Paolo was on the morning of 25 October 2021. I had invited him to give a virtual talk (in French) to the students enrolled in a week-long course I co-organized with the Musée de la civilisation (Québec City). As he was wont to do, he accepted my invitation within 24 hours. Since the course focused on documenting scientific instruments, I asked Paolo to speak about his lifetime's work at the Fondazione Scienza e Tecnica in Florence (more on this below). The students were simply mesmerized. Even challenged by a buggy Microsoft Teams link, Paolo managed to captivate his audience with his knowledge of the field, his bonhomie, and his YouTube videos of experiment reconstructions. He was kind enough to chat the following day with a couple of students who had questions regarding some optical instruments manufactured by Duboscq. This, in a nutshell, summarizes the Paolo Brenni we all knew and admired: generous, unselfish, kind, knowledgeable, ever helpful, funny, and much more.

When I was asked to write this *in memoriam* comment, I didn't know exactly how to approach it. Though I am very familiar with Paolo's work, I wasn't a close collaborator as were some other colleagues in the field. I met Paolo for the first time in the late 1990s, when I was still being introduced to the field (at the Musée Stewart in Montréal) and in charge with Lewis Pyenson of overseeing a work on Jean-Antoine Nollet (1700–1770). Paolo's contribution to it was illuminating and an important chapter in the edited volume.[1] From then on, I met him regularly at the Scientific Instrument Commission (SIC) annual symposium and at various other occasions. I was always struck by his contagious *joie de vivre* and a charming *disinvoltura* with

J.-F. Gauvin (✉)
Université Laval, Quebec city, Canada
e-mail: Jean-Francois.Gauvin@hst.ulaval.ca

[1] Brenni (2002).

G. Bernasconi et al. (eds.), *Astronomical Observatories and Chronometry, 18th-20th Century*, History of Physics, https://doi.org/10.1007/978-3-031-95874-8_17

which he guided a group of experts through an instrument collection—even one he'd discovered for the first time. I was drawn to his stunning expertise; one I would never dream of mirroring. No wonder that, when I was hired at Université Laval in 2018 to revive and lead the museum studies programme, I began teaching a general course on material cultural heritage featuring Paolo's work on scientific instruments. I focused not so much on his vast scholarship of the topic, but on his efforts at preserving, restoring, and caring for scientific instruments. This aspect of Paolo's work was introduced in a book I published recently and from which I borrow heavily in what follows.[2] Written before Paolo passed away, the excerpts I share here demonstrate how central his thinking on scientific instrument conservation-restoration was to me (and to our field).

The International Council of Museums (ICOM), with close to 45,000 members worldwide, adopted a resolution in 2008 advocating for a conservation-restoration terminology that would facilitate communication between various actors, including the public. ICOM defines conservation as "all measures and actions aimed at safeguarding tangible cultural heritage while ensuring its accessibility to present and future generations." Under this general definition, ICOM describes three subtypes of conservation: (1) *Preventive conservation*, or "all measures and actions aimed at avoiding and minimizing future deterioration or loss"; (2) *Remedial conservation*, or "all actions directly applied to an item or a group of items aimed at arresting current damaging processes or reinforcing their structure"; (3) *Restoration*, or "all actions directly applied to a single and stable item aimed at facilitating its appreciation, understanding and use".[3] The latter subtype, furthermore, is explicit about the end result:

> These actions are only carried out when the item has lost part of its significance or function through past alteration or deterioration. They are based on respect for the original material. Most often such actions modify the appearance of the item (ICOM, 2008).

The ICOM definition of conservation-restoration seeks to standardize and give guidelines to museums located across the world. Yet it is fluid and debatable as to whether it can be applied in all cultural contexts. Conservation-restoration doesn't mean the same thing everywhere, and different interpretations and practices do not necessarily lead to the ruin of collections.[4] As discussed below, one key issue is the concept of *authenticity* and how it relates to museum objects (Mairesse & Peters, 2019).

For over thirty-five years, Paolo developed a conservation-restoration practice emulated by most of us in the field. It is common knowledge that he brought back to life the physics cabinet of the Istituto Tecnico Toscano (opened in 1857), now under the auspices of the Fondazione Scienza e Tecnica in Florence.[5] Of the Fondazione's circa 2,000 physics devices from the latter half of the nineteenth century, about

[2] Gauvin (2023).

[3] ICOM (2008).

[4] Njoya Ntieche (2021).

[5] The Fondazione website: http://www.fstfirenze.it/gabinetto-di-fisica/ (accessed 21 October 2024).

50% required only cleaning and minimal intervention to exhibit them. Another quarter necessitated a more thorough conservation-restoration, while the rest were heavily damaged. Several of these were "severed" *(smembrati)*, with parts scattered all over the building. In a laboratory specifically dedicated to the restoration of the Fondazione's instruments, Paolo and Anna Giatti had, over several decades, systematically disassembled, cleaned, repaired, and reassembled all the instruments of the collection. In most cases it was feasible and meaningful to make them operational again (as seen on the Fondazione's YouTube channel showcasing corresponding lecture demonstrations).[6] Paolo and Anna Giatti acknowledged that bringing the instruments back to working order was carried out only when the restoration remained non-invasive, since the chief goal was to preserve these objects of historical value–not to repair them: "restoration [is] aimed above all at preserving (and not repairing!) objects of historical interest." Between 2004 and 2007, the room where the instruments were stored and displayed originally was refurbished and opened to the public.[7]

Contrary to some who would prefer that human intervention on science and technological objects remained parsimonious and kept to a strict minimum,[8] Paolo opted for conservation-restoration (and repair when necessary) that allowed an instrument to regain its lustre lost through time—too often through pure negligence. When, for example, he tackled the restoration of a major meteorological device—Father Angelo Secchi's meteorograph—several lost parts were reconstructed not only to bring back the device's original configuration, but also to facilitate the performance of lecture demonstrations. Paolo clarified, however, that any urge to "invention or interpretation" was avoided. Missing parts were recreated only when solid documentation existed, allowing complete knowledge of their exact shape, size, and characteristics. Sometimes old recipes or nineteenth-century craft methods (the preparation of varnish, for instance) were put to good use during the conservation-restoration of an instrument. This was not an attempt at replicating old techniques, according to Paolo, but simply a fact that they had proven to be effective in the past.[9] More recently, working on a refractor built by Joseph von Fraunhofer and delivered to the Observatory of Tartu (Estonia) in 1824, Paolo and his team deployed a similar know-how in restoring the instrument: the old and chipping varnish on the large wooden beams supporting the telescope was reconditioned; some missing mahogany fragments were replaced to fill the gaps; the parts made in brass and bronze, much oxidized, were cleaned and revarnished (unfortunately, the original nineteenth-century lacquer was

[6] YouTube channel: https://www.youtube.com/user/florencefst?feature=watch (accessed 21 October 2024). See also Brenni 2022.

[7] This formidable contribution to the field over three decades was cavalierly put to an end by the Fondazione in January 2020, upsetting people in Florence. See the short eulogies by Galluzzi and Miniati (2022), Brenni (2009).

[8] He writes that "The maxim of restoration should be to reduce all measures to the absolute minimum necessary" Kühn (1989, p. 405).

[9] When historical and technical information is lacking, it is better not to attempt bringing the object back to working order, a decision Brenni and his colleagues made when restoring an air pump. See Serra et al. (2017) See also Brenni (1993a, b, 2000, 2006).

removed a decade before); only the telescope optics, properly cleaned in 1993, did not require any treatment.[10]

The school of thought supported by Paolo is comparable to the conservation and restoration practices adopted for musical instruments, whose function as sound producers also grants them aesthetic and epistemic values when their operation is brought back to life.[11] What he proposed was located squarely in the middle of a spectrum advocating, at one end, repair, without regard to cost, to restore a device's operation and, at the other end, a circumscribed and parsimonious conservation that defers to the dust of time and other dubious historical traces (often simple dirt, resulting from conservation neglect). However, it was possible (and eminently practical) to slide toward one end or the other of this restoration spectrum, depending on the state of an instrument. Take, for example, three astronomical instruments Paolo worked on at the Palermo Astronomical Observatory in Sicily, Italy. For the Ramsden Circle, he cleaned it and made a small repair (the wooden channel protecting the plumb line, which was missing); for the Pistor & Martins Meridian Circle, after stripping the instrument and removing the parts added in the 1950s, Paolo decided not to go forward with a reconstruction since too many original parts were missing; lastly, for the Merz telescope, he realized it was impossible to remove the components also introduced in the 1950s without compromising the stability of the instrument, choosing instead to leave them on and lacquer them in black so they were easily distinguished from the original ones. Three instruments, three different treatments and solutions depending on the state of the instrument.[12]

This middle-of-the-road stance is widely accepted and recognized today by those who are determined to safeguard the more recent scientific and technological objects, particularly those from the nineteenth and twentieth centuries. We find it, for example, in the Gautier meridian circle of the Rio de Janeiro Observatory (Paolo was an advisor)[13] and in the public display project of the Cockcroft-Walton particle generator formerly built for the Institut de physique nucléaire de Lyon.[14] In the latter case, the hulking size, the composite materials employed to manufacture it, the incomplete and dispersed documentation, and the complexity of its operation compel one to rethink the overall value of such a scientific instrument, including its heritage and public display. Currently stored in fragments, it will be viewable by the public once the conservation-restoration project is completed.[15]

Studying Paolo's conservation-restoration methodological approach prompted me to look further into the past (and the present) to get a glimpse at the historical arch of a savoir-faire that turned into a genuine field of research over time.[16] In concluding his historical and conservation analysis of the old Cockcroft-Walton particle generator

[10] Brenni (2012).

[11] Barclay (2001); Pollens (2015).

[12] Brenni et al. (2001).

[13] Granato and Le Guet Tully (2019).

[14] Geindreau (2016).

[15] Derolez (2017).

[16] Bewer (2010).

in Lyon, Rémy Geindreau argued that a conservator, holding on too closely to the Brandian tradition, "will be marginalized and misunderstood" if s/he does not try "to endorse the questions raised by the new conservation-restoration theories" and practices upheld by Paolo, for instance, regarding scientific instruments.[17] Cesare Brandi (1906–1988) is often seen as the founder of the modern conservation-restoration theory. By establishing the Istituto Centrale per il Restauro (ICR) in Rome in 1939, Brandi laid the groundwork for the field's current professionalization. In the first page of his influential *Teoria del restauro* (1963) Brandi categorically disconnects "repair" (*riparazione*) from "conservation" (*restauro*). Whether, in the former case, the goal is to bring the industrial object (built by artisans or mass-produced) back to its original working conditions ("*restituzione in pristino*"), in the latter the recovery of the artwork's original function "even if it is part of the restoration, represents ultimately but a secondary or concomitant component, never the primary and fundamental one that considers the work of art as a work of art."[18] From this conclusion naturally derives an ontology divided into two classes: the "organic or industrial object" ("*organico-funzionale*") as a "unit of the total" *("unità del totale")* and the artwork as a "unit of the whole" ("*unità dell'intero*"). For the latter, Brandi argues it is critical to recognize the work of art as such ("*riconoscimento dell'opera d'arte come tale*") insofar as the restoration aims to recover the potential unity of the work of art ("*ristabilimento dell'unità potenziale dell'opera d'arte*"). The artwork is not composed of distinct parts–the sum of its parts forming a total; its form is conceived rather as indivisible, making up a unique figurative and constitutive unit of a whole: the image *per se*. The impact on the field of conservation-restoration is twofold following this consideration: (1) an artwork physically damaged continues "to exist *potentially* as a whole in each of its fragments" ("*dovrà continuare a sussistere potenzialmente come un tutto in ciascuno dei suoi frammenti*"); (2) if the same artwork appears to have been materially divided, "we have to search for the original potential unity [*potenziale unità*] contained in each of the fragments, proportionally to the initial form that survives in them."[19] From the concept of "unit of the whole", we deduce the three chief conservation-restoration's deontological principles, i.e., *stability* in the choice of material employed on the artwork being restored, *reversibility* of actions taken with the latter, and *legibility* of an artwork from a historical point of view.[20]

Putting scientific and technological devices back into working order does not mean ignoring all ethical principles of conservation. While being unsure of Paolo's assessment regarding Brandi's conservation-restoration philosophy, he certainly would have agreed with Elizabeth Pye, Emerita Professor of Archaeological and Museum

[17] An excellent example of this "forced" hybridisation for scientific and technological objects is given Mirambet-Paris and Mirambet (2011). Geindreau (2016, p. 18).

[18] Brandi (1977, p. 4).

[19] See also Brandi and Cordaro (1994); Brandi (1977, pp. 16–17).

[20] Note, however, that a restoration's reversibility is often questioned. See Oddy (1999, pp. 1–5). For a historical and philosophical context of conservation-restoration, see Price et al. (1996). For a critical analysis of Brandi's work, see Muñoz Viñas (2015).

Conservation at the University College London's Institute of Archaeology, when she recognized that the conservation of museum objects requires a rigorous ethic, though it should not be excessively rigid or inflexible. The operation of complex scientific apparatuses and machines benefit both museum visitors and researchers, who can learn how they work and discover the artisanal or industrial know-how that produced them. If representation is the quintessence of a painting, according to Pye, the quintessence of a mechanical device is its function. The museum must offer visitors the experience of a moving mechanism, considering it implies special conservation-restoration methods and the risk of permanent damage.[21] Though it is not complete "heresy" to encourage repair or restoration of mechanical devices as a goal in itself, it has been, nevertheless, a contentious issue for decades.[22] The safe solution is to call for an in-depth study of the object's status and its general state of conservation. It is better if the object is part of a circumscribed and well-defined scientific and cultural project (like the Fraunhofer, Ramsden, Merz or Gautier projects described above). Furthermore, it must benefit from the expertise of an interdisciplinary team made up of historians, restorers, technicians, collection managers, designers, and engineers, who will ensure detailed specifications (list of interventions and maintenance measures) and the object's full documentation.[23]

But why is it important to reconstruct the functionality of objects? I hope Paolo would have agreed with me here when I state that we live (for some time now) in a world that reduces everything to art and aesthetics—what Jean Clair calls, as regards the Musée du Quai Branly-Jacques Chirac, a universal *reductio ad aestheticam*.[24] It does not mean, however, we must keep cultivating it. Why does this assertion matter? It matters because material cultural studies have repeatedly demonstrated the importance of the shifting and fluid, even unstable use of objects in all socio-cultural,

[21] A rare guide for the conservation-restoration of scientific instruments supports bringing them back into working order. The author argues that "the purpose of a collection is to show current and future generations what the objects on display were, both in their presentation and in their technical functioning." Biraud (1987); Pye (2006).

[22] Early in the history of the Science Museum, not unlike what happened at the Conservatoire national des arts et métiers (CNAM) in Paris, it was decided to "destroy" objects either by revealing their internal mechanisms (sectioning) or by switching them on (operating). This apparently controversial destruction of historical clues and traces, almost impossible to imagine in museums today, is nonetheless one of the Science Museum's signature identities. It also illustrates the central paradox of science and technology museums: the need to preserve objects *and* demonstrate their operation. This paradox is closely linked to how we think about the conservation and restoration of scientific objects. Mann 1989 argues that this paradox between conservation and operation keeps museum curators in a state of "schizophrenia": "Because they [curators] have been labouring under an inappropriate ethic derived from other types of museum, they have been living in a schizophrenic state in which their actions contradict their beliefs. They are not dishonest, thoughtless and uncaring, but merely confused." (p. 384) See also Mann (1994). For the CNAM, see André (2000) and Jacomy (1995).

[23] Rolland-Villemot (2001); Monger (1988); Tomsin (2007).

[24] The philosopher Bernard Deloche (2010) draws a similar conclusion. Françoise Choay (2007) dismisses the quai Branly as a "Disneyland of exoticism." On the origins of the quai Branly museum, Price (2007). On the generalization of aesthetism, Lipovetsky and Serroy (2013); Clair (2007, p. 110).

historical, and scientific fields. Hide this lesson behind a veil of aestheticism, shroud it in a learned narrative about emergence and transience and the object—particularly the museum artifact—loses its multiple meanings. *Its function is destroyed.* The late Jean-Pierre Cometti, who taught philosophy and aesthetics at the École supérieure d'art in Avignon, reminds us that "it is not the material and physical features that define an object—and consequently its identity—but the uses for which it lends itself."[25] In an illuminating book, Cometti proposes a fundamental ontological shift regarding the field of conservation-restoration. No longer a question of exclusively examining the material properties of objects (a strict ontology of the object), one needs to focus his or her "attention to operation and operating conditions." This shift, however, does not alter the Brandian "unit of the whole." The object must continue to be considered in its extensiveness. The interest shown in the object's separate components must be "subordinated to the interest shown in the whole as a significant totality." Here technology "has no priority," because if this were the case, i.e., if the intervention on a specific part "is conceived independently [within the framework of the *unità del totale*], where what matters is the production of an effect by all appropriate means," we would fall back—relapse—straight into the notion of repair.[26]

Starting from the precept of the unit of the whole, Cometti examines the connections federating the idea of identity, integrity, and authenticity tied to an object. He suggests that the notion of identity, which "better supports an unvarnished treatment [*traitement désauratisé*]," is preferable to that of authenticity for attesting to the practice of conservation-restoration. Citing the famous thought experiment involving the ship of Theseus, Cometti recalls that its identity never ceases to be that of the *ship of Theseus*, even when most of the wooden planks and beams composing it were removed and replaced over time. What allows us to affirm it is still and always the same ship? The "guarantee of identity or sameness, if you prefer, lies in a greater or lesser *continuity*, which is that of uses and functioning [*des usages et d'un fonctionnement*]." To avoid "restoring by imagining," according to the Brandian formula, it is necessary to locate in the present moment the identity of an object—the ship of Theseus at a precise time t_1 after its inaugural journey. From this derives the notions of integrity (the ontologically defined identity) and authenticity (the historically situated identity), both of which are specified by the very object–its identity at an instant t_n. As a result, the "identity of a thing is presupposed by the very idea of a thing, its integrity guarantees its genuineness [*en apporte la garantie*], and the authenticity indicates the conformity of its state at a given time, for example, to a previous state assumed to be original and constitutive of its identity."[27] By emphasizing a

[25] Cometti (2015, p. 71).

[26] Cometti (2015, pp. 31–39).

[27] The concept of time is crucial for Brandi: "Non ci sarà allora bisogno di maggiore insistenza per affermare che l'unico momento legittimo che si offre per l'azione di restauro è quello del presente stesso della coscienza riguardante, in cui l'opera d'arte è nell'attimo ed è presente storico, ma è anche passato e, a costo, altrimenti, di non appartenere alla coscienza umana, è nella storia" Brandi (1977, pp. 26–27). On the metaphysics of identity and the ship of Theseus, Ferret (1996). See also Lipovetsky (2021); Castriota (2019). Cometti (2015, pp. 56–74).

philosophy of use, Cometti wants to shift the discourse of conservation-restoration toward a practice which, insofar as it respects the principle of the unit of the whole, seems to favour putting into working condition science and technology devices and machines (to name but these, at the centre of our study). The author readily recognizes that the museum environment "tends to ontologize what is conceived in contexts of action modalities." Whether they are mechanical devices or "recalcitrant" ethnological objects, "the point of view of operation [*le point de vue du fonctionnement*] leads precisely to consider things differently and, as a result, to dismiss the aesthetic point of view."[28]

In a short article focused on the circulation of astronomical instruments in the eighteenth and nineteenth centuries, Simon Schaffer (2011)[29] suggests that instruments "in a state of disrepair" should be exhibited next to others considered more prestigious and dazzling. The former, he contends, have as much to say—perhaps even more to teach—about the circulation of knowledge and the epistemic and sociological value of repair, use, and maintenance, when delving into the material culture of science. The state in which an object is found at a time *t* remains in and of itself a convincing indication of its historical and material trajectory. This is one of the main features connected to the field of conservation-restoration. I believe Paolo would have agreed with this assessment. But he also strongly believed that carefully restoring scientific instruments (and making them operational when possible) provided relevant information about the object, its materiality and its use that otherwise would be impossible to acquire. No wonder why, a long time ago, Gerard L'E. Turner introduced Paolo to Stuart Talbot as the "Man with Golden Hands."[30] Paolo's contributions to our field went beyond his written scholarship (formidable as it is). He helped safeguard individual artefacts of great value and salvage from oblivion entire collections by making them known, operating them, documenting their historical significance, and demonstrating how to care for them. This is what, in my own way, I wanted to celebrate in this short and incomplete overview of Paolo's incredible conservation-restoration achievements. You are dearly missed, *caro amico*.

References

André, L. (2000). Scientific instrument restoration: problems and practice at the Musée des arts et métiers and in France. In: L. Manetti (Ed.), *Proceedings of the workshop held in Florence, 14–15 December 1998* (pp. 23–30). Le Lettere.

Barclay, R. (2001). Instruments, conservation, restoration, copying of. *Grove Music Online*. Retrieved September 12, 2022, from https://www-oxfordmusiconline-com.ezp-prod1.hul.harvard.edu/grovemusic/view/; https://doi.org/10.1093/gmo/9781561592630.001.0001/omo-9781561592630-e-0000046526

[28] Cometti (2015, pp. 118–119).

[29] Schaffer (2011).

[30] Dawes and Talbot (2022).

Bewer, F. G. (2010). *A laboratory for the art: Harvard's Fogg museum and the emergence of conservation in America, 1900–1950*. Yale University Press.
Biraud, G. (1987). *La restauration et la conservation des appareils scientifiques de collection*. Édition Biraud.
Brandi, C. (1977). *Teoria del restauro*. Piccola Biblioteca Einaudi.
Brandi, C., & Cordaro, M. (Eds.). (1994). *Il restauro: teoria e pratica*. Editori Riuniti.
Brenni, P. (1993a). Strumenti scientifici: Il restauro del meteorografo Secchi. *OPD Restauro, 5*, 99–107.
Brenni, P. (1993b). Il meteorografo di padre Angelo Secchi. *Nuncius, 8*, 197–247.
Brenni, P. (2000). Better than new? Scientific instrument restoration in Italy. In L. Manetti (Ed.), *Proceedings of the workshop held in Florence,*14–15 December 1998 (pp. 51–56). Le Lettere.
Brenni, P. (2002). Jean Antoine Nollet and physics instruments. In: J. F. Gauvin, & L. Pyenson (Eds.), *The art of teaching physics: The eighteenth-century demonstration apparatus of Jean-Antoine Nollet* (pp. 11–27). Les Éditions du Septentrion.
Brenni, P. (2006). Quelques considérations sur la restauration des instruments scientifiques. *CoRé, 17*, 8–11.
Brenni, P. (2009). *Il gabinetto di fisica dell'Istituto tecnico toscano: Guida alla visita*. Edizioni Polistampa.
Brenni, P. (2012). The Fraunhofer's refractor of Tartu (Dorpstat) and its restoration. *Bulletin of the Scientific Instrument Society, 113*, 2–7.
Brenni, P. (2022). Filming nineteenth-century physics demonstrations with historical instruments. In: E. Cavicchi, & P. Heering (Eds.), *Historical scientific instruments in contemporary education* (pp. 34–49). Brill.
Brenni, P., Chinnici, I., & Foderà Serio, G. (2001). The restoration of three large telescopes of the Palermo astronomical observatory. *Bulletin of the Scientific Instrument Society, 71*, 11–16.
Castriota, B. (2019). Authenticity, identity, and essentialism: Reframing conservation practice. In: F. Mairesse, & R. F. Peters (Eds.), *What is the essence of conservation? Materials for a discussion* (pp. 39–48). ICOFOM.
Choay, F. (2007). Un nouveau Luna Park était-il nécessaire? *Le Débat, 147*, 57–64.
Clair, J. (2007). *Malaise dans les musées*. Flammarion.
Cometti, J. P. (2015). *Conserver/restaurer: L'œuvre d'art à l'époque de sa préservation technique*. Gallimard.
Dawes, H., & Talbot, S. (2022). Paolo 'man with golden hand.' *The Bulletin of the Scientific Instrument Society, 152*(March), 11–12.
Deloche, B. (2010). *Mythologie du musée: De l'uchronie à l'utopie*. Le Cavalier Bleu.
Derolez, S. (2017). La patrimonialisation des objets scientifiques contemporains et leurs contextes de valorisation: Cas de l'accélérateur de particules Cockroft-Walton. *Culture & Musées, 29*, 147–150.
Ferret, S. (1996). *Le bateau de Thésée: Le problème de l'identité à travers le temps*. Les Éditions de Minuit.
Galluzzi, P., & Miniati, M. (2022). Eulogies. *The Bulletin of the Scientific Instrument Society, 152*, 4–5.
Gauvin, J. F. (2023). Instruments of knowledge: Finding meaning in objects, habits, and museums. In *Instruments of knowledge* (pp. 173–206). Brill.
Geindreau, R. (2016). Étude préalable à la restauration d'un objet technico-scientifique contemporain: Le générateur Cockcroft-Walton de l'Institut de Physique Nucléaire de Lyon. *CeROArt*, EGG 5. Retrieved October 21, 2024, from http://journals.openedition.org/ceroart/4933
Granato, M., & Le Guet Tully, F. (2019). Les principes de la restauration d'instruments scientifiques: Le cas du cercle méridien Gautier de l'Observatoire de Rio de Janeiro. In Situ 10. Retrieved October 21, 2024, from http://journals.openedition.org/insitu/3865
ICOM. (2008). *Terminology for conservation*. Retrieved October 21, 2024, from https://www.icom-cc.org/en/terminology-for-conservation

Jacomy, B. (1995). Du cabinet au conservatoire: Les instruments scientifiques du Conservatoire des arts et métiers à Paris. *Journal of the History of Collections, 7*, 227–233.

Kühn, H. (1989). The restoration of historic technological artefacts, scientific instruments and tools. *The International Journal of Museum Management and Curatorship, 8*, 389–405.

Lipovetsky, G. (2021). *Le sacre de l'authenticité*. Gallimard.

Lipovetsky, G., & Serroy, J. (2013). *L'esthétisation du monde: Vivre à l'âge du capitalisme artiste*. Gallimard.

Mairesse, F., & Peters, R. F. (Eds.). (2019). *What is the essence of conservation? materials for a discussion*. In: *ICOFOM*.

Mann, P. R. (1989). Working evidence and the destruction of evidence in the science museum. *The International Journal of Museum Management and Curatorship*, 369–387.

Mann, P. R. (1994). The restoration of vehicles for use in research, exhibition, and demonstration. In: A. Oddy (Eds.), *Restoration: Is it acceptable?* (pp. 131–138). British Museum, Department of Conservation.

Mirambet-Paris, A., & Mirambet, F. (2011). La conservation-restauration du patrimoine technique et industriel dans le cadre de la loi sur les Musées de France, une mission impossible? *La Lettre De l'OCIM, 135*, 27–35.

Monger, G. (1988). Conservation or restoration? *The International Journal of Museum Management and Curatorship, 7*, 375–380.

Muñoz Viñas, S. (2015). Who is afraid of Cesare Brandi? Personal reflections on the Teoria del Restauro. *CeROArt* (June). Retrieved October 21, 2024, from https://journals.openedition.org/ceroart/4653

Njoya Ntieche, G. (2021). Traditional versus contemporary conservation methods in West Cameroon's community museums. *Museum International, 73*(1–2), 156–163.

Oddy, W. A. (1999). *Reversibility—does it exist?* The British Museum Press.

Pollens, S. (2015). *The manual of musical instrument conservation*. Cambridge University Press.

Price, N. S., Talley, M. K., & Vaccaro, A. M. (Eds.). (1996). *Historical and philosophical issues in the conservation of cultural heritage*. The Getty Conservation Institute.

Price, S. (2007). *Paris primitive: Jacques Chirac's museum on the Quai Branly*. The University of Chicago Press.

Pye, E. (2006). Challenges of conservation: Working objects. *Science Museum Group Journal, 6*. Retrieved October 21, 2024, from http://journal.sciencemuseum.ac.uk/browse/issue-06/challenges-of-conservation/

Rolland-Villemot, B. (2001). Le traitement des collections industrielles et techniques, de la connaissance à la diffusion. *La Lettre De l'OCIM, 73*, 13–18.

Schaffer, S. (2011). Easily cracked: Scientific instruments in states of disrepair. *Isis, 102*, 706–717.

Serra, L., Giatti, A., & Brenni, P. (2017). The pneumatic pump of the mariano college in Bergamo. In S. Esposito (Ed.), *Atti del XXXVI convegno annuale SISFA (Naples, 2006)* (pp. 371–379). Pavia University Press.

Tomsin, P. (2007). Les patrimoines mobiliers scientifique et technique: Spécificités de leur restauration, de leur conservation et de leur valorisation. *CeROArt*, 1. http://journals.openedition.org/ceroart/370

Jean-François Gauvin is a professor of the history of science and museum studies at Université Laval in Québec city (Canada). He earned his PhD at Harvard University and has worked in museums and with scientific instruments for close to 30 years. His latest book is *Instruments of Knowledge. Finding Meaning in Objects, Habits, and Museums* (Brill, 2023).

MIX
Papier aus verantwortungsvollen Quellen
Paper from responsible sources
FSC® C105338

If you have any concerns about our products,
you can contact us on
ProductSafety@springernature.com

In case Publisher is established outside the EU,
the EU authorized representative is:
Springer Nature Customer Service Center GmbH
Europaplatz 3, 69115 Heidelberg, Germany

Printed by Libri Plureos GmbH
in Hamburg, Germany